Opportunistic Networks

Mobility Models, Protocols, Security, and Privacy

Edited by

Khaleel Ahmad

Nur Izura Udzir

Ganesh Chandra Deka

CRC Press is an imprint of the
Taylor & Francis Group, an **informa** business

A CHAPMAN & HALL BOOK

CRC Press
Taylor & Francis Group
6000 Broken Sound Parkway NW, Suite 300
Boca Raton, FL 33487-2742

Printed on acid-free paper

International Standard Book Number-13: 978-1-138-09318-8 (Hardback)

Library of Congress Cataloging-in-Publication Data

Names: Ahmad, Khaleel, editor. | Udzir, Nur Izura, editor. | Deka, Ganesh Chandra, 1969- editor.
Title: Opportunistic networks : mobility models, protocols, security, and privacy / editors, Khaleel Ahmad, Nur Izura Udzir, Ganesh Chandra Deka.
Description: Boca Raton : Taylor & Francis, a CRC title, part of the Taylor & Francis imprint, a member of the Taylor & Francis Group, the academic division of T&F Informa, plc, 2018. | Includes bibliographical references and index.
Identifiers: LCCN 2018021208| ISBN 9781138093188 (hardback : alk. paper) | ISBN 9780429453434 (ebook)
Subjects: LCSH: Ad hoc networks (Computer networks)
Classification: LCC TK5105.77 .O676 2018 | DDC 004.6/85--dc23
LC record available at https://lccn.loc.gov/2018021208

Visit the Taylor & Francis Web site at
http://www.taylorandfrancis.com

and the CRC Press Web site at
http://www.crcpress.com

Printed and bound in Great Britain by
TJ International Ltd, Padstow, Cornwall

Contents

Preface vii
Editors ix
Contributors xi

1 **Foundations of Opportunistic Networks** 1
Musaeed Abouaroek and Khaleel Ahmad

2 **Opportunistic Resource Utilization Networks and Related Technologies** 11
Mai A. Alduailij and Leszek T. Lilien

3 **Buffer Management in Delay-Tolerant Networks** 25
Sweta Jain

4 **Taxonomy of Mobility Models** 59
Jyotsna Verma

5 **Taxonomy of Routing Protocols for Opportunistic Networks** 85
Khaleel Ahmad, Muneera Fathima, and Khairol Amali bin Ahmad

6 **Congestion-Aware Adaptive Routing for Opportunistic Networks** 117
Thabotharan Kathiravelu and Nalin Ranasinghe

7 **Vehicular Ad Hoc Networks** 133
Sara Najafzadeh

8 **Energy Management in OppNets** 159
Itu Snigdh and K. Sridhar Patnaik

9 **Network Coding Schemes** 171
Amit Singh

10 **Taxonomy of Security Attacks in Opportunistic Networks** 193
Gabriel de Biasi and Luiz F. M. Vieira

11 **Pervasive Trust Foundation for Security and Privacy in Opportunistic Resource Utilization Networks** 213
Ahmed Al-Gburi, Abduljaleel Al-Hasnawi, Raed Mahdi Salih, and Leszek T. Lilien

12 **Future Networks Inspired by Opportunistic Networks** 229
Anshul Verma, Mahatim Singh, K. K. Pattanaik, and B. K. Singh

13 **Time and Data-Driven Triggering to Emulate Cross-Layer Feedback in Opportunistic Networks** 247
Rintu Nath

14 Applications of DTN 259
Rahul Johari, Prachi Garg, Riya Bhatia, Kalpana Gupta, and Afreen Fatimah

15 Performance Evaluation of Social-Aware Routing Protocols in an Opportunistic Network 267
Makshudur Rahman and Md. Sharif Hossen

16 Hands-On ONE Simulator: Opportunistic Network Environment 279
Anshuman Chhabra, Vidushi Vashishth, and Deepak Kumar Sharma

Index 305

Preface

In recent decades, networking has moved from the stationary model of wired communication toward a new and dynamic environment with various attributes, for example, nomadic computing, pervasive computing, virtual home environment, and the anytime-anywhere model of mobile communication, where information is exchanged between wireless devices and made available to mobile users. Opportunistic networks provide pervasive networking to exploit any pairwise contact in its neighbor to identify next hops towards the destination. It allows mobile users to share data without any existing network infrastructure. The key observation of opportunistic computing is that the environment around (mobile) users features a steadily increasing set of heterogeneous resources available on fixed and mobile devices with wireless networking capabilities.

Opportunistic Networks: Mobility Models, Protocols, Security, and Privacy is designed to serve as a textbook for undergraduate and postgraduate courses. Beside this, the aim of this book is to provide research scholars, scientists, and network engineers with an expert guide to the fundamental concepts, mobility models, architecture, protocols, security issues, and state-of-the-art research developments in opportunistic networks. This book has been composed in a straightforward and understandable way to enable students to grasp important concepts easily.

In this book, each of the sixteen chapters covers a unique topic in detail: Foundations of Opportunistic Networks, Opportunistic Resource Utilization Networks, Buffer Management in Delay-Tolerant Networks, Taxonomy of Mobility Models, Taxonomy of Routing Protocols, Congestion Aware Adaptive Routing for Opportunistic Networks, Vehicular Ad Hoc Networks (VANETs), Energy Management in OppNets, Network Coding Schemes, Taxonomy of Security Attacks in Opportunistic Networks, Pervasive Trust Foundation for Security and Privacy in Opportunistic Resource Utilization Networks, Future Networks Inspired by Opportunistic Networks, Time and Data-Driven Triggering to Emulate Cross-Layer Feedback in Opportunistic Networks, Applications of DTN, Performance Evaluation of Social-Aware Routing Protocols in an Opportunistic Network, and Hands-On ONE Simulator.

Editors

Dr. Khaleel Ahmad, Ph.D., is currently an assistant professor in the School of Computer Science & Information Technology at Maulana Azad National Urdu University, Hyderabad, India. Prior to this he has worked at prestigious universities and institutions of national repute. He holds a Ph.D. in Computer Science & Engineering and an M.Tech. in Information Security. His research areas are opportunistic networks, cyber security, cryptography, and cloud computing. He has over 30 published papers in refereed national/international journals and conferences (viz. Elsevier, ACM, IEEE, and Springer), and ten book chapters (CRC Press/Taylor & Francis Group, IGI Global, and IGNOU New Delhi). He has also delivered expert talks, guest lectures in Central University of Haryana, Telangana University, International Conference and chaired the session in an international conference in New Delhi. He is also a life member of various international/national research societies viz. ISTE, CRSI, ISCA, IACSIT (Singapore), IAENG (Hong Kong), IAOE (Austria), ISOC (USA), etc. Besides this, he is associated with many international research organizations as an editorial board member and reviewer.

Nur Izura Udzir, Ph.D., is an academic staff member at the Faculty of Computer Science and Information Technology, Universiti Putra Malaysia (UPM) since 1998. She received her B.Sc. (1995) and M.Sc. (1998) from UPM, and her Ph.D. in Computer Science from the University of York, UK (2006). Her areas of specialization are access control, intrusion detection systems, secure operating systems, coordination models and languages, and distributed systems. She is currently the head of the Department of Computer Science and a member of the Information Security Group (which she led in 2008–2013) at the faculty. She is a member of IEEE Computer Society, Malaysian Society for Cryptology Research (MSCR), the Society of Digital Information and Wireless Communications (SDIWC), and currently serves as a committee member of the Information Security Professionals Association of Malaysia (ISPA.my). Dr. Udzir has supervised and co-supervised over 30 Ph.D. students and over 15 MA (by research) students. She has written a book, *Pengenalan kepada Pengaturcaraan C++* (*Introduction to C++ Programming*) (Prentice Hall, 2001), and has published over 60 articles in journals and as book chapters, and over 80 international conference proceedings, thus earning a H-index of 8 with 232 citations in Scopus (H-index 14 and 718 citations in Google Scholar) as of March 2016. In addition, she has won various awards for her contributions in academic and research, including six Best Paper Awards at international conferences and the MIMOS Prestigious Award 2015 for the supervision of her student's doctoral thesis. She has also delivered keynote speeches and expert talks, and also as an invited speaker at various international conferences.

Ganesh Chandra Deka, Ph.D., is the deputy director (training) in the Directorate General of Training, Ministry of Skill Development and Entrepreneurship (formerly DGE&T, Ministry of Labor & Employment), Government of India. He is a member of the IEEE and IETE, an associate member of the Institution of Engineers, India, and the editor-in-chief of the *International Journal of Computing, Communications, and Networking.*

His research interests include information and communications technology (ICT) in rural development, e-governance, cloud computing, data mining, NoSQL databases, and vocational education and training. He has edited three books on cloud computing and published more than 40 research papers. He earned a Ph.D. in Computer Science from Ballsbridge University.

Contributors

Musaeed Abouaroek
Department of Computer Science and IT
Maulana Azad National Urdu University
Hyderabad, India

Khaleel Ahmad
School of Computer Science and IT
Maulana Azad National Urdu University
Hyderabad, India

Mai A. Alduailij
College of Computer and Information Sciences
Princess Nourah Bint Abdulrahman University
Riyadh, Saudi Arabia

Ahmed Al-Gburi
Department of Computer Science
Western Michigan University
Kalamazoo, Michigan

Abduljaleel Al-Hasnawi
Department of Computer Science
Western Michigan University
Kalamazoo, Michigan

Khairol Amali bin Ahmad
Department of Electrical and Electronics Engineering
National Defence University of Malaysia
Kuala Lumpur, Malaysia

Riya Bhatia
University School of Information, Communication and Technology, Guru Gobind Singh Indraprastha University
Delhi, India

Anshuman Chhabra
Department of Electronics and Communication Engineering
Netaji Subhas Institute of Technology
New Delhi, India

Gabriel de Biasi
Department of Computer Science
Federal University of Minas Gerais
Minas Gerais, Brazil

Muneera Fathima
School of Computer Science and IT
Maulana Azad National Urdu University
Hyderabad, India

Afreen Fatimah
University School of Information, Communication and Technology
Guru Gobind Singh Indraprastha University
Delhi, India

Prachi Garg
USI, C&T
GGSIP University
Delhi, India

Kalpana Gupta
Centre for Development of Advanced Computing(C-DAC)
Noida, India

Md. Sharif Hossen
Department of Information and Communication Technology
Comilla University
Comilla, Bangladesh

Sweta Jain
Department of Computer Science and Engineering
Maulana Azad National Institute of Technology
Bhopal, India

Rahul Johari
University School of Information, Communication and Technology
Guru Gobind Singh Indraprastha University
Delhi, India

Thabotharan Kathiravelu
Department of Computer Science
University of Jaffna
Jaffna, Sri Lanka

Leszek T. Lilien
Department of Computer Science
Western Michigan University
Kalamazoo, Michigan

Sara Najafzadeh
Department of Computer Engineering
Yadegar-e-Imam Khomeini (RAH) Shahre Rey Branch
Islamic Azad University
Tehran, Iran

Rintu Nath
Scientist F, Vigyan Prasar
Department of Science and Technology
Ministry of Science and Technology
Noida, India

K. Sridhar Patnaik
Department of Computer Science and Engineering
Birla Institute of Technology Mesra
Ranchi, India

K. K. Pattanaik
ABV-Indian Institute of Information Technology and Management
Gwalior, India

Makshudur Rahman
Department of Computer Science and Engineering
Port City International University
Chittagong, Bangladesh

Nalin Ranasinghe
University of Colombo School of Computing
Colombo, Sri Lanka

Raed Mahdi Salih
Department of Computer Science
Western Michigan University
Kalamazoo, Michigan

Deepak Kumar Sharma
Department of Information Technology
Netaji Subhas Institute of Technology
New Delhi, India

Amit Singh
Department of Computer Science
Central University of Rajasthan
Rajasthan, India

B. K. Singh
Department of Computer Science and Engineering
National Institute of Technology
Jamshedpur, India

Mahatim Singh
Department of Computer Science
Institute of Science, Banaras Hindu University
Varanasi, India

Itu Snigdh
Department of Computer Science and Engineering
BIT Mesra
Ranchi, India

Vidushi Vashishth
Department of Information Technology
Netaji Subhas Institute of Technology
New Delhi, India

Anshul Verma
Department of Computer Science
Institute of Science, Banaras Hindu University
Varanasi, India

Jyotsna Verma
Department of Computer Science
Central University of Rajasthan
Rajasthan, India

Luiz F. M. Vieira
Department of Computer Science
Federal University of Minas Gerais
Minas Gerais, Brazil

1

Foundations of Opportunistic Networks

Musaeed Abouaroek and Khaleel Ahmad

CONTENTS

1.1 Introduction..1
1.2 Needs of Opportunistic Networks..2
1.3 Migration from MANET to Opportunistic Networks 2
1.4 Sensor Networks..3
1.4.1 Examples of Sensor Network Applications 3
1.5 Mobile Ubiquitous LAN Extensions (MULEs)..4
1.6 ZebraNet 4
1.7 Shared Wireless Infostation Model (SWIM) 5
1.8 Mobile Ad Hoc Network (MANET)...5
1.8.1 Advantages of an Ad Hoc Network..6
1.9 Mobile Social Networks (MSNs)...6
1.10 Vehicular Ad Hoc Networks (VANETs) 7
1.10.1 Applications of VANETs..7
1.10.2 Characteristics of a VANET...7
1.11 Delay\Disruption-Tolerant Networking (DTN) 8
1.11.1 Network Topology Classification 8
1.11.2 Routing Strategy Classification...8
1.11.3 Replication and Semantic Classification...8
1.12 Freenet 9
References...9

1.1 Introduction

Opportunistic networks (OppNets) are a subclass of delay-tolerant networks (DTNs) where communication opportunities (contacts) are discontinuous in nature, so an end-to-end path may not exist. The link performance in OppNets is highly dynamic in nature. Therefore, the TCP/IP protocol will not work in such an environment because the end-to-end path may only exist for a short period of time (Ritu & Sidhu, 2014).

Each OppNet originates from a seed, which is an arrangement of nodes engaged together at the phase of the initial OppNet deployment. The seed is pre-designed (and can, subsequently, be seen as a system in its own right). In the most extreme case, it can at least involve a single node. The seed forms into a bigger system by extending the request to join the OppNet to remote devices, networks, clusters, or different systems which it may contact. Any new node that transforms into a completely developed OppNet member, which is called a helper node, might be allowed to welcome external nodes. By welcoming "free" nodes, the OppNets can be especially competitive economically. Before a seed OppNet can develop, it must find its own particular arrangement of potential partners accessible to it. As an example of a discovery, a PC can be found by an OppNet once the OppNet recognizes a subset of Internet addresses (Internet protocol [IP] addresses) situated in its geographical area. Another case of revelation could include an OppNet node scanning the range for radio signals or beacons and gathering enough information to have the capacity to contact their senders (Lilien, Kamal, Bhuse, & Gupta, 2006).

1.2 Needs of Opportunistic Networks

Smartphones and other mobile devices are widely used nowadays. To facilitate smooth communication between them, a service provider, as a central authority, has a specific topology and a path between nodes is predetermined. This is a traditional scenario, but in an OppNet the same scenario is not applicable since the path between nodes may not exist. An OppNet has no infrastructure; this can be considered an advantage in some situations where the network infrastructure may collapse due a disaster or emergency. In these situations, there should be a connection to call a rescue team, as an example of OppNet capabilities. Other scenarios where OppNets can operate are inaccessible areas and areas where there is a sparse population, in which building infrastructure is difficult and expensive. Telecommunication providers can also benefit when there is more data loading in particular; if the load causes a slow response from the network, the provider can send their content information to some devices in the particular area and those devices feed content information to other devices opportunistically (Trifunovic et al., 2017).

1.3 Migration from MANET to Opportunistic Networks

In the last few years, a lot of research has been conducted to design mobile ad hoc network (MANET) technology. A MANET is a kind of network where the node connects to other nodes within the network range wirelessly. In MANETs there are no network infrastructures, so the nodes exchange the packets through intermediate nodes between them. MANETs are dependent networks; thus, no centralized service interrupts the communication. MANETs are characterized by self-configuration, highly dynamic network topology, severe resource constraints, and shared wireless medium. The OppNet has come into view because of the challenges faced by MANETs and some other issues faced by itself. In MANETs, the connection between nodes that are in need of communication should be always available through a common internetwork, which is not usually possible in common scenarios. In such cases, and because of some environmental challenges, some users switched off their devices or moved out of range, thus, making the MANET protocol inapplicable for the OppNet. In an OppNet, mobility disconnections, partitions, etc., are considered as norms rather than exceptions. So, in an OppNet, mobility is used to provide connection among disconnected groups instead of being seen as a drawback to be solved (Ritu & Sidhu, 2014).

The OppNet is a wireless ad hoc network. It is the evolution of the MANET, which supports delay tolerance. The message forwarding process in OppNets accrues when the node finds an opportunity to forward it. The node cannot send the message until it finds another node that has entered into the same network range. The source node searches for the neighbor node to send its message and the neighbor node does the same process until the message reaches the destination node (Martín-Campillo, Crowcroft, Yoneki, & Martí, 2013).

OppNets are considered as a very appropriate choice in emergency situations as they have no infrastructure and central services. The OppNet uses the process of store-carry and forward to deliver the message between two nodes. Thus, the connection between these nodes is set up automatically. So this explains the ability of delay tolerance in the OppNet. The OppNet has a very significant role during emergency situations as it is guaranteed to deliver the message without any loss in disaster areas (Sugihara & Gupta, 2011).

1.4 Sensor Networks

A sensor network is a collaborative result of hundreds or thousands of sensor nodes. Each sensor node contains four essential parts, such as a sensing unit, a processing unit, a radio unit, and a power unit. Together, the units can fit into a space as small as a matchbox or even a smaller module. As the sensor nodes have restricted sensing and computational abilities, they can travel only within short distances. These nodes spread out and arrange themselves to accomplish a common task (Akyildiz, Su, Sankarasubramaniam, & Cayirci, 2002).

Wireless sensor networks (WSNs) are the networks which are spatially distributed over the range and gather information from the physical world. These are used for observing environmental factors like pressure, temperature, moisture, etc., and sending this data to the sink or destination node. WSNs have proven beneficial in a number of applications in the area of traffic surveillance, military applications, weather forecasting, landslide detection, fire detection, etc. They are the backbone of emerging technologies such as the Internet of Things (IoT), the cyber-physical system (CPS), etc. The most interesting contribution of WSNs is in healthcare. WSNs in healthcare itself is a topic of research that has gained much popularity these days. The potential of sensing the information from physical entities makes WSNs a hot topic for research (Jadhav & Satao, 2016).

1.4.1 Examples of Sensor Network Applications

Intrusion detection and tracking: Sensors are conveyed along the fringe of a battlefield to detect, classify, and track intruding personnel and vehicles.

Weather monitoring: Weather monitoring is an important aspect; dedicated nodes are used which can predict weather parameters like temperature, wind velocity, pressure, amount of rainfall, humidity, etc.

Indoor surveillance: These are used to provide security in art galleries, hospitals, shopping malls, and other facilities.

Traffic analysis: These can monitor traffic or a congested part of a city to help people reach their destination easily (Akyildiz, Su, Sankarasubramaniam, & Cayirci, 2002).

1.5 Mobile Ubiquitous LAN Extensions (MULEs)

MULEs are a kind of data collector; they are used to collect data from static sensor nodes in WSNs. Data mules, when compared to multi-hop transmitting approaches, remarkably reduce energy utilization at sensor nodes, yet have a disadvantage related to latency of data delivery. Movement planning issues of data mules comprise three parts: deciding the path, measuring speed change over time, and scheduling when to collect data from each node. These three aspects are not independent from each other, and enhancing them is difficult, as shown by the NP-hardness (Sugihara & Gupta, 2011).

MULEs receive data from the sensor nodes; they buffer it and then send it to the wired access points (APs) within range. A significant power saving occurs at the sensors as they just transmit it over a short range. The main advantage from our perspective is the capability of large power savings at the sensors because of communication that now happens over a short range. New radio technologies like ultra-wideband (UWB), which works at exceptionally low power with an extensive data burst limit, are suited for MULE communication. The main drawback of this approach is increasing latency since sensor nodes must wait for a MULE to approach before the exchange can happen (Shah, Roy, Jain, & Brunette, 2003).

1.6 ZebraNet

ZebraNet is a type of mobile sensor network that has limited network coverage and high-energy sensors, such as Global Positioning System (GPS). Because of this, it operates in an isolated part of the hardware design space.

Application:	Data Logging, Application Protocol
Impala Middleware:	Operating System, Network Services
System Firmware:	Peripheral, Clock, and Low-Level Energy Management
Hardware:	Physical Chips, Power Supplies, Battery Charger, and Solar Array

In ZebraNet, once the data samples are acquired they are accumulated and analyzed. Zebras must be fairly mobile i.e., only a few are collared, and they must spread across a distance of several kilometers, so data accumulation becomes difficult. To transfer data in the network, nodes will communicate in a pairwise fashion. The main objective is to get position logs back to the biologists so that single-hop transmissions are enough to pass data to other collars. In such a case, latency can be seen, but it is ignored. Since collars come into contact very often, a dedicated system like pairwise communications is used instead of multi-hop transmissions. A manned base station intermittently comes into contact with a zebra, so that it can download data from all the zebras. Flash memory is used to store position logs, which compensate latency of data proliferation.

To support connectivity in such a sparse system, the radio range should be greater than one kilometer. However, radio communications consume a lot more energy than the regular short-range radios utilized in sensor networks (Zhang, Sadler, Lyon, & Martonosi, 2004).

ZebraNets are controlled by the Impala middleware layer. Impala permits the combination of planned and voluntary events, and through this application GPS detection and radio correspondence times are predefined. A few protocols are conceivable; however, in our deployment, the information is proliferated through the network using a coding protocol. This enables the base station to obtain data from all the collared zebras by experiencing them as one. In January 2004, we conveyed seven nodes on zebras in Kenya. In view of the outcomes from the reset (RST) sending of the system, we are now making improvements. The power utilization of this system, including occasional data sampling and correspondence time, is estimated with an oscilloscope (Zhang, Sadler, Lyon, & Martonosi, 2004).

1.7 Shared Wireless Infostation Model (SWIM)

Shared Wireless Infostation Model (SWIM) is a union of Infostations with the ad hoc networking model. Information dissemination in SWIM resembles the spreading of a disease, where the nodes in the network act as carriers of an ailment, and data sharing with other nodes acts as the spread of the disease from a contaminated to a well person. Finally, once the node has reached any Infostation, it is conveyed and then the packet is 'offloaded' and disposed of from the network. In our analogy, the disease epidemic is similar to a contaminated carrier reaching a 'healing' center, in which the carrier is cured of the disease. Storing the identity of the packets that were offloaded to an Infostation is essential so that they will not be allowed in the future, i.e., developing 'immunity' to the disease.

By allowing the packet to spread over all mobile nodes, the delay until one of the replications reaches an Infostation can be remarkably reduced. However, it has a drawback, i.e., spreading the packets to other nodes will consume much network capacity. But again, we will be faced with the capacity–delay tradeoff. We have come up with an approach to control this tradeoff by regulating the parameters of packet spreading; for instance, by managing the likelihood of packet communication between two adjoining nodes (practically equivalent to the likelihood of contamination (infection) in the epidemic model), the communication range of every node (closely resembling the infection distance in the epidemic model), or the number and appropriation of the Infostations (same as the number and location of hospitals in the epidemic model) (Small & Haas, 2003).

1.8 Mobile Ad Hoc Network (MANET)

MANETs are self-managing wireless networks. These nodes are infrastructure-less, routed to each other, and contain temporary and dynamic wireless networks. Ad hoc is a Latin word. For this purpose, it means every device in a MANET has the ability to change its place to another place freely and its links to other devices will be changed frequently (Mehta, Nupur, & Gupta, 2015).

Moreover, the data transmission in a MANET among mobile devices can be accrued directly without any dependency on any centralized service unit. Some services are added into transmission standards like Bluetooth, which follow a store-carry and forward principles. Over the last few years, we have seen a drastic increase in mobile phone applications; it seems that centralized services are outstripping decentralized approaches. We will discuss current areas of application for MANETs, as well as the parameters that are barriers to their diffusion (Stieglitz & Fuchß, 2011).

A MANET is an independent scheme that works in mobile nodes without any pre-defined infrastructure communication system. Also, it uses a wireless connection between its devices. These nodes are able to connect wirelessly with each other directly, within the range of transmission. Thus, there will be a multi-hop of intermediate nodes that forward the packets from the source node to the destination node. To ensure communication success, intermediate nodes must co-operate to hand over the packets between the source and destination nodes (Kaushik & Kaushik, 2012).

1.8.1 Advantages of an Ad Hoc Network

1. It allows access to information and services; geographical location does not matter.
2. It works independently from a centralized network (i.e., it is a self-organizing network), the nodes operate as a router, and it is cheaper than a wired network.
3. It is scalable and able to accommodate a lot of nodes.
4. It is flexible.
5. It is robust, as it has no centralized administration.
6. The network can be built up at any time and any place (Mehta, Nupur, & Gupta, 2015).

1.9 Mobile Social Networks (MSNs)

Mobile Social Networks (MSNs) are used to provide a communication environment between people in an attractive way. MSNs take the benefits of wireless networks (the mobile Internet) and OppNets (the ad hoc networks). Thus, MSNs can be located in the mobile device and carried out by the user. So a human interaction with a mobile device, i.e., human–computer interaction (HCI) might be important to MSNs. Based on mobile distributed systems, MSNs are treated as an integral computation of mobile computing and social computing in several aspects (Hu, Chu, Leung, Ngai, Kruchten, & Chan, 2015).

In Mobile Social Networks (MSNs), users co-operate to set up the connection between them as they all have the same needs in the absence of network infrastructure. Since these networks do not need any pre-defined network infrastructure, they can be set up in many critical situations, including battlefields, huge disaster recoveries, and wide-area sensor networks. In such environments, the network connectivity in MSNs is usually highly dynamic in nature due to the physical obstructions, uncertain node mobility, and limited radio range. In such a highly dynamic network, connectivity and uncertainty make data dissemination in MSNs a challenging issue (Fan, Du, Gao, Chen, & Sun, 2010).

Recent technological development has given the mobile devices the ability to create, store, and share data in a decentralized way and in the unusual conditions like high speed, high temperature and different location the communications capabilities are added. As there is content available for the same interests among a group of people, various opportunistic nodes might be available in the form of an OppNet to share the content among these people in their daily lives. The OppNet is embedded into mobile devices to transmit the social network content between mobile devices in an ad hoc manner. The OppNet is used to carry out the contents among mobile device users in the same geographical location. Mobile users have the ability to exchange data through short-range communications like Bluetooth or Wi-Fi Direct through mobile devices. Since opportunistic data sharing is independent of any network infrastructure, it minimizes the data traffic through the Internet. It can also allow data exchange in remote areas or disaster scenarios (e.g., earthquake, tsunami) where infrastructure may be damaged or not available (Hu, Chu, Leung, Ngai, Kruchten, & Chan, 2015).

1.10 Vehicular Ad Hoc Networks (VANETs)

A VANET is a kind of network that came about with the idea of setting up a network of vehicles for the same purpose and situation of use. VANETs are considered one of the most reliable networks that are used to connect between vehicles in an environment such as highways and urban areas. The major goal of a VANET is to establish network communication between a number of vehicles independently from any controller or base station. VANETs have many applications that make it helpful for emergency situations, for example, sharing information to an ambulance to save a human life in the case of an accident. Along with these useful applications, VANET vehicles face many problems such as lack of infrastructure. This gives the vehicles the big responsibility of maintaining communication according to the VANET's requirements (Rehman, Khan, Zia, & Zheng, 2013).

1.10.1 Applications of VANETs

1. *Real-time traffic*: The data can be stored at a Road Side Unit (RSU) so it can be used whenever required.
2. *Co-operative message transfer*: Slow/Stopped vehicles share messages and co-operate with each other.
3. *Co-operative collision warning*: It warns drivers of any dangers ahead, so they can correct their route.
4. *Vision enhancement*: In bad weather, it guides drivers with clear information on vehicles and obstacles.

1.10.2 Characteristics of a VANET

1. *Dynamic topology*: The topology of the VANET environment changes continuously due to the high mobility of vehicles. The connection between two vehicles traveling in opposite directions at an average speed remains for a short time. This connection time is even shorter in a freeway/highway environment, where vehicle speeds are higher.
2. *Predictable mobility patterns*: Pre-defined routes are available in the VANET environment through which a vehicle travels. Thus, the network designers can predict the mobility patterns in the network.
3. *Network topology*: Due to the high mobility and random speeds of nodes, network topology changes frequently.
4. *Unbounded network sizes*: VANETs can vary in size from one city or several cities to one or more countries.

5. *Protection of nodes*: VANET nodes are physically better protected than others and can reduce the effect of infrastructure attack.
6. *Low energy consumption*: VANETs have no issues with energy consumption; they use very little energy.

The connection association between the vehicles in a VANET has visit separations in view of the high development of the hubs and continuous change in geography.

1.11 Delay\Disruption-Tolerant Networking (DTN)

Delay-tolerant networking (DTN) is a technique of setting up the network environment for the purpose of locating the technical issues in diverse networks. These technical issues may cause a lack of permanent network connection. Examples of such networks are those operating in mobile networks or extreme terrestrial environments (Ali, Qadir & Baig, 2010).

Delay-tolerant networks (DTNs) are mobile networks that may never have an end-to-end contemporaneous path. The characteristics of DTN are different from those of traditional ad hoc networks. The routing protocols of DTN follow the store-carry-forward mechanism to minimize the probability of packets dropping, tolerate delays, and improve delivery performance in terms of delivery ratio and delivery latency (Benhamida, Bouabdellah, & Challal, 2017).

1.11.1 Network Topology Classification

According to the availability of network topology information, DTN protocols can be classified into deterministic, stochastic, and coding-based schemes. In the *deterministic class,* the systems assume that information on the network topology is known a priori. When the network behavior is random, the routing protocol is *stochastic*. It depends on decisions regarding where and when the message should be forwarded. The decision varies from forwarding to any contacts within range to using information such as historical data or mobility patterns. The *basic idea of coding-based* approaches is to encode a large number of messages into one message using a linear combination (network coding) or to encode one original message to a large number of coding blocks (erasure coding).

1.11.2 Routing Strategy Classification

This category classifies DTN solutions according to the routing strategy into social-based and opportunistic-based protocols. In social-based protocols, each node selects relay nodes to forward messages using social metrics such as between community, centrality, social similarity, etc. These metrics define the common interest and proximity between nodes. However, opportunistic-based routing protocols select relay nodes based on factors such as mobility patterns, historical information about the node contacting the destination, and the probability of contacting the destination again.

1.11.3 Replication and Semantic Classification

The multicast-based transmission delivers the message to a group of interested receivers while unicast-based transmission delivers the message to its unique destination.

In unicast-based transmission, anycast destination cannot be initialized since it can be any one of the nodes within the membership group (Benhamida, Bouabdellah, & Challal, 2017).

1.12 Freenet

Freenet is a distributed content sharing system, in which users have the ability to insert and retrieve. Known for its peer-to-peer anonymous network, it aims to provide anonymity for both content publishers and retrievers. No node in Freenet has any information about other nodes except its immediate neighbors. Techniques like hop-by-hop transmitting of user messages, and rewriting the (source) address of the messages of every node, are applied in Freenet to provide user anonymity. Freenet provides two operational modes: Darknet and OppNets. In Darknet, only trusted friends are allowed to connect with each other, whereas in OppNet anyone can get connected to Freenet. Those nodes have to query each other for the purpose of storing and retrieving data, which are named after location-independent keys. Every node manages its own local data store and ensures its availability to other nodes in the network for the sake of reading and writing data; as well as a dynamic routing table which contains the addresses of other nodes and also the keys that they are thought to hold. Freenet is a peer-to-peer anonymous content sharing system. Every node has to maintain a part of its hard disk space to contribute the content sharing in the Freenet network. The node in Freenet has the ability to dynamically join or leave the network at any time as it is considered a peer-to-peer system. The Freenet circular range [0,1] is associated with every node within the range, where location [0] and location [1] are the same. The nodes choose their location in the range randomly when they first join the Freenet network. To ease the connection of an arbitrary node to the Freenet network, a group of seed nodes are provided by the Freenet network (Tian, Duan, Baumeister, & Dong, 2017).

References

Akyildiz, I. F., Su, W., Sankarasubramaniam, Y., & Cayirci, E. (2002). A survey on sensor networks. *IEEE Communications Magazine*, 40(8), 102–114.

Ali, S., Qadir, J., & Baig, A. (2010, October). Routing protocols in delay tolerant networks – A survey. In *6th International Conference on Emerging Technologies (ICET)*, Islamabad, Pakistan, 70–75.

Benhamida, F. Z., Bouabdellah, A., & Challal, Y. (2017, April). Using delay tolerant network for the internet of things: Opportunities and challenges. In *8th International Conference on Information and Communication Systems (ICICS)*, Irbid, Jordan, 252–257.

Fan, J., Du, Y., Gao, W., Chen, J., & Sun, Y. (2010, November). Geography-aware active data dissemination in mobile social networks. In *IEEE 7th International Conference on Mobile Adhoc and Sensor Systems (MASS)*, San Francisco, CA, 109–118.

Hu, X., Chu, T. H., Leung, V. C., Ngai, E. C. H., Kruchten, P., & Chan, H. C. (2015). A survey on mobile social networks: Applications, platforms, system architectures, and future research directions. *IEEE Communications Surveys & Tutorials*, 17(3), 1557–1581.

Jadhav, P., & Satao, R. (2016). A survey on opportunistic routing protocols for wireless sensor networks. *Procedia Computer Science*, 79, 603–609.

Kaushik, S., & Kaushik, M. (2012). Analysis of MANET security, architecture and assessment. *International Journal of Electronics and Computer Science Engineering (IJECSE, ISSN: 2277-1956)*, 1(2), 787–793.

Lilien, L., Kamal, Z. H., Bhuse, V., & Gupta, A. (2006). Opportunistic networks: The concept and research challenges in privacy and security. *Proceedings of the WSPWN*, Miami, FL, 134–147.

Martín-Campillo, A., Crowcroft, J., Yoneki, E., & Martí, R. (2013). Evaluating opportunistic networks in disaster scenarios. *Journal of Network and Computer Applications*, 36(2), 870–880.

Mehta, J. S., Nupur, S., & Gupta, S. (2015). An overview of MANET: Concepts, architecture & issues. *International Journal of Research in Management, Science & Technology*, 3(2), 98–101.

Rehman, S., Khan, M. A., Zia, T. A., & Zheng, L. (2013). Vehicular ad hoc networks (VANETs) – An overview and challenges. *Journal of Wireless Networking and Communications*, 3(3), 29–38.

Ritu, M. & Sidhu, M. K. (2014). Routing protocols in infrastructure-less opportunistic networks. 4(6). Available online at: http://ijarcsse.com/Before_August_2017/docs/papers/Volume_4/6_June2014/V4I6-0599.pdf.

Shah, R. C., Roy, S., Jain, S., & Brunette, W. (2003). Data mules: Modeling and analysis of a three-tier architecture for sparse sensor networks. *Ad Hoc Networks*, 1(2–3), 215–233.

Small, T., & Haas, Z. J. (2003, June). The shared wireless infostation model: A new ad hoc networking paradigm (or where there is a whale, there is a way). In *Proceedings of the 4th ACM International Symposium on Mobile Ad Hoc Networking & Computing*, Annapolis, MD, 233–244.

Stieglitz, S., & Fuchß, C. (2011). Challenges of MANET for mobile social networks. *Procedia Computer Science*, 5, 820–825.

Sugihara, R., & Gupta, R. K. (2011). Path planning of data mules in sensor networks. *ACM Transactions on Sensor Networks (TOSN)*, 8(1), 1.

Tian, G., Duan, Z., Baumeister, T., & Dong, Y. (2017). A traceback attack on freenet. *IEEE Transactions on Dependable and Secure Computing*, 14(3), 294–307.

Trifunovic, S., Kouyoumdjieva, S. T., Distl, B., Pajevic, L., Karlsson, G., & Plattner, B. (2017). A decade of research in opportunistic networks: Challenges, relevance, and future directions. *IEEE Communications Magazine*, 55(1), 168–173.

Zhang, P., Sadler, C. M., Lyon, S. A., & Martonosi, M. (2004, November). Hardware design experiences in ZebraNet. In *Proceedings of the 2nd International Conference on Embedded Networked Sensor Systems*, New York, NY, 227–238.

2

Opportunistic Resource Utilization Networks and Related Technologies

Mai A. Alduailij and Leszek T. Lilien

CONTENTS

2.1 Introduction ... 11
2.2 Structure and Operation of Oppnets ... 12
2.3 Design and Implementation of the Oppnet Virtual Machine (OVM) ... 13
2.4 An OVM-Based Healthcare and Wellness Monitoring Application ... 14
2.5 A Review of Oppnet-Related Technologies ... 16
2.5.1 Resource-Sharing Technologies ... 16
2.5.2 Connectivity-Based Technologies ... 18
2.5.3 Specialized and Other Networks ... 20
2.6 Previous Oppnet Work ... 21
Key Terminology & Definitions ... 22
References ... 22

2.1 Introduction

A wide variety of heterogeneous devices and systems ("entities") coexist but are often unable to collaborate. This is particularly wasteful when an underutilized entity is a neighbor of a very busy system but the entity is unable to help the system due to, say, software incompatibility.

We propose a solution to this problem, namely, *opportunistic resource utilization networks* or *oppnets* (Lilien et al., 2006a, 2007, 2010; Kamal et al., 2008). Oppnets rely on "helpers" employed by them to expand in an opportunistic and ad hoc manner, in the process acquiring more resources. Please note that the oppnets are a superset of other kinds of *opportunistic networks* (defined by other researchers), in which opportunism is usually limited to a subset of oppnet capabilities (e.g., to only opportunistic communication or opportunistic message forwarding).

The following characteristics taken together distinguish oppnets from other opportunistic systems: (1) support for the *helper* paradigm—as the basis for ad hoc search for more resources; (2) opportunistic use of *all kinds* of computing resources—not limited to, e.g., communication resources; (3) *ad hoc* operation for most of the oppnet lifetime (as the *expanded* oppnet); (4) *universality* of the helper nodes—regardless of the make or function of a helper's system or device, and regardless of its communication media, protocols, etc.; and (5) *lack of* third-party *mediators*—since interactions among oppnet-enabled entities occur without third parties.

Our goal is ensuring that any oppnet-enabled entity can easily communicate with any oppnet, become its helper, and help the oppnet by providing its resources to the oppnet. We contrast *oppnet-based* entities—which have oppnet capabilities from their deployment—with systems or devices that become *oppnet-enabled* in an ad hoc fashion. It should be noted that oppnet nodes can become helpers for other oppnets.

Another goal is proposing a *standard* middleware to facilitate both building oppnet-based systems and modifying oppnet-oblivious entities into oppnet-enabled ones. In the latter case, interoperation of entities (regardless of their make or means of communication) occurs ad hoc.

In this chapter, we contrast opportunistic resource utilization networks with other opportunistic networks, and provide key terminology and definitions. We discuss the structure and operation of oppnets, including the classification of oppnet helpers. We present the design and implementation of oppnets using the Oppnet Virtual Machine (OVM)—including the motivation for the OVM, its definition, and an overview of the OVM primitives. We describe an OVM-based healthcare and wellness monitoring application, show a use scenario for it, and indicate how OVM primitives are used in the application. Finally, we review oppnet-related technologies and previous oppnet work.

2.2 Structure and Operation of Oppnets

Oppnets are a pervasive computing paradigm and universal middleware, capable of breaking barriers preventing the collaboration of diverse entities. An oppnet starts as a collection of *seed* nodes, which constitute the initial pre-designed (not ad hoc) *seed oppnet*; it can be as small as a single node but must include a *controller*, which consists of a subset of seed nodes. When needed, the controller can decide to expand its oppnet by asking foreign entities to assist the oppnet by joining it. In this way, the initial seed oppnet grows into an *expanded oppnet* (and an expanded oppnet can keep growing into an even bigger expanded oppnet). An entity that accepts an invitation to assist an oppnet makes an obligation to keep on helping for as long as the oppnet needs it and becomes a *helper*. All of a helper's resources are integrated into the oppnet. When a helper completes its tasks, it is released from its obligation by the node that invited it to join the oppnet, and it is free to return to its original duties.

There are two *types of helpers* with regard to (w.r.t.) their *capabilities*: (1) *regular helpers*—which are powerful, able to perform all oppnet activities, including inviting and integrating other helpers; and (2) *lightweight helpers* a.k.a. *lites*—which have limited oppnet capabilities, including restricted oppnet communication capabilities (Lilien et al., 2007, 2010).

There are two *types of helpers* w.r.t. their *refusal rights* (Lilien et al., 2010): (1) *reservist helpers* that can be ordered to assist an oppnet requesting help, and are obliged to comply (they can be subject to appropriate penalties if they refuse to help); and (2) *voluntary helpers* a.k.a. *ad hoc helpers* that must be asked for help, and are free to refuse (except, possibly, in cases when an oppnet requesting help faces a life-or-death situation; this will be elaborated upon in a moment).

Before an oppnet can grow, it must find a set of potential reservists or voluntary helpers available to it. Finding reservists requires looking them up in a *reservist directory*. In contrast, finding voluntary helpers is a true discovery, involving scanning communication spectra for signals or beacons, and collecting enough information to contact their senders.

It is obvious that any candidate for a helper can be asked to join in any situation. It should also be obvious that any candidate can be ordered to join in life-or-death situations. (It is analogous to citizens being required by law to assist with their property—e.g., vehicles—and labor in saving lives or critical resources.)

Using reservists requires further explanation. First of all, we envision different application-related categories of reservists, based on their declared application areas. Reservists sign up for one or more application areas (or just for individual application). For example, an "emergency reservist" is obliged to help (and can be ordered to help) any oppnet deployed for an emergency, while an "entertainment reservist" is obliged to help (and can be ordered to help) any oppnet running an entertainment application.

Once the reservists sign up, they are "trained" for an "active duty." Training the reservists requires providing them with "oppnet-enabling" facilities for assisting oppnets in their discovery and for using the reservists by oppnets. For example, a standard OVM middleware (discussed in the following section) is installed on the reservists. Such "training" makes reservists fully prepared for their oppnet duties.

2.3 Design and Implementation of the Oppnet Virtual Machine (OVM)

We designed oppnets as the middleware that consists of a collection of *OVM primitives* (modules). An oppnet includes OVM primitives from the moment of its deployment. In contrast, an entity that is not an oppnet can become *oppnet-enabled* by downloading and installing a needed and customizable subset of OVM primitives; after the installation, the entity becomes ready to be a *helper* for any oppnet.

Any oppnet is oppnet-enabled (since it includes the OVM). Hence, OVM ensures that oppnets and oppnet-enabled entities can communicate, and oppnets can acquire resources from other oppnet-*based* entities or oppnet-*enabled* entities in an opportunistic and ad hoc manner.

We strived to find the optimal set of OVM primitives in the sense that they are a minimal set of non-overlapping (or, at least, minimally-overlapping) modules. The optimal set of OVM primitives is implemented for four separate node categories: control center nodes, seed nodes, regular helpers, and lightweight nodes. For example, regular helpers can accommodate a large subset (even the full set) of the OVM primitives while lightweight nodes need only a small subset of the OVM primitives. Furthermore, nodes in the same category can select different subsets of OVM primitives, allowing further customization. In this way we avoid "one size fits all" problems, including reduced security levels and waste of resources.

We designed and developed the oppnet OVM middleware in a modular (non-monolithic) fashion, and structured it into *OVM primitives*; each primitive is an atomic element of the oppnet middleware that implements a well-defined activity of an oppnet (Alduailij and Lilien, 2015). Table 2.1 lists, as an example, the set of OVM primitives implemented for only the regular helper nodes.

TABLE 2.1

List of OVM Primitives for the Regular Helper Nodes

No.	OVM Primitive Name	OVM Primitive Function
1	NODE_addNode	Add a node to oppnet
2	NODE_discover	Discover services of a certain device
3	NODE_evalAdmit	Evaluate a device and admit it into oppnet if the device meets criteria for admittance
4	NODE_isMember	Checks if a device is already an oppnet node (oppnet member)
5	NODE_joinOppnet	Join oppnet
6	NODE_listen	Listen to incoming connections, receive and save messages in buffer
7	NODE_processMsg	Process a message from buffer
8	NODE_release	Release a helper when no longer needed
9	NODE_remNode	Remove a node from oppnet
10	NODE_report	Report information to an oppnet device
11	NODE_reqHelp	Request help from candidate helpers
12	NODE_reqRelease	Request the inviting node to be released
13	NODE_runApp	Execute application indicated by authorized oppnet seed or helper node
14	NODE_scan	Scan communication spectrum to detect devices that could become candidate helpers
15	NODE_selectTask	Select a task from the task queue to execute
16	NODE_sendData	Send data (e.g., task list) to another oppnet device
17	NODE_validate	Validate the credentials of the inviting node, and check the ability to help

2.4 An OVM-Based Healthcare and Wellness Monitoring Application

An *oppnet application* is an application implemented with OVM primitives. When it needs additional resources, it can contact (via OVM primitives) potential helpers—that is, oppnet-enabled entities (incl. entities in other oppnets)—for *application-level resource acquisition*. The acquisition of communication resources is the foundation for the acquisition of other resources.

We tested our OVM—the set of OVM primitives—by developing an oppnet application for healthcare and wellness monitoring (Alduailij and Lilien, 2015). We assumed the scenario illustrated in Figure 2.1. When a person loses consciousness outdoors, his Body Sensor Network (BSN) Wristband (shown as CC in Figure 2.1) detects an emergency situation; the signs include: low temperature, low blood pressure, and collapsing. Wristband is both the oppnet controller and the only seed node in this case. Wristband initiates an oppnet expansion by scanning for potential helpers using NODE _ scan (Item 14 in Table 2.1). By using NODE _ discover (Item 2 in Table 2.1), Wristband discovers Tablet T1 (indicated in Figure 2.1 as H1) as a potential helper. If Tablet T1 provides services that allow it to accomplish any of the oppnet tasks, then Wristband uses NODE _ isMember (cf. Table 2.1) to check if Tablet T1 is already a member of its oppnet. If this is not the case, then Wristband uses NODE _ reqHelp (cf. Table 2.1) to send to it a request for help.

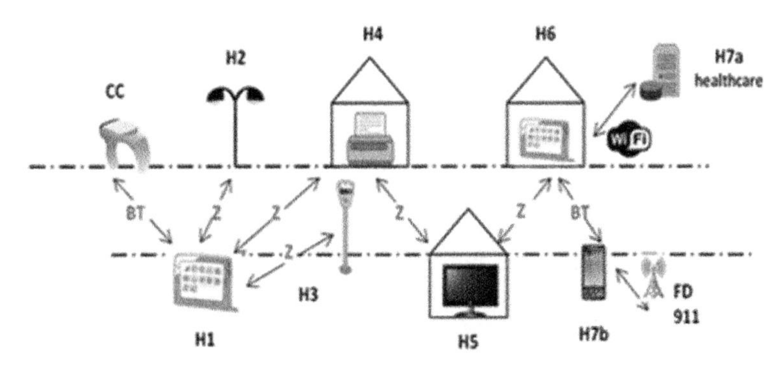

FIGURE 2.1
The healthcare and wellness monitoring scenario.

In the meantime, Tablet T1 is listening for any requests for help using `NODE _ listen`. When it receives the request for help, it uses `NODE _ processMsg` to process the message. Then, it uses `NODE _ validate` to validate the credentials of the inviting node and to find out if it is able to help. The decision to join the oppnet can be the result of two situations: (1) Tablet T1 is idle; or (2) Tablet T1 is busy, but it knows that the tasks that it is asked by the oppnet to do are far more important (e.g., involve a life-or-death situation). Tablet T1 decides to join, uses `NODE _ joinOppnet` to do so; the primitive, among others, sends a join acknowledgment to Wristband (which invited Tablet T1).

Wristband uses `NODE _ evalAdmit` to evaluate Tablet T1 in terms of having adequate capabilities, trustworthiness, etc., and admits Tablet T1 into its oppnet as Helper H1 (since in our scenario the evaluation result is positive). Wristband uses `NODE _ addNode` to add H1 to its list of helpers and then uses `NODE _ sendData` to send to H1 specification of the tasks that it expects H1 to complete.

Upon receiving the messages, Helper H1 checks if it can perform the most urgent task on the task list received from Wristband, which is calling the emergency number 9-1-1. However, H1 (a tablet) is incapable of making the call due to the lack of a cellular phone module. Helper H1, in the possession of a passer-by, uses its AI capabilities and takes the initiative to display on its tablet screen the help-request message in a human-readable form, hoping that its owner will see the message. This is done by running `NODE _ selectTask` and `NODE _ runApp`.

Due to space limitations, from now on we no longer indicate primitives used by oppnet entities.

Since Helper H1 could not perform the most urgent task, it tries to perform at least one of the five less urgent ("unhurried") tasks. H1 is not connected to a Wi-Fi network, but its Bluetooth (BT) and ZigBee networks are on. H1 looks for BT devices but finds none within its reach. So, next H1 searches for ZigBee devices. It finds three ad hoc candidate helpers: Streetlight (H2 in Figure 2.1), Parking Meter (H3), and Printer (H4). Helper H1 then sends requests for help to all three candidate helpers. The candidate helpers validate the credentials of the inviting node and check their ability and their willingness to help. Eventually, all three candidate helpers accept the invitation to join the oppnet. However, after evaluation, Helper H1 chooses only Printer since—in H1's evaluation—Printer has more capabilities than Streetlight and Parking Meter. Printer becomes Helper H4 (it is a "lite" helper, with very limited resources).

Upon receiving its tasks, Helper H4 prints the help message (hoping that a human will see it), and searches for devices on the ZigBee network (the only one it is connected to).

H4 finds a reservist: DigitalTV in a nearby house. DigitalTV becomes Helper H5 (it is also a "lite" helper). It plays the video message (received from Wristband via H1 and H4) to attract TV viewers' attention.

Since the goal of notifying 9-1-1 is not realized yet, Helper H5 searches for devices on the ZigBee network (the only network it is connected to) and discovers Tablet T2 (H6 in Figure 2.1) in a neighbor's house. In due process, Tablet T2 becomes Helper H6.

Helper H6 can use not only ZigBee but also Wi-Fi and BT. H6 is able to upload via Wi-Fi into a healthcare database (H7a) the help message created by the Wristband. Via BT, H6 finds Smartphone (H7b), which becomes Helper H7b. H7b delivers to the 9-1-1 service the SMS message containing the text of the help message created by the Wristband.

Since the overall goal (notifying 9-1-1) is now achieved, H7b sends a release request to its inviting node, Helper H6, indicating completion of its task. Helper H6 (Tablet T2) removes Helper H7b from the list of its helpers, and then sends a release message to it. The process of releasing helpers is repeated until Wristband (CC) receives the last release request from Helper H1 (Tablet T1). Wristband releases H1, and then terminates the oppnet-based application.

Please note that in a luckier scenario, the goal (notifying 9-1-1) could have been achieved earlier, maybe even by Helper H1 (if it had, e.g., cellphone capabilities to contact 9-1-1).

2.5 A Review of Oppnet-Related Technologies

The oppnet-related technologies discussed in this section are categorized as resource-sharing technologies, connectivity-based technologies, and specialized and other networks.

2.5.1 Resource-Sharing Technologies

This subsection discusses the differences between resource-sharing technologies and the oppnet middleware in a way that indicates the novelty of the OVM-based oppnet middleware.

Peer-to-Peer (P2P) Systems: P2P systems allow for the sharing of content—such as files—among devices or networks that interact via an "appropriate communication and information channels" (Subramanian and Goodman, 2005). Such resource-sharing and application-level communication is achieved by direct sharing among "peers" without depending on centralized servers (Kamal, 2008).

FlashLinq (Scalable, 2015) is a P2P platform developed to complement traditional cellular-based services. The technology advances a concept known as *proximal communication*, which enables devices to discover each other and communicate at a high speed without intermediary infrastructure. FlashLinq can create a "neighborhood-area network," where fixed and mobile peer applications can interact directly.

Although the OVM-based oppnet middleware shares many characteristics with P2P systems, such as the decentralized sharing of resources, our approach is distinguished by the following key features: (1) it allows the collaboration of devices in order to achieve a common goal rather than meeting individual node requirements; (2) it is designed to allow the acquiring of all kinds of resources—such as communication, sensing, storage, data, etc.—rather than just communication or data; and (3) it distinguishes four types of nodes (i.e., CC, seed, regular helpers, and lites) that have different capabilities and duties.

Grid Computing: This term refers to the hardware and software infrastructure that enables the aggregation of distributed computing resources to achieve a common goal (Abbas, 2004). It can use a set of open standards and protocols to gain access to applications and data through application-level communication while sharing processing power, storage capacity, and a vast array of other computing resources over the Internet (Siddiqui and Fahringer, 2002).

Globus Toolkit (GT) (Foster, 2005) is a service-oriented distributed computing toolkit encompassing applications and infrastructures. The main components addressed by GT are: security, resource access, resource management, data movement, and resource discovery. These components enable a broader "Globus ecosystem" of tools and components that build on or interoperate with core GT functionality to provide a wide range of useful application-level functions. These tools can be used to develop a wide range of grid infrastructures supporting distributed applications.

Although the OVM-based oppnet middleware shares many characteristics with some grid computing systems (such as the ability to grow by joining, and the use of "foreign" resources), our approach is distinguished from grid computing systems by the following key features: (1) it is designed to deal with a wide range of application areas while grid computing systems are concentrated on computationally-intensive operations; (2) it relies on helpers joining a system and remaining a part of it until released while grids allow sites to join and leave as they choose (Kamal, 2008); and (3) its entities collaborate in highly heterogeneous environments while grid entities work in a more homogenous environment.

Spontaneous Networks (SNs): SNs are a subset of ad hoc networks created when a group of people meet for a collaborative activity using heterogeneous wireless mobile devices (Feeney et al., 2001). They are limited in space and time by the space and time of the meeting. They consist of diverse devices connected by a variety of wireless technologies (including Wi-Fi, infrared, BT).

Rewadkar and Karve (2014) propose an SN based on a wireless ad hoc network that enables devices to communicate with each other without the availability of a central server (as in a client-server communication) and without a prior configuration (as in P2P networks). In this SN, laptops are placed close to each other at a particular place for a short period of time to access available services and resources within the network without an Internet connection. *Energy Efficient Self Configure Secure Protocol (EESCSP)* is proposed to create and manage the SN, which is self-configured to integrate services and resources into the network.

Many features of SNs are shared by the OVM-based oppnet middleware, for example, heterogeneity of devices, collaboration among devices without a central server, and the use of diverse communication technologies. However, our approach has three key features that are not found in SNs: (1) opportunistic growth of oppnets not limited by space or time as severely as in SNs; (2) the administrative capabilities and role of the seed/CC nodes; and (iii) categorization of nodes into four different node categories.

Internet of Things (IoT): IoT aims at sharing resources via heterogeneous devices by minimizing the communication gap between heterogeneous devices (Hersent et al., 2012; Ning et al., 2013; Ukil et al., 2012). To reach this goal, some of the proposed solutions require that the *Original Equipment Manufacturers (OEMs)* agree to produce new products capable of interacting with devices of other OEMs, regardless of their make or model. The idea is that collectively OEMs can agree on an open, universal development framework/standard and ecosystem.

Other IoT initiatives focus their efforts on developing IoT applications that allow devices to interact, discover, sense, and actuate with minimal human interaction. These applications communicate through different technologies (such as Wi-Fi, Ethernet, or IEEE 802.15.4 low-rate wireless personal area networks); they can form a mesh or utilize the Internet to scale up.

AllJoyn (2014) is an open-source project (initiated by Qualcomm), which aims to create a platform and services that enable products and applications to interoperate regardless of their make or operating system. AllJoyn aims to connect devices like televisions, home appliances, home entertainment devices, and automobiles through Wi-Fi to create smart homes or vehicles.

In spite of all the current IoT research efforts (AllJoyn, 2014; IoT Toolkit, 2014; Ning et al., 2013), there is no single IoT platform that incorporates all the functionality proposed by the OVM-based oppnet middleware. The major shortcomings of the current IoT solutions (well visible in the oppnet context) include the following: (1) most, if not all, IoT systems rely on devices that are pre-configured in a deterministic manner; (2) some IoT platforms do not fully utilize the ability of some devices to discover other devices, are unable to admit them into their network, are not capable of interacting with them in an ad hoc manner, and cannot help each other directly—without the mediation of a directory, a control center, or another third party; (3) most of the IoT solutions are designed for home automation (however, with some tweaking many can lend their capabilities to other applications); (4) some IoT platforms are limited to using only a certain communication protocol, such as Wi-Fi; and (5) some IoT solutions have a built-in support only for manufacturer-specific protocols, and to extend their functionality to devices from other manufacturers users have to write new device drivers.

2.5.2 Connectivity-Based Technologies

This subsection identifies the key features distinguishing our oppnet approach from each of the connectivity-based technologies.

Mobile Ad Hoc Networks (MANETs): MANETs are decentralized networks comprised of a collection of autonomous mobile nodes communicating over wireless communication channels. A MANET topology can change rapidly over time as nodes can join and leave the network at any time (Singh et al., 2012). Since the network is decentralized, a MANET's nodes are responsible for the network's activities such as discovering the network's topology, self-configuration, and routing.

Beddernet (Gohs et al., 2011): is a platform-agnostic MANET framework. The Beddernet architecture is designed to work with different networking protocols. Beddernet was tested to work with BT ad hoc networks or scatternets. Beddernet middleware has been tested on Java and Android devices.

A MANET shares the self-configuration feature with the oppnet middleware as both are decentralized. However, the nodes of the oppnet middleware are capable of realizing more challenging tasks than self-configuration and routing, since helper nodes are responsible for sharing their resources to assist oppnets in achieving their goal. Another distinguishing factor is the commitment of oppnet middleware nodes to keep assisting until the oppnet mission is accomplished, while in MANETs nodes can join and leave a network without any restrictions.

Mesh networks: These networks have a decentralized topology, in which each node is connected to at least two other nodes (TeckTarget, 2015). Such infrastructure eliminates failure in the network due to a *single point of failure (SPoF)*, since if a node fails other nodes can still communicate with each other. Mesh network systems can be wired or wireless, a full mesh or a partial mesh.

Vision Mesh (Zhang and Cai, 2010) is a video sensor mesh network platform used for water conservancy engineering. Vision Mesh is composed of a massive number of image or video sensor nodes from which multi-view image or video is acquired. The OpenCV machine vision library is integrated into Vision Mesh to enable video and image processing.

Although the topology of mesh networks provides reliability in communication between nodes, the expenses of constructing such a topology can be too high. Moreover, the interactions among devices in mesh network systems are pre-configured in a deterministic manner. On the other hand, nodes invited by the oppnet middleware need not be configured a priori, and there is no determinism in the manner in which the joining helper nodes are called to service (Lilien et al., 2007).

Opportunistic Networks: *Opportunistic networks* (e.g., Pelusi et al., 2006; Wang et al., 2005) are, in our view, a proper subset of opportunistic resource utilization networks (oppnets). They can be viewed as a generalization of the MANET paradigm, in which the assumption of having complete paths between data senders and receivers is relaxed. This enables stations to communicate in disconnected environments, in which islands of connectivity appear, disappear, and reconfigure dynamically (Lilien et al., 2007).

Seto et al. (2010) present a mobile platform for body sensor networking. It allows for local processing of data and uses opportunistic sensing strategies, in which the capabilities of onboard sensors and smartphones may be collected and fused with body sensor data.

The OVM-based oppnet middleware (cf. Lilien et al., 2006a, 2010) is significantly different from systems using opportunistic networks (as defined by others, cf. Pelusi et al., 2006; Wang et al., 2005). In the latter, opportunism is limited to communications when devices are within each other's range. In the "other" opportunistic network systems (e.g., Pelusi et al., 2006) there is no notion of utilizing resources of "foreign" nodes in a network to perform a task of the network. In contrast, oppnets enable opportunistic use of all kinds of resources, services, or capabilities (incl. hardware, software, human skills, etc.) that happen to be within the oppnet's reach (Lilien et al., 2006a) through any communication technologies (regardless of the device make or function). Another distinguishing feature is that an oppnet starts as a relatively small network, known as the seed network, which can keep growing into a larger and larger expanded oppnet (Lilien et al., 2010).

There have been some attempts to incorporate more resources—other than communication capabilities—into opportunistic networks (e.g., Seto et al., 2010; Lu et al., 2013; Bleda et al. 2014). However, the utilized resources are limited to a few kinds of resources (such as sensing or computation), and through a limited set of types of communication technologies.

Delay-Tolerant Networks (DTNs): DTNs can be viewed as the superclass of wireless network systems. The main goal of DTNs is to provide connectivity between local/ regional networks when communication is prone to discontinuation and interruptions. By adapting store-and-forward message switching, a DTN can overcome problems causing large delays (Seligman et al., 2007).

Husni and Wibow (2012) build a DTN-based email server capable of sending and receiving emails even if the server is not continuously connected to a network. The proposed email server exploits DTN's ability to overcome problems associated with extreme environments, intermittent connectivity, large or variable delays, and high bit error rates (which are common characteristics for communications with remote areas). The authors propose using facilities and infrastructures such as the public transportation system for building a DTN-based network. A train system was chosen to be the infrastructure for DTN routers using Wi-Fi. A train passing through a village (even not stopping there) can have a mobile Wi-Fi station, which brings to the village new email from the outside world and also collects local email awaiting sending to the outside world.

The main goal for the DTNs is to provide connectivity in the most challenging networking circumstances. On the other hand, the OVM-based oppnet middleware can include not only DTN capabilities but also all their own unique capabilities, starting with gaining resources from "foreign" entities via helpers.

Self-Organizing Networks (SONs): SON automation technology provides cost-efficient deployment, operation, and maintenance of mobile networks (Jorguseski et al., 2014). SON was among the requirements within the *3rd Generation Partnership Project (3GPP) Long Term Evolution (LTE)* standardization (UTRA, 2015). SON functions can be categorized into: (1) self-configuration; (2) self-optimization; and (3) self-healing.

Tonguz and Viriyasitavat (2013) propose a biologically-inspired self-organizing network system whereby certain vehicles serve as *Road Side Units (RSUs)*. Vehicles that act as temporary RSUs act as a communication bridge for other networks by making occasional brief stops. The authors believe that using vehicles as RSUs could improve message reachability and network connectivity between vehicles while avoiding the cost of deploying stationary RSUs.

Although SONs share with the oppnet middleware the ability to form, organize, and manage nodes in the absence of a core center (Prehofer and Bettstetter, 2005), SONs are not focused on the concept of acquiring resources via helpers, which is the core feature of oppnets.

2.5.3 Specialized and Other Networks

In this section, we discuss the differences between the oppnet middleware and some specialized networks, namely *Wireless Sensor Networks (WSNs)*—a monitoring and control technology—and *Ambient Networks (ANs)*—a 3G cellular technology.

Wireless Sensor Networks (WSNs): WSNs are specialized systems that consist of autonomous, resource-constrained devices embedding sensors, processors, and transceivers to monitor and control a physical environment (Akyildiz and Vuran, 2010). WSN nodes communicate over wireless communication technologies, usually 2.4 GHz radios based on the IEEE 802.15.4 standard.

Capella et al. (2014) develop and deploy a WSN for continuous in-line monitoring of nitrate concentration in a river in Eastern Spain. The authors also implement policies to improve the features of the whole system. The improvements include optimizing the times at which measurements are to be carried out, and sampling frequency being altered according to the system evolution, the user preferences, and the application features.

The major differences between the OVM-based oppnet middleware and WSNs are: (1) the oppnet middleware allows integration of heterogeneous devices that can have powerful computing capabilities (are not just resource-constrained nodes); and (2) the oppnet middleware allows nodes to communicate over diverse communication channels (not just over a single frequency channel, as is typical for WSNs (Lilien et al., 2007)).

We must note that whole WSNs or any subset of WSN nodes can be admitted into an oppnet as helpers.

Ambient Networks (ANs): ANs are a networking paradigm for use beyond 3G mobile systems. ANs were developed as a part of the European Union Sixth Framework Programme (European Commission, 2015). They aim to provide existing and new services over any access technology and any network type. To attain such services, ANs enable on-demand and transparent cooperation between heterogeneous networks, with little or no pre-configuration or off-line agreement (Belqasmi et al., 2008).

Vodel et al. (2010) present a communication concept that aims to create a lightweight radio standard that combines the advantages of different communication technologies (such as cognitive radios and ANs) while overcoming the design limitations of a single radio standard.

ANs resemble the OVM-based oppnet middleware in their ability to integrate heterogeneous devices to provide various services. However, there are two major differences between oppnets and ANs: (1) ANs are completely predesigned and all their facilities are built-in, whereas the operations of oppnets are mostly ad hoc without prior configuration; and (2) ANs are global networks intended to replace the Internet, while the OVM-based oppnets are intended to build local/wide area networks to serve a wide range of applications (Lilien et al., 2007).

2.6 Previous Oppnet Work

Lilien et al. (2007) and Kamal et al. (2008) propose the *OVM* for oppnets and define its basic primitives. The original OVM primitives were evaluated for completeness and consistency only via an intellectual analysis, without any actual simulation, emulation, or implementation (Lilien et al., 2010).

We critically evaluated the original OVM primitives, and decided to make the following modifications:

1. Some of the original primitives were eliminated or modified to avoid redundancies.
2. A number of new primitives were added to provide more modular and non-overlapping OVM functionality.
3. The whole set of OVM primitives was redesigned using object-oriented principles (such as modularity, polymorphism, and inheritance).

Kamal et al. (Lilien et al., 2007; Kamal et al., 2008) present OVM as the API framework for oppnets, describing the design and implementation of the oppnet testbed named MicroOppnet v.2.2. MicroOppnet v.2.2 was developed as a proof of concept for oppnets, not as an OVM validation platform. Hence, despite using the OVM concept, it has a monolithic structure, with the code not using OVMs (the code instead of being divided into separate OVM primitives is written without such separation). Moreover, MicroOppnet v.2.2 is rudimentary in its resource utilization opportunism as sensing is the only resource utilized.

In contrast, in our work we simulate, emulate, and implement different components of the OVM-based oppnet middleware. Our code is not monolithic but modular, divided into the set of OVM primitives developed by us.

Key Terminology & Definitions

Controller or a control center (CC): It consists of an arbitrary subset of seed nodes that controls and manages its oppnet (but its tasks can be delegated to other nodes, if desirable). Among others, CC can decide to start an oppnet's expansion (growth).

Helper: This is an entity that can assist an oppnet in its activities by accepting an invitation to help the oppnet. Before it joins an oppnet, a future helper is a *candidate helper.* A candidate helper can possess diverse capabilities and "skills" potentially useful for the oppnet (such as specialized software, or communication, computing, storing, sensing, or actuating capabilities). A helper selected from among (a potentially broad set of) candidate helpers and integrated into an oppnet becomes an *actual helper.*

Oppnet-based entities: These are entities (e.g., systems or applications) that include OVM primitives *by design,* and are capable of growing their oppnet by integrating helpers; oppnet growth starts with oppnet-based entities searching for potential helpers.

Oppnet-enabled entities: These are entities (e.g., systems or applications) that—after downloading and installing a required subset of OVM primitives *in an ad hoc manner*—are capable of becoming oppnet helpers, providing resources to oppnet-based entities.

Oppnet Virtual Machine (OVM): This is a collection of primitives (modules or building blocks) for application-level communication and resource acquisition. A subset of OVM primitives can be used by any oppnet-based entity or oppnet-enabled entity, as its middleware supports all oppnet activities.

Opportunistic networks: As defined by other researchers, these are a proper subset of oppnets, because their opportunism is usually limited to a subset of oppnet capabilities, such as only opportunistic communication capabilities, or opportunistic message forwarding.

Opportunistic resource utilization networks or oppnets: These are a new paradigm for specialized ad hoc networks. The salient feature of oppnets is their use of "helpers" to expand in an opportunistic and ad hoc manner. The helpers can be either predefined or ad hoc, and assist in the acquisition of various resources.

Seed oppnet and expanded oppnet: An oppnet starts as a *seed oppnet* a.k.a. a *seed,* which is a pre-designed (not ad hoc) collection of nodes that initiate oppnet activity. A *seed* can be as small as a single node. A seed oppnet grows into an *expanded oppnet* by inviting, admitting, and integrating so-called helpers. In turn, an expanded oppnet can keep growing into a larger and larger expanded oppnet as long as it is beneficial (can get more resources to do its job faster or better).

References

A. Abbas, *Grid Computing: A Practical Guide to Technology and Applications,* Charles River Media, Hingham, MA, 2004.

I. Akyildiz and M. Vuran, *Wireless Sensor Networks,* Wiley, Hoboken, NJ, 2010.

M. Alduailij and L. Lilien, "A collaborative healthcare application based on opportunistic resource utilization networks with OVM primitives," *Intl. Conf. on Collaboration Technologies and Systems (CTS),* Atlanta, GA, June 2015, pp. 426–433.

"AllJoyn", online, last accessed on Mar. 25, 2014, available at: https://www.alljoyn.org/z.
F. Belqasmi, R. Glitho, and R. Dssouli, "Ambient network composition," *IEEE Network*, IEEE Press, Piscataway, NJ, vol. 22(4), Jul.–Aug. 2008, pp. 6–12.
A. Bleda, R. Maestre, A. Jara, and A. Skarmeta, "Ambient assisted living tools for a sustainable aging society," *Modeling and Optimization in Science and Technologies*, Springer, NY, 2014, pp. 193–220.
J. Capella, A. Bonastre, R. Ors, and M. Peris, "A step forward in the in-line river monitoring of nitrate by means of a wireless sensor network," *Sensors and Actuators B: Chemical*, vol. 195, May 2014, pp. 396–403.
European Commission, "The sixth framework programme in brief. Dec. '02 Edition," online, last accessed on Apr. 10, 2015, available at: https://ec.europa.eu/research/fp6/pdf/fp6-in-brief_en.pdf.
L. Feeney, B. Ahlgren, and A. Westerlund, "Spontaneous networking: An application oriented approach to ad hoc networking," *IEEE Communications*, IEEE Press, Piscataway, NJ, vol. 39(6), Jun. 2001, pp. 176–181.
I. Foster, "Globus toolkit version 4: Software for service-oriented systems," *Network and Parallel Computing*, Springer, Berlin – Heidelberg, Germany, 2005, pp. 2–13.
R. Gohs, S. Gunnarsson, and A. Glenstrup, "Beddernet: Application-level platform-agnostic MANETs," *Distributed Applications and Interoperable Systems*, Springer, Berlin – Heidelberg, Germany, 2011, vol. 6723, pp. 165–178.
O. Hersent, D. Boswarthick, and O. Elloumi, *The Internet of Things: Key Applications and Protocols*, Wiley, 2012.
E. Husni and A. Wibowo, "Delay tolerant network based e-mail system using trains," *Asian Internet Eng. Conf. (AINTEC)*, Bangkok, Thailand, Nov. 2012, pp. 17–22.
"IoT Toolkit," online, last accessed on Mar. 25, 2014, available at: http://iot-toolkit.com/.
L. Jorguseski, A. Pais, F. Gunnarsson, A. Centonza, and C. Willcock, "Self-organizing networks in 3GPP: standardization and future trends," *IEEE Communications*, IEEE Press, Piscataway, NJ, vol. 52(12), Dec. 2014, pp.28–34.
Z. Kamal, "A proof of concept for oppnets and its resource utilization techniques with QoS constraints," Ph.D. Dissertation, Western Michigan University, Kalamazoo, MI, 2008.
Z. Kamal, L. Lilien, A. Gupta, Z. Yang, and M. Batsa, "New UMA paradigm: Class 2 opportunistic networks," *Unlicensed Mobile Access Technology: Protocols, Architectures, Security, Standards and Applications*, ed. by Y. Zhang et al., Auerbach Publications, Taylor and Francis, Boca Raton, FL, 2008, pp. 349–392.
L. Lilien, A. Gupta, Z. Kamal, and Z. Yang, "Opportunistic resource utilization networks: A new paradigm for specialized ad hoc networks," *Computers & Electrical Engineering*, Elsevier B.V., Amsterdam, The Netherlands, vol. 36(2), Mar. 2010, pp. 328–340.
L. Lilien, A. Gupta, and Z. Yang, "Opportunistic networks for emergency applications and their standard implementation framework," *IEEE Intl. Performance, Computing, and Communications Conf. (IPCCC)*, New Orleans, LA, Apr. 2007, pp. 588–593.
L. Lilien, Z. Kamal, V. Bhuse, and A. Gupta, "Opportunistic networks: The concept and research challenges in privacy and security," *Intl. Workshop on Research Challenges in Security and Privacy for Mobile and Wireless Networks (WSPWN)*, Miami, FL, Mar. 2006a, pp. 131–147.
L. Lilien, Z. Kamal, and A. Gupta, "Opportunistic networks: Challenges in specializing the P2P paradigm," *17th Intl. Conf. on Database and Expert Systems Applications (DEXA)*, Kraków, Poland, Sept. 2006b, pp. 722–726.
R. Lu, X. Lin, and X. Shen, "SPOC: A secure and privacy-preserving opportunistic computing framework for mobile-healthcare emergency," *IEEE Transactions on Parallel and Distributed Systems*, IEEE Press, Piscataway, NJ, vol. 24(3), Mar. 2013, pp. 614–624.
H. Ning, H. Liu, and L. Yang. "Cyberentity security in the Internet of Things," *IEEE Computer*, IEEE Press, Piscataway, NJ, vol. 46(4), Apr. 2013, pp. 46–53.
L. Pelusi, A. Passarella, and M. Conti, "Opportunistic networking: Data forwarding in disconnected mobile ad hoc networks," *IEEE Communications*, IEEE Press, Piscataway, NJ, vol. 44(11), Nov. 2006, pp. 134–141.

C. Prehofer and C. Bettstetter, "Self-organization in communication networks: Principles and design paradigms," *IEEE Communications*, IEEE Press, Piscataway, NJ, vol. 43(7), July. 2005, pp. 78–85.

D. Rewadkar and S. Karve, "Energy efficient self-configured secure protocol (EESCSP) for wireless spontaneous ad-hoc network," *Intl. Conf. on Control, Instrumentation, Communication and Computational Technologies (ICCICCT)*, Kanyakumari, India, Jul. 2014, pp. 792–799.

"Scalable peer-to-peer platform advances proximal communications," *Product News Network*, online, last accessed on Oct. 15, 2015, available at: http://bi.galegroup.com.libproxy.library.wmich.edu/global/article/GALE|A249044616/61a5dfb7daaf6d5fd4dec664284503f1?u=lom_wmichu.

M. Seligman, K. Fall, and P. Mundur, "Storage routing for DTN congestion control," *Wireless Communications and Mobile Computing*, vol. 7(10), Dec. 2007, pp. 1183–1196.

E. Seto et al., "Opportunistic strategies for lightweight signal processing for body sensor networks," *3rd Intl. Conf. on Pervasive Technologies Related to Assistive Environments*, Samos, Greece, June. 2010, pp. 56–61.

M. Siddiqui and T. Fahringer, *IBM Solutions Grid for Business Partners - Helping IBM Business Partners to Grid-Enable Applications for the Next Phase of e-Business on Demand*, IBM Corporation, Austin, TX, 2002.

G. Singh, N. Kumar, and A. Verma, "Ant colony algorithms in MANETs: A review," *Journal of Network and Computer Applications*, Academic Press, London, UK, vol. 35(6), Nov. 2012, pp. 1964–1972.

R. Subramanian and B. Goodman, *Peer-to-Peer Computing: The Evolution of a Disruptive Technology*, Idea Group Publ., Hershey, PA, 2005.

TeckTarget, "Mesh network topology (mesh network) definition," online, last accessed on Jul. 5, 2015, available at: http://searchnetworking.techtarget.com/definition/mesh-network.

O. Tonguz and W. Viriyasitavat, "Cars as roadside units: A self-organizing network solution," *IEEE Communications*, IEEE Press, Piscataway, NJ, vol. 51(12), Dec. 2013, pp. 112–120.

A. Ukil, S. Bandyopadhyay, J. Joseph, V. Banahatti, and S. Lodha, "Negotiation-based privacy preservation scheme in Internet of Things platform," *First Intl. Conf. on Security of Internet of Things (SecurIT)*, Kollam, India, Aug. 2012, pp. 75–84.

UTRA, "Evolved Universal Terrestrial Radio Access (E-UTRA) and Evolved Universal Terrestrial Radio Access Network (EUTRAN); overall description; stage 2, 3GPP standard TS 36.300, v. 12.1.0," online, last accessed on Aug. 3, 2015, available at: http://www.3gpp.org/dynareport/36300.htm.

M. Vodel, M. Caspar, and W. Hardt, "A capable, lightweight communication concept by combining Ambient Network approaches with Cognitive Radio aspects," *17th IEEE Intl. Conf. on Telecommunications (ICT)*, Apr. 2010, Doha, Qatar, pp. 430–434.

Y. Wang, S. Jain, M. Martonosi, and K. Fall, "Erasure-coding based routing for opportunistic networks," *ACM Conf. of the SIG on Data Communication (SIGCOMM)*, Philadelphia, PA, Aug. 2005, pp. 229–236.

M. Zhang and W. Cai, "Vision mesh: A novel video sensor networks platform for water conservancy engineering," *3rd IEEE Intl. Conf. on Computer Science and Information Technology (ICCSIT)*, Jul. 2010, Chengdu, China, pp. 106–109.

3

Buffer Management in Delay-Tolerant Networks

Sweta Jain

CONTENTS

3.1 Introduction 25
3.2 Issues Related to DTN Routing 27
3.3 Buffer Management in DTNs 28
 3.3.1 Local Buffer Management Policies 29
 3.3.2 Fuzzy-Based Buffer Management Techniques 43
 3.3.3 Buffer Management in Social-Based Routing Algorithms 43
 3.3.4 Global Buffer Management Policies 44
 3.3.5 Traffic Differentiation Schemes for Delay-Tolerant Networks 49
3.4 Summary and Discussions 52
3.5 Scope for Future Work 53
References 54

3.1 Introduction

Delay-tolerant networks (DTNs) are an emerging class of wireless networks where the next node in the path toward the destination is not known in advance and is chosen dynamically as and when contact opportunities occur. Hence, they are also popularly known as opportunistic networks. The sparse network density and long delays between node meetings generally lead to intermittent node connectivity. Even in such challenged network environments, where it is difficult to establish end-to-end paths, they can perform data transmission (Marano & Socievole, 2012).

Conventional ad hoc routing protocols may not be applicable in such extreme environments due to their requirement of the presence of a continuous end-to-end path between communicating nodes for the entire period of communication, which is not the case in DTNs due to frequent link disruptions. Hence, a new routing mechanism called a store-carry-forward strategy (Cerf Burleigh, Hooke, Torgerson, Durst, & Scott, 2007) is employed by these networks for data communication between the nodes. As the next hop node may not be immediately available, the node buffers the messages and carries them with itself until it meets another node. Upon the occurrence of a new contact opportunity, it forwards the message and repeats this process of message forwarding until the message eventually reaches its destination. As the delays between node meetings in DTNs are unpredictable and may be very long, the DTN nodes must have sufficient storage capacity for storing messages in their buffers for long periods of time. Thus, the problem of buffer management has aroused special interest in the field of DTNs.

Two major issues to achieve data delivery in the challenging network environments of DTNs include (Kim & Shin, 2011): a routing strategy to determine whether to forward a message on node encounter or not; and a buffer management policy which comprises of a scheduling decision regarding the sequence of messages to be forwarded when a contact opportunity arises and a dropping decision which selects the message(s) to be dropped when the node buffer overflows (Jain, Chawla, Soares, & Rodrigues, 2014).

Routing protocols in DTNs tend to create and spread multiple copies of a message to increase the delivery probability, although at the cost of network resources. Long-term storage combined with message replication may cause buffer overflowing and result in frequent message drops (Krifa, 2008). At the same time, as the contact opportunities are of short duration it restricts the number of messages that may be exchanged successfully. Hence, buffer scheduling and dropping policies help to make the best use of limited network resources, which results in improved performance of DTN routing protocols (Soares, Farahmand, & Rodrigues, 2010; Tang, Chai, & Weng, 2012).

The work in Sun & Cao (2013) identifies some of the major differences between mobile ad hoc networks (MANETs) and DTNs. In MANETs, the routing protocol tries to establish a contemporaneous end-to-end path between the source (A) and destination (C) nodes. This path is symmetric in nature, i.e. the path from A to C is similar to the path from C to A as shown in Figure 3.1.

Routing in DTNs, however, make use of contact opportunities between nodes, occurring due to node mobility, visible in Figure 3.2. The routing behavior in DTNs is asymmetric, i.e. the path from A to C may be different from the path from C to A due to intermittent node connectivity. The delivery delay in DTNs may be larger than that in MANETs due to the asymmetrical routing behavior. Moreover, the transmission reliability may be low in DTNs due to limited encounter duration and lack of contemporaneous end-to-end connectivity. Achieving successful data delivery in challenged DTN environments is extremely difficult. With the lack of network topology information and limited availability of node

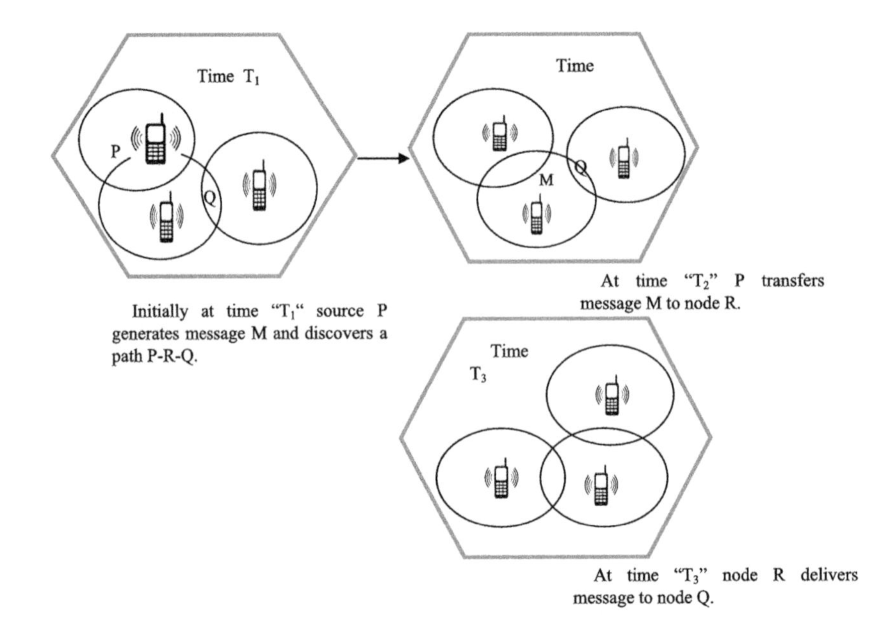

FIGURE 3.1
Routing in MANETs.

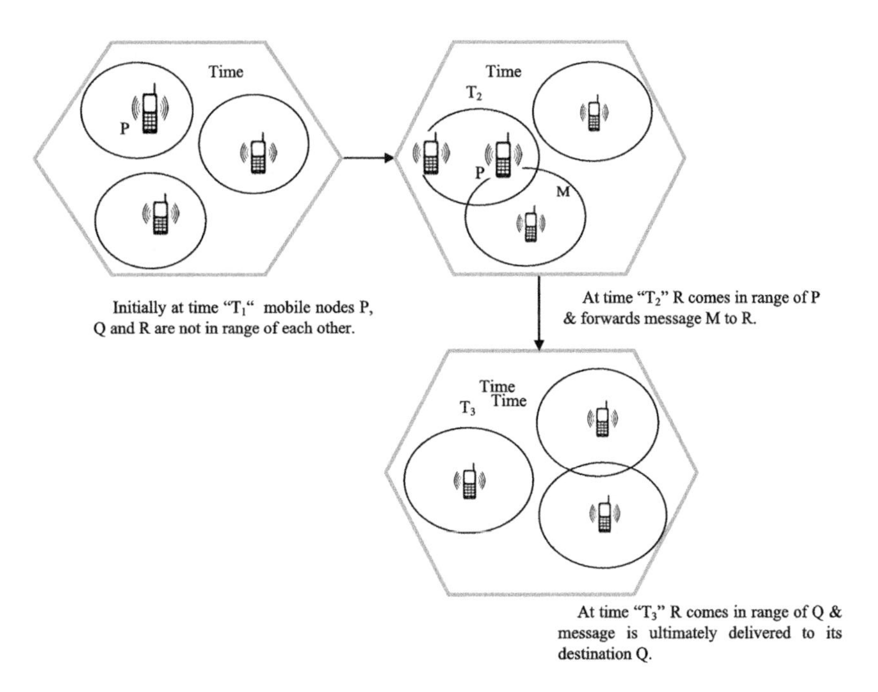

FIGURE 3.2
Store-carry-forward routing in DTNs.

and network resources, the problem in DTNs is further elevated than in mobile ad hoc networks. Some of the major challenges to routing in DTNs can be summarized as node mobility and limited resources such as bandwidth, buffer space, and energy of its nodes (Sun & Cao, 2013).

3.2 Issues Related to DTN Routing

Four main issues related to DTN routing are (Gupta, 2013):

1. *Forwarding Policy*: The first and foremost issue in DTN routing is the forwarding policy, in which a node decides whether to forward a stored message in its buffer to the encountered node or not. It may happen that the source node or the intermediate node carrying the message may not meet the destination for some time, but if it can predict the next meeting time with the destination then this may help it in taking better routing decisions. If the current node meets a node that has a higher probability of meeting the destination in the near future, then the node may forward the message to that particular node. However, it is difficult to determine the meeting probability between nodes if the network topology is dynamic.
2. *Replication Policy*: The second important issue is the replication policy, i.e. how many times a message is replicated in the network. Multiple replicas of a message are propagated over alternative network paths so as to increase its delivery probability (Spyropoulos, Turletti, & Obraczka, 2009). The main concern is to which nodes to replicate the message, and how to distribute the copies of a message to

different encountered nodes and so on. Creating too many copies of a message may cause buffer overflow at intermediate nodes and subsequently drop of messages.

3. *Dropping Policy*: The third issue is related to the buffer dropping policy. DTNs use the store-carry-and-forward routing paradigm, whereby nodes even after forwarding messages to other nodes carry those messages in their buffer for long periods of time for future dissemination. Long-term storage along with message replication strategy may quickly fill up the limited buffer of nodes resulting in message drops. Deciding which message to drop when the buffer overflows is a critical decision.
4. *Scheduling Policy*: The fourth issue is concerned with the buffer scheduling policy. As the period of contacts between nodes is very short and network bandwidth is also limited, only a limited amount of data may be exchanged during node meetings (Makhlouta, Harkous, Hutayt, & Artail, 2011). Estimating the contact capacity and determining the number of messages that can be exchanged in a given contact opportunity may help in better utilization of already limited bandwidth. Determining the order in which messages may be sent when a contact opportunity occurs is an important issue.

3.3 Buffer Management in DTNs

Buffer management is an important issue in DTNs due to the use of message buffering and replication strategy. Without buffer management, implementation of most of the routing schemes proposed for DTNs may not be effective due to frequent message drops or limited contact opportunities. It involves two major decisions that must be taken carefully: (a) which message(s) to drop in case of buffer overflow and the arrival of new messages, and (b) which message(s) to schedule for transmission when a contact opportunity arises. The goal of an efficient buffer management policy can be summarized as follows.

- The primary and necessary goal must be to improve the delivery ratio of the network.
- However, its secondary goal is to minimize the resource consumption; hence, the buffer management policy must also aim to reduce the overhead, i.e. minimize the number of relays to which a message is forwarded. Although it is desirable to minimize delay, maximizing the delivery rate, even at the cost of higher delay, is acceptable.

Buffer management policies can be classified on the basis of the type and the amount of information about the messages they use (Jain, Chawla, Soares, & Rodrigues, 2014). They can be mainly classified as: (a) local policies which make decisions on the basis of local information about messages available at nodes, such as their arrival time, age, time-to-live (TTL), size, hop count, forward count of message, etc.; and (b) global policies which try to find out optimal solutions on the basis of complete network-wide information about messages. Section 3.3.1 presents a state-of-the-art survey of the local-information-based buffer dropping and scheduling policies, and Section 3.3.4 discusses the global buffer management policies.

3.3.1 Local Buffer Management Policies

a. The earliest work on buffer management was proposed by Davis, Fagg, & Levine (2001), where four buffer dropping policies were evaluated:
   - *Drop-Random*: This policy drops randomly chosen messages from the buffer. This is a blind strategy which does not take into account the message status of messages in the buffer.
   - *Drop-Least-Recently-Received*: This policy removes the message that arrived first among all the messages that reside in a node's buffer. The reason being that such a message must have seen a lot of forwarding opportunities during the past encounters. Therefore, it can be safely dropped as a large number of replicas of such messages have been created in the network and hence their probability of delivery is higher as compared to other messages in the buffer.
   - *Drop-Oldest*: This policy drops the message that has been in the network for the longest time. The same logic applies to this strategy as in the case of drop-least-recently-received.
   - *Drop-Least-Encountered*: This policy drops the message whose destination the current node has least delivery probability. This dropping policy is based on the meeting probability of two nodes, which is estimated on the basis of their past encounters. For each time step, node A updates the meeting time for another node C, with respect to co-located node B, using the following rules as mentioned in Equation 3.1(a–c):

$$M_{t+1}(A,C)=\begin{cases}\lambda M_t(A,C) & \text{if none co-located}\\ \lambda M_t(A,C) & \text{if } C=B\\ \lambda M_t(A,C)+\alpha M_t(B,C) & \text{for all } C\neq B\end{cases} \quad (3.1a\text{–}c)$$

   where:

   $M_t(A,C)$ is the meeting value at time t

   $\alpha=0.1$ is a parameter that decides how much portion of B's meeting value is to be added with A

   $\lambda=0.95$ is the decay rate of the meeting value

   The value $M_t(A,C)$ is initially zero for all node pairs.

   When two nodes A and B encounter each other, then they sort the messages stored in their buffers according to the *relative ability* of two nodes to deliver the message to the destination, i.e. according to $M_t(A,C)-M_t(B,C)$, where C is the destination of the message. With this, it ensures that messages are always forwarded to nodes that have a higher chance of meeting the destination.

   Simulation results showed that drop-least-encountered performs better than drop-oldest as it takes into account the meeting probability of nodes while taking message dropping decisions and the message that has the least probability of reaching its destination is dropped first.

b. Although Davis, Fagg, & Levine, (2001) used meeting probability only for taking dropping decisions at nodes, MaxProp (Burgess, Gallagher, Jensen, & Levine, 2006) used an estimate of meeting likelihood for taking routing decisions as well.

It was basically designed for vehicular networks. In MaxProp, each node estimates the meeting likelihood for all other nodes in the network. Initially, at the start of the network, all nodes set this likelihood metric to $1/(n+1)$ where n is the total number of nodes in the network. On each successive encounter, the meeting likelihood metric is incremented and all values are renormalized using the incremental average. The nodes exchange their meeting likelihood metric with other nodes they encounter, and this helps in determining the optimal routes in the network. MaxProp uses two parameters, hop counts and path likelihood to nodes, to decide the transmission and dropping priority of messages stored in a node's buffer. The path likelihood is calculated and updated using the historical data of encounters between nodes, as discussed previously.

If only the delivery likelihood metric is used, as in drop-least-encountered (Davis, Fagg, & Levine, 2001), the messages with a high chance of reaching their destination will get forwarding opportunities at nodes, while those with a low likelihood of meeting their destination would never get a chance to spread. However, MaxProp avoids such a situation by giving more preference to newer messages in the network by giving them more forwarding opportunities. To do this, it divides the node buffer into two parts, one containing packets that have a hop count less than a predefined threshold t (hops which are sorted according to hop count), and the other containing packets that have a hop count more than t (hops which are sorted according to the delivery likelihood metric). In case of buffer overflow, MaxProp drops packets on the basis of hop count; hence, those that have been propagated in the network for a long time are dropped first. MaxProp also deletes those messages whose acknowledgements are received; this frees buffer space for new messages in the network.

The main limitation of using hop count as a parameter for message prioritization is that it only reflects the number of times a message has been forwarded on a particular path. Hence, the hop count of a message does not give a proper estimate of the distribution of the message among other nodes, which might have a copy of the message but do not lie on this forwarding path.

c. Probabilistic Routing Protocol using History of Encounters and Transitivity (PRoPHET) (Lindgren, Doria, & Schelen, 2003) is a probabilistic routing protocol which works under the assumption that nodes in a network do not move randomly but follow certain meeting patterns. The protocol utilizes the history of node meetings and the concept of transitivity to estimate the delivery probability between two nodes.

Each node initializes the delivery predictability metric $P(A, B)$ for every other node B in the network to a constant value P_{init}. When two nodes A and B meet each other, they update their delivery predictability metric as:

$$P(A,B)_{new} = P(A,B)_{old} + \left(1 - P(A,B)_{old}\right) * P_{init} \tag{3.2a}$$

If the nodes do not meet for a long time (say K time units) this delivery predictability metric is reduced by some aging factor γ.

$$P(A,B)_{new} = P(A,B)_{old} * \gamma^{K} \tag{3.2b}$$

The concept of transitivity is also used to update the delivery predictability metric, i.e. if node A meets node B frequently and node B meets another node C frequently then node B could be a good relay for node A to transfer its messages to node C.

$$P(A,C)_{\text{new}} = \max\left(P(A,C)_{\text{old}}, P(B,C), *P(A,B)_{\text{new}} * \beta\right) \tag{3.2c}$$

where β is a constant.

Thus, the nodes maintain the delivery predictability metric for each node they encounter and use it for taking forwarding decisions on encounter opportunities.

The performance of the PRoPHET routing protocol has been assessed using different combinations of queuing policies and scheduling strategies in Phanse & Lindgren (2006). The following queuing policies have been proposed and evaluated by Phanse & Lindgren:

- *Most Forwarded (MOFO)*: This dropping policy takes into account the number of times a message has been replicated by a node; it drops the message that has been forwarded the most as this message has already been spread in the network and has a high probability of reaching its destination.
- *Shortest Lifetime (SHLI)*: This dropping policy drops the message whose remaining lifetime, i.e. remaining time-to-live is shortest. This is based on the idea that a message with a low remaining TTL (RTTL) does not have enough time to reach the destination and will expire soon and therefore it can be easily dropped. This policy will prevent the messages that are about to expire from reaching their destination, even if they are near their destination.
- *Most Probable (MOPR)*: This dropping policy is based on the delivery predictability metric; it uses a new count called Forwarding Predictability (FP), which is initialized to zero; for each message replication, FP is incremented by the delivery predictability metric P of the receiving node for the destination of the message. This policy is similar to MOFO but it drops the message with the largest value of FP.

$$\text{FP} = \text{FP}_{\text{old}} + P \tag{3.3}$$

- *First in First Out (FIFO)*: This policy drops the message that is at the front of the queue, i.e. which entered the node queue first.
- *Least Probable (LEPR)*: This policy drops the message whose destination the current carrying node has lowest delivery predictability. However, if the source has a small delivery probability for the destination of a message, this message will never get forwarded and will not be spread in the network, further reducing its chance of reaching the destination.

The comparative analysis of these strategies showed that MOFO displays the best performance whereas LEPR gives the worst performance, as LEPR always drops messages that have the least delivery probability. It may happen that if a source has a lower delivery probability for some destination then the messages destined to these nodes may get dropped at the source itself without being spread in the network, resulting in an overall reduced delivery ratio. Therefore, a comprehensive buffer dropping policy is needed which integrates routing metric with message parameters to take appropriate buffer management decisions, as some resources must have already been spent on such messages which are being dropped.

A number of scheduling strategies whose purpose is to determine the order of messages to be scheduled on node encounters have also been presented in Phanse and Lindgren, (2006).These strategies prioritize messages stored in a node's buffer in terms of the delivery predictability metric of the PRoPHET routing protocol as defined in Lindgren, Doria, & Schelen (2003). All the scheduling policies give priority to messages for which the encountered node has a higher delivery predictability than the node currently carrying it.

- *GRTR*: This scheduling strategy is similar to the forwarding criterion used in basic PRoPHET. The current carrier of the message, node A, forwards a message M having destination D to encountered node B if and only if $P(B, D) > P(A, D)$.
- *GRTRSort*: This scheduling strategy schedules the messages in descending order of difference between the delivery predictability metric of two nodes, i.e. $P(B, D) - P(A, D)$, and forwards the message only if $P(B, D) > P(A, D)$.
- *GRTRMax*: This scheduling strategy is similar to GRTRSort but it first forwards those messages for which node B has highest delivery predictability.
- *COIN*: This strategy is not used with PRoPHET but in conjunction with the Epidemic routing protocol. Rather than simply forwarding the message as in Epidemic, this method performs a COIN test to decide whether the message is to be forwarded or not. For instance, a message is forwarded only if $X > 0.5$, where $X \varepsilon U(0, 1)$.

The simulation results confirmed that GRTRSort in conjunction with the MOFO queuing policy gives the best performance.

d. Less Probable Sprayed (LPS) and Least Recently Forwarded (LRF) are two newly developed hybrid buffer management policies (Naves, Moraes, & Albuquerque, 2012). These policies have been evaluated over two traditional DTN routing protocols: Epidemic and PRoPHET. LPS is an enhancement of LEPR; like LEPR, LPS also selects the message with lowest delivery probability for dropping in case of congestion but only if its predefined minimum number of replicas have already been spread in the network. To estimate the number of replicas of a message it simply uses a counter field appended to the message header. For each message forwarding, this counter field is incremented by one. This helps to overcome the premature dropping of a message that has a lower delivery probability which is faced in LEPR (Phanse & Lindgren, 2006). As LPS requires contact probability metric between two nodes for taking dropping and scheduling decisions it can be used only with DTN routing protocols which use such a metric.

On the other hand, LRF selects the least recently forwarded message for dropping on buffer overflow, based on the reasoning that such a message has already travelled several hops in the past and now newer messages should get a forwarding opportunity. LRF can be used with any DTN routing protocol as it only requires the last forwarding time of a message by a node. Through simulation results, authors show the superior performance of LPS and LRF over other buffer dropping policies such as FIFO, MOFO, and LEPR in terms of higher delivery ratio and reduced overhead, with LRF providing the best performance among all compared schemes.

e. Zhang et al. studied the effect of three traditional buffer management strategies on the Epidemic routing protocol (Zhang, Neglia, Kurose, & Towsley, 2007), namely: drop-tail, drop-head, and source-prioritized drop-head. In drop-tail, a node always drops the message at the end of the buffer. In drop-head, a node drops the message at the front of the buffer. Source-prioritized drop-head is similar to drop-head except that it gives priority to messages generated by nodes themselves as source. In case of buffer overflow, a node first drops those messages for which it acts as a relay, then if necessary drops source messages.

   The simulation results demonstrate that source-prioritized drop-head gives the best performance among the three dropping policies and that it is beneficial to give priority to source packets.

f. Prioritized Epidemic (PREP) (Ramanathan, Hansen, & Basu, 2007) is a modified version of Epidemic routing which consists of both the routing module and the bundle prioritization module. The routing module is responsible for maintaining the topology information via link state advertisements and determining the optimal paths between nodes. The bundle prioritization scheme determines the dropping and schedule order of the messages stored in a node's buffer.

   Each node maintains a low watermark and high watermark for buffer occupancy. As a node's buffer occupancy crosses the high water mark, the node starts dropping the bundles, and when the buffer occupancy reaches the low water mark the node stops dropping the bundles. This mechanism prevents the continuous calling of the dropping procedure. Bundles that have a hop count less than the hop count threshold are selected first for dropping and those that have the largest shortest path cost to the destination are dropped in priority. If the buffer level is greater than the low water mark and there are no such bundles then the bundles are selected in a random fashion and deleted. This preserves more copies of bundles near their destination and copies farther from the destination are deleted in case of resource crisis.

   Upon occurrence of a transmission opportunity, a node again divides its buffer into parts, downstream and upstream bins. If an encountered node has a smaller path cost to the destination of the bundle than the node currently carrying the message, then that bundle is placed in the downstream bin; otherwise, it is placed in the upstream bin. Each of these bins is then sorted using Radix sort on the basis of the remaining Time-to-live property and the age of the bundles. Bundle transmission starts from downstream bin followed by the upstream bin.

g. N-drop policy (Li, Zhao, Liu, & Liu, 2009) takes dropping decisions on the basis of the number of times a message has been forwarded. When a condition of buffer overflow arises, then those messages which have been forwarded N or greater than N number of times by the node are dropped. If none of the messages have a forward count greater than threshold value N, then messages are dropped according to FIFO policy. The value of N is chosen according to the buffer state of the node. As the buffer size of nodes increases, the threshold value N increases, i.e., more replicas of a message may be created and distributed in the network.

h. Vasco et al. have also evaluated and presented a number of buffer management policies (Soares, Farahmand, & Rodrigues, 2010). They have conducted this study for an application of DTNs called vehicular delay-tolerant networks. The dropping and the scheduling order of messages have been considered on the basis of a number of factors: the position of the message in the queue, its remaining time-to-live, and its number of message copies. Based on these factors, several dropping and scheduling policies have been compared, namely: FIFO, Random, Ascending Remaining Lifetime, Descending Remaining Lifetime, Ascending Replicated Copies, and Descending Replicated Copies. The scheduling policies analyzed by the authors are as follows:

- In FIFO scheduling, the messages are ordered on the basis of the time of arrival in their buffer, i.e. on first come first served basis.
- In Random scheduling, the messages stored in the buffer are selected randomly for transmission upon the occurrence of a contact opportunity.
- In Remaining-Lifetime-based (RL-based) scheduling, the scheduling policy orders messages stored in the node buffer on the basis of the TTL of the messages. The TTL of a message denotes the amount of time a message is valid in the network; after the expiration of the TTL, the message is discarded from the network. Two scheduling policies have been evaluated: in the first case, messages are ordered in descending order of TTL, and in the second case, messages are ordered in ascending order of TTL.
- In Replicated-Copies-based (RC-based) scheduling, the messages are ordered according to the number of times a message has been replicated in the network since its inception. Again, two variants have been evaluated: in one scheduling policy, the messages are ordered in decreasing order of the replication count of a message, and in the other policy, messages are ordered in increasing order of the replication count of messages.

The dropping policies are also based on the same criterion.

- In the Drop-Head dropping policy, the message at the head of the buffer is dropped, i.e. the message that came first in a node's buffer is discarded.
- In the Random dropping policy, on buffer overflow, a randomly selected message from the node's buffer is dropped.
- In the Remaining Lifetime dropping policy, the messages are dropped on the basis of remaining TTL of messages. Two variations of this policy are considered; in the first, the variant message with the smallest remaining TTL is given highest priority for dropping (RL Ascending, denoted RL ASC), and in the second, the variant message with the highest remaining TTL is given highest priority for dropping (RL Descending, denoted RL DESC).
- In Replicated-Copies-based dropping policy, the dropping priority of messages is decided on the basis of the replication count of messages. Again, two variations of this policy have been evaluated; either the message with the smallest replication count is dropped first (RL Ascending Order) or the messages with the highest replication count is dropped first (RL Descending Order) on buffer overflow.

From the simulation results, authors observed that an RL-DESC-based dropping and scheduling policy, which gives higher priority to messages that have higher RTTL for scheduling results, on average reduced latency compared to all other policies. Whereas RL ASC, which gives higher priority to messages that have lower RTTL, trying to deliver them before expiring leads to an increased average delay of messages. Although the RC ASC policy requires slightly more time to deliver bundles than the RL DESC policy, it achieves a higher delivery ratio as it provides more forwarding opportunity to messages that have not been spread well in the network. FIFO scheduling only gives priority to messages at the head of a buffer, which limits the transmission opportunities of other messages in a node's buffer and hence registers a lower delivery ratio and a large average delay compared to all other policies. Hence, the authors concluded that the combination of a scheduling policy based on ascending replicated copies and a dropping policy based on descending replicated copies displayed the best performance in comparison to other combinations. This study highlighted the various factors that affect dropping and scheduling decisions in delay-tolerant networks, but, as discussed previously, a better approach, which considers multiple message parameters, is required.

i. A study on a variety of dropping policies for a DTN project called "Haggle" has been presented (Bjurefors, Gunningberg, Rohner, & Tavakoli, 2011). Haggle is a data-centric architecture where mobile nodes opportunistically connect to each other using any available communication technology such as Wi-Fi, Bluetooth, etc. Each node in Haggle architecture is assumed to have a buffer management system which stores a collection of data objects along with the metadata, such as user interest, associated with them. The dropping policies proposed in this work are mainly based on user interests and replication count. They are listed as follows:
   - *Least Interested (LI)*: This policy drops the data object in which the least number of neighbors of a node are interested. This scheme reduces content diversity but it also results in increased delivery ratio.
   - *Most Interested (MI)*: This policy drops the data object in which most neighbors of a node are interested. This scheme adversely affects the delivery ratio as it removes those data objects in which most nodes are interested.
   - *Max Copies (MAX)*: This policy drops that particular data object which has been replicated a predefined number of times (max).
   - *Most Forwarded (MF)*: This policy drops that data object from the node buffer for which the node has created the highest number of replicas.
   - *Least Forwarded (LF)*: This policy drops that data object from the node buffer for which the node has created the lowest number of replicas.
   - *Random*: This policy drops any randomly chosen data object from the node buffer.

   Two extreme cases have also been compared:
   - *Infinite Buffer*: This is the best case, where each node is assumed to have an infinite buffer and can, therefore, accommodate an infinite number of data objects without dropping any data object.
   - *No Buffer*: This is the worst case, where each node stores data objects of its own interest and not of others' interests.

   The authors concluded that MF strategy outperforms all other strategies in terms of all major network performance parameters, i.e. delivery ratio, delay, and overhead.

j. Rohner, Bjurefors, Gunningberg, McNamara, & Nordstrom (2012) state that not all data items are equally important for all nodes in the network and this realization can make a dissemination system do more justice to the network resources. A host can determine data items that are relevant to it or not by matching its own interests with the item's metadata and then, based on matching scores, it can then decide which data items to accept and store. To make the best utilization of short contact opportunities, the authors suggest the concept of relevance-driven data dissemination. Relevance scores are used to decide what items are to be selected for transfer and in which order to forward them. It is assumed that each data item has a set of attributes D associated with it, known as metadata, and each node also contains a set of interests I corresponding to the data attributes. Interests are forwarded in the network either by direct contact or through spreading them via third parties. Data items are said to be related to a node if $R = D \cap I$ is non-empty. A score value is used to measure the strength of node's interest in a data item.

$$\text{Score} = \frac{\sum_{a \in R} \omega_a}{\sum_{a \in I} \omega_a} \tag{3.4}$$

where ω_a are respective weights of data items in D and I. This score is then used by a dissemination system to decide the order of data items to be transferred. Using this concept, five data-centric ordering strategies have been proposed and analyzed:

Score Local: The data items are ordered according to the relevance scores of data items computed with respect to the receiver of data. The most relevant data item is exchanged first and hence this strategy results in the transfer of data items that are relevant in a particular locality.

Score Global: The aggregated relevance scores, with respect to all the nodes known at a particular time, are used for data ordering and hence it results in the exchange of data items which are globally relevant.

Random Local: Only data items that are locally relevant are exchanged but in random order.

Random: This strategy randomly selects data items for transfer irrespective of their relevance score.

LIFO Local: Last in first out strategy is used for exchanging only locally relevant data items.

Score local, Random local, and LIFO local are "local strategies", as they select items for transfer based only upon the receiver's interests. Score global and Random, however, attempt to take advantage of the complete system and therefore use opportunistic contact to transfer data that is relevant not only to its recipient but to other nodes in the network as well.

k. A number of buffer dropping policies based on message size have been proposed by S. Rashid et al. in their works (Ayub & Rashid, 2010), (Rashid & Ayub, 2010), (Rashid, Ayub, Soperi, Zahid, & Abdullah, 2011), (Rashid, Abdullah, Soperi, Zahid, & Ayub, 2012), and (Rashid, Ayub, Soperi, Zahid, & Abdullah, 2013).

Drop Largest (DLA) (Ayub & Rashid, 2010) suggests dropping the message of largest size from the buffer in case of buffer overflows until space is created for the incoming message. This policy will create the largest amount of space for incoming messages and relatively the number of message drops may be less. This dropping policy is also used in Rashid, Ayub, Soperi, Zahid, & Abdullah (2013).

Message Drop Control Source Relay (MDC-SR) (Rashid, Ayub, Soperi, Zahid, & Abdullah, 2013) as the name suggests, tries to control the number of message drops by placing an upper bound threshold on the number of messages that a node can buffer; if this threshold is crossed then a new incoming message is discarded except in the case where the message is destined to this particular node. This prevents unnecessary message drops at the receiving node and also saves bandwidth. In a case where a message is destined to the node itself and it does not satisfy the threshold condition, DLA policy is used.

In Threshold Drop (Rashid & Ayub, 2010), an upper bound and a lower bound is defined on the message size and only the messages which lie on this threshold range are selected for dropping

In Equal Drop (Rashid, Ayub, Soperi, Zahid, & Abdullah, 2011), if a message arrives at a node and the buffer is full then the message that is the same size as incoming message is selected for dropping.

In Mean Drop (Rashid, Abdullah, Soperi, Zahid, & Ayub, 2012), on buffer overflow, only those message(s) whose size is greater than or equal to the mean of the size of messages currently residing at the node are selected for dropping.

The dropping decisions in these policies are guided by only a single message parameter, i.e. the size of the message, while other important message parameters such as the number of replicas, its remaining TTL have been totally ignored. However, it may happen that the message(s) selected for dropping may be relatively newer in the network or may have a high chance of reaching its destination in next few hops.

l. The work in Tang, Chai, & Weng (2012) utilizes the average contact frequency (ACF) between nodes for taking buffer scheduling decisions. It is based on the observation that if two nodes have met in the recent past then they have a possibility of meeting again in the near future. The ACF *fij* is defined as the number of encounters between node i and node j during a predefined length of time T; hence, ACF is calculated as:

$$\mathrm{ACF} = \frac{N_{ij}}{T} \tag{3.5}$$

Here, N_{ij} is the number of contacts between i and j during T time.

When two nodes A and B meet each other, then for each message i in *A's* buffer that has the destination D that is not there in *B's* buffer, A will compare the respective ACF value $\mathrm{ACF}_{AD}(i)$ and $\mathrm{ACF}_{BD}(i)$ in the ACF table of node A and node B and then forward all those messages for which B has a higher ACF to the destination of message. The messages will be scheduled in descending order of $\mathrm{ACF}_{BD}(i)$. On buffer overflow, the message with the largest number of replicas is dropped first.

m. In the Enhanced Buffer Management Policy (EBMP) (Kim & Shin, 2011), two message-utility-based buffer dropping policies have been proposed. The first utility function, denoted as $EBMP^{delivery}$, combines the age of a message (AGE) with its estimated total number of replicas (ETR) and is defined in Equation 3.6.

$$EBMP_i^{delivery} = \frac{1}{ETR_i} + \frac{1}{\log(AGE_i)} \tag{3.6}$$

The age of a message is calculated as the difference between current time and the time of the creation of the message. The reason for giving higher dropping priority to older messages is due to the fact that they have spent a long time in the network and a large number of copies of these messages have already been created in the network, thereby increasing their chances of delivery to the destination. This also gives more chances to relatively new messages that have seen fewer transmission opportunities.

The second utility function, denoted by $EBMP_{delay}$, combines the estimated replica count (ETR_i) and remaining TTL ($RTTL_i$) of the message is defined in Equation 3.7:

$$EBMP_i^{delay} = \frac{1}{ETR_i} + \log(RTTL_i) \tag{3.7}$$

As the message properties age and the remaining TTL are counted in seconds and their value is in several hundreds and thousands, log function is applied over them to limit their impact on the utility function and to give proper weightage to the replica count property of the message.

In both of the utility functions, replication count is used. In the first utility function, which additionally considers $1/\log(age_i)$, the ETR of a message plays a more dominating role than in the second utility function, which adds $\log(RTTL_i)$ to $1/ETR_i$. Hence, the first utility function aims to increase the delivery ratio of the network by dropping messages that have a large value of estimated replica count, while the second utility function aims to reduce the average delivery delay in the network by dropping messages that have a lower value of RTTL.

The work also presents a new mechanism to estimate the replication count of the messages in the network. For this, each node in the network maintains two variables for each message it holds in its buffer: Estimated Replica (ER) count and My Forward (MF) count. ER is used for storing the estimated number of replicas of a message and MF is used to store the count of the number of new replicas of a message created by a node itself. Whenever two nodes meet each other, they update the ER value of messages stored in their buffer by exchanging the my_forward values of the corresponding messages.

Let ER_i^A denote the estimated total number of replicas of message i as estimated by node A and MF_i^A denote the forward count of message i which means the number of message replicas created by node A itself. The source node A of message i initializes ER_i^A to 1 and MF_i^A to 0. Two cases may arise:

In the first case, the node A meets another node B which does not carry message i and forwards it message i. In this case, node A increments both *my forward* count MF_i^A, and estimated replica count ER_i^A for message i, by one. The node B

then derives its estimated replica count ER_i^B for message i from ER_i^A and sets its *my forward* count MF_i^B as zero.

$$ER_i^A = ER_i^A + 1 \qquad ER_i^B = ER_i^B \tag{3.8a}$$

$$MF_i^A = MF_i^A + 1 \qquad MF_i^B = 0 \tag{3.8b}$$

In the second case, two nodes A and B having a common message i in their buffers meet each other. In this case, both the nodes exchange their MF values for message i and update their ER values respectively.

$$ER_i^A = ER_i^A + MF_i^B \tag{3.9a}$$

$$ER_i^B = ER_i^B + MF_i^A \tag{3.9b}$$

But this method of replica count estimation may result in false updates in situations where two nodes meet frequently (the probability of which would be higher as nodes are in close proximity to each other), but have not created any new copies of the common messages they are carrying since their last meeting. Hence, there is a high probability that the ER value is falsely incremented due to exchange of MF values. The authors, through simulation analysis, show that local buffer management policies based on local message properties can also give comparable performance to global buffer management policies, which use global message information.

n. Y. Liu et al. have proposed a buffer dropping and scheduling scheme called the Message Transmission Status Based Buffer Management Scheme (MTSBS) in (Prodhan, Das, Kabir, & Shoja, 2011). The messages are prioritized on the basis of their replication count and dissemination speed. The dissemination speed of a message denotes how fast the message is distributed in the network and is calculated as the ratio of the number of hops that a message has travelled to its RTTL. Firstly, message replica count is used to prioritize messages and in case of a tie, i.e. if two messages have the same replica count, then dissemination speed is used as the tiebreaker.

   Their buffer management policy is inspired by a famous theory of finance called the *law of diminishing marginal utility*, which states that "as a user increases consumption of a product, there is a decline in the marginal utility that user derives from consuming each additional unit of that product" (Prodhan, Das, Kabir, & Shoja, 2011). Hence, it is not advantageous to create additional copies of a single message that has already been spread in the network, as it will restrict the buffer space available for other less spread messages in case of resource crisis. Hence, their buffer management policy suggests giving priority to messages that have a low replica count and a slow speed of dissemination a chance for transmission in the network.

o. The work in Rashid, Ayub, & Abdullah (2015) presents a protocol named TTL-based routing (TBR), which implements buffer scheduling and dropping in the Spray and Wait routing protocol. This protocol orders the messages to be forwarded in a contact opportunity according to a priority function P_{fk} as defined in Equation 3.10(a) and the messages to be dropped when the node buffer overflows according to the priority function P_{dk} as defined in Equation 3.10(b). Here, s_k denotes the size of message k, TTL_k denotes time-to-live, h_k denotes hop count, and L_k denotes the number of copies of the message. The disadvantages of using hop count as an estimate of replica count of the message have already been discussed.

$$P_{fk} = \frac{1}{h_k * TTL_k * s_k} \tag{3.10a}$$

$$P_{dk} = \frac{L_k}{s_k} \tag{3.10b}$$

p. Rashid et al. propose a weight-based buffer management policy (called WBD) (Silva, Nunes, Miniy, & Loureiro, 2017), where messages are assigned weights based on their size (MS), remaining time-to-live (RTTL), their stay-time in the node buffer (MSTQ), its hop count value (HC) and number of replicas (RC). These five message parameters are combined using the formula mentioned in Equation 3.11

$$\text{Weight} = \frac{1}{\text{MS}_i} + \frac{1}{\text{RTTL}_i} + \frac{1}{\text{MSTQ}_i} + \text{HC}_i + \text{RC}_i\text{L} \tag{3.11}$$

After weight computation, the messages stored in a node buffer are divided into two parts based on a threshold value: a High Weight Message List (HWML) and a Low Weight Message List (LWML). In case of buffer overflow, only messages from HWML are selected for dropping in descending order of their weight value, and in case of message transmission opportunity messages from LWML are scheduled for transmission in ascending order of their weight values. It also uses an acknowledgment mechanism to delete messages that have already been delivered to their destinations from other nodes' buffers and update this list of delivered messages on node encounters.

q. ST-Drop (Chen, Yao, Zong, & Wang, 2017) is another local buffer management policy which relies only on locally available information to estimate the time and space coverage of messages in the network. It is based on the idea that a message with a greater time and space coverage is more likely to have been delivered, so it can be dropped first. Each message has associated with it two variables: a space coefficient S_c to measure the space coverage of the message in the network, and a time coefficient T_c to measure the time coverage of the message in the network. These two parameters are then combined using a simple function to compute the space-time coverage ST_c of a message.

$$ST_c = S_c * T_c \tag{3.12}$$

The space coverage of a message is measured in a way similar to the MOFO scheme (list items p,q,r,s) by counting the number of times a message was forwarded by a node. To estimate the time coverage of a message it combines the concept of FIFO and SHLI. The reason for combining these two is that using FIFO alone may result in a recently arrived message being dropped while retaining a message that has short TTL left (which may be automatically dropped after some time due to TTL expiration); while using SHLI alone, a message may reside in the node's buffer unnecessarily for a long time due to its high TTL value even if its delivery probability is low. Thus T_c is computed as the ratio of the message stay-time in the node buffer to its TTL, thereby penalizing messages that have short TTL. Thus the space-time coverage of a message is computed using only its local information. On buffer overflow, the ST-Drop drops those messages that have the highest value of ST_c. It does not drop messages with zero ST_c value as these are newer messages in the network and should get a chance to get forwarded in the network. ST-Drop has been evaluated in conjunction with three routing protocols, namely Epidemic, PRoPHET, and Bubble Rap. The simulation results show that ST-Drop works best with the BubbleRap algorithm in comparison to the other two routing protocols and helps in improving its delivery ratio.

r. In Pan, Ruan, Zhou, Liu, & Song, (2013), the authors propose a new data transmission probability metric which is based on the average contact frequency and average contact duration between two nodes. This transmission probability is used to take both forwarding and dropping decisions at a node. In addition to the data transmission probability, the remaining TTL and the size of message are also taken into account to decide the dropping priority of a message stored in a node's buffer.

s. Most of the works on buffer management focus either on dropping policy or scheduling policy, overlooking the routing technique. A comprehensive-integrated buffer management (CIM) technique is proposed (Mathurapoj, Pornavalai, & Chakraborty, 2009), which combines information about the state of messages and node delivery history to determine the scheduling and dropping order of messages and delete redundant messages from node buffers. CIM has 4 modules:

Queuing Policy: When two nodes meet each other, they share a state information packet, where they exchange the list of messages held by them, their details, time stamp, location information, and list of frequently contacting nodes. If a message's destination is in the frequently contacted list of the encountered node, then that message is added preferentially in the queue. The messages in the queue are then sorted according to a priority level P_i assigned to them:

$$P_i = \frac{1}{H_{\text{relay}} + N_{\text{forward}}} \tag{3.13}$$

where:

H_{relay} is the number of hops travelled by message i

N_{forward} is the number of times the message has been forwarded by the current carrier

The messages which have travelled long in the network and have been forwarded a large number of times are assigned a lower priority level.

Buffer Replacement Policy: On buffer overflow, a node needs to drop some of its stored messages to accommodate new incoming messages. For this purpose, each node j calculates the utility value for each message i stored in its buffer using a weighted combination of its relative TTL, size, and forward count.

$$W_{ij} = \alpha \frac{\text{TTL}_{\text{min}}}{\text{TTL}_{\text{o}}} + \beta \frac{BS_j}{S_{m_i}} + \gamma \frac{\text{count}_{\text{avg}}}{\text{count}} \tag{3.14}$$

where:

α, β, and γ	are the weight factors for the three message parameters
TTL_{min}	is the remaining TTL of the message
TTL_{o}	denotes the initial TTL value of the message
BS_j	denotes the buffer space of node j
Sm_i	denotes the size of the message i
count_{av}	represents the average number of times a message is forwarded by node j

count denotes the number of times message i has been forwarded by node j

The overall buffer utility of node j is then calculated as:

$$U_j = \frac{\left(\sum_{i}^{M_j} w_{ij}\right)}{M_j} \tag{3.15}$$

where M_j is the total number of messages carried in the node buffer.

If a congested node j receives a new message i, it first computes the buffer utility of message i in its buffer and then compares it with its overall buffer utility; if W_{ij} is less than U_j then the message is discarded, otherwise the message with least value for buffer utility is replaced by the new incoming message i. This replacement continues until enough space is created for the newly arrived message in the node buffer.

Congestion Control: This module mainly tries to control the number of copies of a message disseminated in the network by using a threshold value on the hop count and on the relay time of the message. This helps in preventing uncontrolled flooding of messages in the network which cause congestion and waste network bandwidth. A node stops forwarding a message if it has travelled a predefined number of hops or the time it has spent in the network is greater than a threshold relay time value. This helps to give more forwarding chances to the newer messages in the network.

Redundant Deletion: Lastly, this module prevents further spreading of messages which have already reached their destinations and deletes them from node buffers creating overall more buffer space for other remaining messages in the network. To accomplish this, each node maintains a list of ACKid containing the ID of messages which are known to be successfully delivered. On each node meeting, this list is exchanged between the nodes and is updated for further dissemination.

The authors have compared the performance of CIM with Epidemic routing using random buffer management and Epidemic using FIFO policy and PRoPHET routing protocols. The simulation results show improved performance of CIM both in terms of delivery ratio and message delivery delay due to the improved queuing policy and buffer replacement strategy.

3.3.2 Fuzzy-Based Buffer Management Techniques

Some works, such as Senttawatcharawanit, Yamada, & Haque, (2013), Makhlouta, Harkous, Hutayt, & Artail, (2011), and Jain, Chawla, Soares, & Rodrigues, (2014), have used the fuzzy logic technique for multiple message parameter aggregation and to determine the message priority in the buffer, which is then used for taking the scheduling or dropping decision.

a. Fuzzy-Spray (Senttawatcharawanit, Yamada, & Haque, 2013) is the first work in this area which prioritizes messages stored in a node buffer using fuzzy logic to determine their scheduling order. The two inputs to the fuzzy controller are the forward transmission count (FTC) of a message and its size. The FTC is simply determined by increasing FTC by one each time a node replicates a message to a new node. However, it may happen that two nodes that have common messages may have different values of FTC.
b. Adaptive Fuzzy Spray and Wait (AFSnW) (Makhlouta, Harkous, Hutayt, & Artail, 2011) uses the same buffer scheduling policy as in Mathurapoj, Pornavalai, & Chakraborty, (2009), but their scheduling policy is applied over the Spray and Wait routing protocol, in contrast to the Epidemic routing protocol used in Fuzzy-Spray, due to Spray and Wait's ability to achieve comparable delivery ratio with reduced overhead ratio.
c. Enhanced Fuzzy-based Spray and Wait Routing (EFSnWR) (Jain, Chawla, Soares, & Rodrigues, 2014) enhances the Spray and Wait routing protocol functionality by integrating a dropping and scheduling policy in it. The message rank is determined using the Fuzzy Logic Controller (FLC), which takes as input three message parameters, namely message TTL, size, and the number of replicas using fuzzy logic. While designing the fuzzy rule base, maximum weight age is given to replica count parameter, then TTL, and, lastly, size. But the major advantage of EFSnWR is in its use of an enhanced method for message replica count estimation, which is an improvement over the method proposed in Kim & Shin (2011). Each node, in addition to maintaining the replica count of a message, also maintains a list of nodes to which it has forwarded a message. This list is updated whenever two nodes meet each other for each of the common messages they contain. This helps in proper estimation of the replica count of a message in a distributed manner with the use of only local information.

3.3.3 Buffer Management in Social-Based Routing Algorithms

Social-based routing is an emerging field in DTN routing paradigms whereby the routing decisions are based on the social characteristics of nodes such as its importance in the network, the community to which it belongs, etc. In addition to routing decisions, some of the recent works (Shen, 2013), (Souza, Mota, Galvao, Manzoni, & Cano, 2014), (Zhou, Lin, Zhou, & Liu, 2016), (Liu, Wang, Guo, Lu, & Sun, 2017) and (Balasubramanian, 2007) have also used these social metrics to take dropping and scheduling decisions at a node.

a. T. Senttawatcharawanit et al. are the first ones in this area to exploit the social characteristics of a node for taking buffer dropping decisions (Shen, 2013). This dropping policy exploits information about community structure and node centrality degree to select messages that are least relevant for a node. In case of buffer overflow, a node first drops messages destined to nodes belonging to other communities and does message ordering based on node centrality values.

b. SEDUM (Souza, Mota, Galvao, Manzoni, & Cano, 2014), proposed by Li and Shen is a social utility-based multi-copy routing protocol. It uses a new utility metric that exploits both contact frequency and contact duration between nodes to take routing decisions. In addition to developing a routing protocol, the authors have also used a buffer scheduling policy based on the priority and time-to-live property of messages. The dropping policy of SEDUM is based on the concept of core replicas, where the core replica of a message is a copy of the message created by its source. The dropping policy of SEDUM is such that a core replica cannot be replaced by a non-core replica in the buffer and further non-core replicas are dropped on the basis of their utility value.
c. A community-based buffer management scheme is proposed in Liu, Wang, Guo, Lu, & Sun (2017), in which the message dropping and scheduling decision is taken on the basis of the community property of a node. In case of buffer overflow, a node first selects those messages that have yet not entered their destination community and sorts them according to the closeness degree of the current node to the destination node of the message; the message with the least value of closeness degree is dropped first. The authors also define a new metric to calculate the closeness degree between two nodes in terms of average intermeeting time between the nodes and fluctuations observed in it. When node i encounters node j, messages destined to encountered node j are scheduled first, followed by messages belonging to the community of node j, followed by messages of other communities which are sent in descending order of closeness degree.
d. Socially Aware Congestion Control algorithm (SACC) (Balasubramanian, 2007), proposes a new metric called the Social Congestion Metric (SCM), which combines the social tie of a node with its congestion level to decide whether it should act as a relay or not. This new social metric is then used to take the forwarding decisions; a node forwards a message to only those nodes which have strong links with the destination and are less congested. At the time of buffer overflow, the messages whose destination has the weakest social link to the current node, and thus the lowest delivery probability, are dropped. Their work is quite similar to the dropping policy in Zhou, Lin, Zhou, & Liu (2016).

These works prove the effectiveness of using buffer management policies along with social-based routing protocols in enhancing the network performance. Hence, it may be verified that buffer management policies are equally effective in enhancing the network performance of social-networking-based routing protocols. However, the buffer management policies designed for social-based routing policies are suitable only for specific DTN applications such as pocket switched networks (PSNs). The routing algorithms and the simulation scenarios used for their evaluation are specific to applications which follow some meeting patterns, such as those of human beings meeting in a conference scenario.

3.3.4 Global Buffer Management Policies

The local buffer management policies, although simple to implement, may result in suboptimal decisions as they make use of only locally available information about messages and not complete network-wide information (Krifa, Barakat, & Spyropoulos, 2012). In order to take optimal buffer management decisions, the buffer management policy requires sufficient information about all the messages in the network. They then decide whether to create an additional replica of a message or to delete a copy of the message by considering the trade-off between the local benefit gained by the receiver and the global benefit for the entire network (Rohner, Bjurefors, Gunningberg, McNamara, & Nordstrom, 2012). The global buffer management policies try to optimize one or more network performance metrics, such as delivery ratio or delivery delay, and derive message utility functions based on network-wide information about the messages.

a. Resource Allocation Protocol for Intentional DTN (RAPID) (Krifa, Barakat, & Spyropoulos, 2012) is the first work in this direction. Two utility functions have been defined: one with the objective of maximizing delivery ratio and another for minimizing average delivery delay. At any contact opportunity, a node first forwards those messages which can locally provide the highest increase in the utility function. But to effectively implement these utility metrics, it requires complete information about all the replicas of a given message. However, such a scheme is difficult to implement in DTN scenarios, where delay between node meetings may be long and random; hence, the information about messages may become obsolete and result in suboptimal decisions.
b. Krifa et al. have developed two message utility functions with the purpose of optimizing two specific performance metrics, i.e. maximizing delivery rate (DR) and minimizing average delivery delay (DD) respectively (Krifa, 2008; Elwhishi, Ho, Naik, & Shahida, 2013). These utility metrics utilize the complete status of all the messages in the network, such as their total number of copies in the network at current time, the intermeeting rate of two nodes, etc. The optimal buffer management (OBM) policy that aims to maximize the average DR drops the message i that has the smallest value of utility function defined in Equation 3.16:

$$\left(1-\frac{m_i(T_i)}{L-1}\right)\lambda R_i \exp\left(-\lambda n_i(T_i)R_i\right) \tag{3.16}$$

Here, K denotes the total number of messages in the network, L is the number of nodes in the network, T_i denotes time elapsed since the creation of message and is computed as the difference between current time and creation time of message; $n_i(T_i)$ denotes the number of message copies currently existing in the network of the message i; $m_i(T_i)$ denotes the number of nodes who have seen the message i, (they may not carry a copy of it); λ is the average of the meeting time which is assumed to follow exponential distribution.

Similarly, the OBM policy that minimizes the average DD drops the message i that has smallest value of utility function, defined in Equation 3.17:

$$i_{\min} = \operatorname{argmin}\left[\frac{1}{\lambda n_i(T_i)^2}\left(1-\frac{m_i(T_i)}{L-1}\right)\right] \tag{3.17}$$

Obtaining global information about all messages in real time in DTNs is impractical. Thus, a history-based dropping and scheduling (HSBD) scheme has also been proposed, which uses message utility functions based on the estimates of $n_i(T_i)$ and $m_i(T_i)$. The replication level of a message is updated during node encounters. The per-message utility for maximizing the delivery ratio used in HSBD is defined in Equation 3.18:

$$i_{\min} = \operatorname{argmin}\left[\left(\frac{1}{n_i(T_i)^2 \lambda}\right)(1 - \frac{m_i(T_i)}{L-1}\right] \quad (3.18)$$

Here, $m_i(T)$ and $n_i(T)$ are supposed to be instances of random variables $M(T)$ and $N(T)$. When global information is not available, the HBSD relies on the method of calculating the average delivery rate using all possible values of $M(T)$ and $N(T)$ and then maximizing it. The performance of HSBD is highly dependent on the accuracy of these estimates.

c. Elwhishi et al. (Wanga, Yanga, & Wub, 2015) use Ordinary Differential Equations (ODE) for determining the optimal buffer dropping and scheduling decisions. Each message has a utility value associated with it, and the creation of its new replica or its deletion from a node's buffer affects the overall delivery ratio of the network. Hence, to study this effect on the network performance the authors have modeled the message propagation process using the fluid flow limit model and designed ordinary differential equation to find the message utility. The reason for selecting the ODE model over the Markov Chain model is due to its scalability with the network size, whereas the computational complexity of the Markov Chain becomes very high as the number of network nodes increases. The ODE solution uses estimates of two global parameters for finding message utility, namely: the number of nodes currently carrying a copy of the message and the number of nodes who have seen a copy of the message. Each node calculates the message utility for each message stored in its buffer and then analyzes the effect of either dropping a message or replicating it, and the message with the least utility value which results in maximum increase in the delivery ratio is selected for dropping and vice versa.

   To maximize the average delivery ratio is to drop message $i_{\min,}$ which satisfies the following: P_{fi} is the probability of forwarding message i to every encountered node, which can be estimated as $P_{fi} = n_i(t)\ m_i(t)$.

$$i_{\min} = \operatorname{argmin}\left[\begin{array}{c}\left(1 - \frac{m_i(T_i)}{N-1}\right)^2 \left(\frac{N}{N - n_i(T_i) + n_i(T_i).e^{\beta N R_i P_f}}\right)^{m_i(T_i)+1} \\ \left[e^{\beta N R_i P_f}\left(\beta R_i n_i(T_i) + \frac{m_i(T_i)}{N}\right) - \frac{m_i(T_i)}{N}\right]\end{array}\right] \quad (3.19)$$

   They assume that all messages in the network are of the same size, and the method used to collect the network parameters does not achieve accurate values. In addition, they do not consider the impact of bandwidth on the delivery ratio.

   The buffer management policy has been evaluated and compared over Epidemic and two-hop forwarding, which are two popular DTN routing protocols.

d. Wanga, Yanga, & Wub (2015) propose a non-heuristic solution to the message scheduling and dropping policy for Spray and Wait routing protocol. The authors say that both RTTL and number of copies of a message are two important factors that affect buffer management decisions and combining them in a heuristic way may not always lead to optimal solutions. As the messages residing in a node buffer may be of variable size, the buffer management problem can be seen as a 0/1 knapsack problem having messages with different utility value and sizes. This can then be solved as an optimization problem with the objective of maximizing the overall delivery ratio of the network. This optimization problem is then solved using a dynamic programming technique. The work presents both a theoretical centralized version as well as a practical distributed version of their buffer management strategy to take into account the dynamic changing environment of DTN.

The buffer management decision is taken by calculating the effect of either dropping or scheduling a message on the delivery ratio. The major advantage of their scheme over other past works, such as Krifa, Barakat, & Spyropoulos (2012) Elwhishi, HoNaik, & Shahida (2013), is that it also takes into account the number of message replicas that may be created during the remaining TTL of a message while calculating message utility. Secondly, they also take into account the fact that a message transmission may also be aborted due to limited bandwidth. The following message utility function is calculated:

$$U_i = \left[N - m_i(T_i)\right] N^{\frac{1-\epsilon_i}{\epsilon_i}} e^{-\lambda N R_i} \frac{1}{\epsilon_i}\left(1 - e^{-\epsilon \lambda N R_i}\right)\Big[n_i(T_i) - n_i(T_i) e^{-\epsilon \lambda N R_i} + N e^{-\epsilon \lambda N R_i}\Big]^{\frac{-1-\epsilon_i}{\epsilon_i}} \tag{3.20}$$

This utility function is based on determining the probability of successful delivery of the message based on the assumption that the intermeeting time between nodes follows exponential distribution. The messages are scheduled according to their utility per-unit value, i.e. U_i/M_i. This is to take into account the fact that a bigger message requires more buffer space and more transmission time if bandwidth is limited. For distributed implementation of this buffer management scheme, local estimates of the parameters $n_i(T)$ and $m_i(T)$ are used. Hence, each node collects and maintains the history information of some of the messages it has seen in the past.

e. A joint relay selection and buffer management scheme has been proposed (Le, Kalantarian, & Gerla, 2016) with the objective of maximizing the delivery ratio of the network. This work addresses three main issues faced in the message delivery process in resource-constrained DTNs: selection of relay node, message scheduling order, and message dropping order on buffer overflow.

The relay selection is based on the concept of social ties, whereby the social tie between two nodes is determined using their frequency of encounters and recency of contact. The social-tie value of node i with node j at any time t_{base}, denoted by $Ri(j)$, is calculated:

$$R_i(j) = \sum_{k=1}^{n} F\left(t_{\text{base}} - t_{jk}\right) \tag{3.21}$$

where t_{base} denotes the current time and t_{jk} denote the time of kth meeting of nodes i and j. Here, $F(x)$ is a function of the intermeeting time between two nodes.

A node is said to be a good relay for a message if it has a higher social tie with the destination of the message than the current carrier of the message. However, it may happen that an encountered node has zero value for the social tie with the destination. Hence, authors suggest another metric called *social delivery potential*, which is simply an aggregation of the social-tie value of a node and its neighbors collectively; thus, a node is a good relay even if it has a high social tie with the destination via its neighbors.

The existing routing schemes based on delivery predictability or the social-tie metric tend to select popular nodes in the network or nodes that have a high value of delivery predictability or social-tie strength with the destination. This leads to congestion at these nodes, causing buffer overflow and frequent message drops, which results in a reduction in the overall delivery ratio of the network. Hence, authors suggest the use of both the social-tie strength and the queue length of a node for the selection of relay nodes. This will help in even distribution of traffic across the network, and prevent the overloading of highly-connected nodes. The reason for selecting queue length is because it reflects a node's connectivity. Hence, a node is selected as a relay node only if it has high social delivery potential to the destination and it is less congested than the current carrier of the message.

Next, in order to prioritize message scheduling and dropping order, a utility function has been developed that takes into account complete network-wide information about the messages, the nodes, and their meeting patterns. The utility function has been designed for heterogeneous node mobility, i.e. each pair of nodes may have different intermeeting rates and contact duration rates, which is different from most of the previous works (Krifa, Barakat, & Spyropoulos, 2012); (Elwhishi, Ho, Naik, & Shahida, 2013); (Wanga, Yanga, & Wub, 2015). It is assumed that node intermeeting times and contact duration follow exponential distribution, having rates λ and θ respectively. The nodes collect global network information through direct contacts or through other nodes they meet. Each node records certain metadata about other nodes in the network, such as their contact statistics with other nodes, the list of messages stored in their buffers, and last update time of the list etc.; each node also maintains information about all the messages in the network. Based on this global information, the following utility value for maximizing the delivery ratio of the network is developed:

$$U_i = -e^{-\lambda_i R_i - \theta_i H_i} \sum_{n=1}^{\infty} \left[\left(1 - e^{-\theta_i H_i}\right)^{n-1} \sum_{k=0}^{n-1} \frac{(\lambda_i R_i)^k}{k!} \right] \tag{3.22}$$

where:

R_i is the remaining TTL of message i,

λ_i is the average intermeeting rate between nodes who contain a copy of message i and its destination

θ_i is the average contact duration rate between nodes who contain a copy of message i and its destination

H_i denotes the contact time required for successful delivery of message i

To take into account the variable size of messages, the optimization problem is formulated as a 0/1 knapsack problem and solved using dynamic programming.

f. Iranmanesh (2016) uses a multi-objective utility function for taking buffer management decisions, which combines both the utility function for maximizing the delivery ratio and the utility function for minimizing delivery delay. The utility function for maximizing delivery ratio and minimizing delivery delay are derived in a manner similar to Elwhishi, Ho, Naik, & Shahida (2013), using the meeting probability of nodes during the remaining TTL of a message; however, the authors also consider the change in delivery ratio with respect to the TTL of a message and the change in the number of replicas of a message. The authors have combined the proposed buffer management scheme with encounter-based routing protocol, which is a quota-based routing protocol and uses encounter frequency between the nodes for message replica distribution (Nelson, Bakht, & Kravets, 2009).

$$UF_i = \alpha * \varphi(Delivery\_\mu_i) + \beta * \varphi(Delay\_\mu_i) \quad (3.23)$$

where $\varphi(Delivery\_\mu_i)$ and $\varphi(Delivery\_\mu_i)$ are delivery and delay utilities combined using weighting factors α and β respectively to combine the impact of delivery ratio and delivery delay. Since the values of delivery ratio utility and delivery delay utility lie over different domains, they have been normalized before combining them.

3.3.5 Traffic Differentiation Schemes for Delay-Tolerant Networks

The concept of traffic differentiation has been widely used on the Internet to distinguish between different types of traffic that are generated by different application sources and that have different quality of service (QoS) requirements in the network. It allows the network to distinguish packets on the basis of their priority class and favor higher priority class messages over others. Differentiated services (DiffServ) and integrated services (IntServ) are two commonly used techniques on the Internet (Wikipedia). While DiffServ classifies traffic on the basis of their class of service (CoS), giving preferential treatment to higher priority traffic class packets so as to meet their QoS requirement, the best effort model gives equal priority to all types of application data generated in the network and hence is unable to meet the QoS requirements of different applications.

DTNs are also designed to support multiple applications which may have different performance requirements, but network resources such as storage and bandwidth are a constraint in DTNs (Jain and Chawla, 2017). Three priority classes have been defined for bundles in DTN architecture (Cerf, Burleigh, Hooke, Torgerson, Durst, & Scott, 2007): bulk class is used for low priority traffic, normal class is used for medium priority traffic, and expedited class denotes high priority traffic. The main purpose of these priority classes is to define the relative scheduling and dropping order of messages stored in a node's buffer. The expedited messages that have the highest priority should be scheduled first at any contact opportunity, followed by normal and bulk messages.

The local and global buffer management schemes discussed in Sections 3.3.1 and 3.3.4 do not prioritize messages on the basis of their class of service and hence give no assurance for quality of service. Therefore, considering the traffic class of messages is also important, along with other message properties that are necessary to meet the QoS requirements of different applications.

a. ORWAR (Opportunistic DTN Routing with Window-aware Adaptive Replication) (Sandulescu and Tehrani, 2008) proposes a message ordering scheme based on the priority class of messages for the Spray and Wait routing protocol. On any transmission opportunity, messages are transmitted in the order of their utility/size values, where the utility of a message is determined by its priority class. Messages belonging to the higher priority class are assigned highest utility value. This ordering helps to give preference to higher priority class messages. In order to give more transmission opportunities to messages belonging to higher priority class, more replicas are allotted to this class of messages, i.e. $L+\Delta$ as compared to the Spray and Wait routing protocol where all messages irrespective of their utility value are assigned a fixed number of message copies, i.e. L. On the contrary, a lesser number of message copies are assigned to lower priority class messages, i.e. $L-\Delta$, which directly affects their delivery probability. In order to avoid bandwidth wastage due to partially transmitted messages, an estimate of contact time is used to determine the amount of data that may be transmitted during an encounter opportunity. On buffer overflow, messages belonging to the lowest priority class are always dropped and this severely affects the delivery probability of these messages, resulting in an overall reduced delivery ratio of the network. Although ORWAR uses the priority class of messages, it misses some of the important message parameters, such as the current message replica count, delivery probability, etc.
b. Two very popular traffic differentiation schemes have been proposed for vehicular delay-tolerant networks (VDTNs) in Soares, Farahmand, & Rodrigues (2011), namely Priority Greedy (PG) and Custom Service Time (CST) scheduling. Their work is based on the priority CoS model as defined in the DTN architecture (Cerf, Burleigh, Hooke, Torgerson, Durst, & Scott, 2007). It assumed that the class of service of a message is assigned by the source node and does not change over time.

   In PG scheduling, at any contact opportunity, first expedited class messages are scheduled, followed by normal class messages and, lastly, bulk class messages. In case of the arrival of a new message and buffer overflow condition, the node always drops the lowest priority traffic class message from its buffer. The main disadvantage of this approach is that it drastically affects the delivery of bulk and normal traffic class messages, as all the network resources are dominated by the expedited traffic class messages. It may happen that bulk and normal traffic class messages are not spread at all in the network due to limited contact durations and buffer storage. Hence, in an attempt to ensure fairness among the three traffic classes and prevent starvation of lower class messages, another scheduling algorithm called CST was considered.

   In CST scheduling scheme, firstly the contact time between nodes is estimated and is then used to decide the amount of data that may be transmitted in that contact time. The estimated contact time is shared among the three traffic classes with the highest service time allocated to messages from expedited class and so on. Along with bandwidth, the buffer storage of nodes is also divided among the three traffic class queues in the same proportion as bandwidth division. On arrival of a new message, the node puts the message in the respective traffic class queue based on its priority class. If the corresponding queue is full, it drops the message from that particular traffic class queue so that messages of all traffic classes may be stored in a node's buffer and get a chance for forwarding.

These schemes consider only the remaining TTL for scheduling messages within a particular traffic class queue, while further works (Jain and Chawla, 2017; Shin and Kim, 2012; Matzakos., Spyropoulos, & Bonnet, 2015) have shown that the use of an appropriate message ranking mechanism with each traffic class queue can result in an improved delivery ratio of each traffic class queue. Multiple message properties along with their proper aggregation are essential to achieving appropriate ranking of messages for buffer scheduling.

c. In Shin and Kim (2012), the Class of Service-based Scheduling and Dropping (CoSSD) strategy uses three additional message properties, namely an estimation of the number of message copies, denoted by ETR; elapsed time, denoted by ET; and remaining TTL, denoted by RT, along with message priority to rank messages. The messages from the expedited class, which is considered as the highest traffic class, are scheduled first, followed by normal and bulk, and within each class ETR is used to order messages. In addition to scheduling, two buffer dropping policies are used in conjuncture for deciding which message to drop in case of buffer congestion. The former combines the priority class of the message with the message utility function, which aims to optimize delivery ratio and is similar to the one defined in OBM (Krifa, 2008). The latter dropping policy combines the ETR value of the message with the time spent by the message in the network, i.e. ET. Hence, on buffer overflow, it selects those messages whose large number of copies already exist in the network and are comparatively older even if they belong to a higher priority class as their probability of reaching the destination is relatively high. As ETR is an important parameter for taking both the dropping and scheduling decisions, estimating its correct value is very important for the schemes to be effective. The authors use their previously proposed method (Kim and Shin, 2011) for the estimation of the number of replicas of a message, which has several shortcomings as discussed in Section 3.3.1(m).
d. The message prioritization problem of DTNs is presented as a constrained optimization problem in Matzakos, Spyropoulos, & Bonnet (2015) and a distributed gradient-descent-based solution has been proposed for it. Each bundle belonging to a particular class is assumed to have a QoS constraint associated with it, which in this work is taken to be the minimum accepted delivery probability. While solving constrained optimization problems, gradient-descent-based algorithms may use appropriate penalty functions for each violated constraint. Thus, if the predicted delivery probability of a bundle is less than its class requirement, then its utility value is incremented by a term proportional to the delivery probability deficit. This ensures that the utilities of bundles that do not satisfy their QoS constraint will always be higher than the utilities of bundles that do satisfy them and hence such messages always get preference over others, resulting in convergence to feasible solutions. In order to implement this distributed algorithm, the bundles stored in a node's buffer are divided into two dynamic groups: the first group contains all bundles whose predicted delivery is below their QoS threshold; the second group consists of the bundles which are above their threshold. The bundles of the first group are scheduled first, followed by bundles belonging to the second

group. Within each group, messages are further divided into subgroups according to their priority class, and within each subgroup the utility function U_i as defined in Krifa, Barakat, & Spyropoulos (2012) is used for ordering of the messages. This ensures that a subgroup attributed to a higher QoS class will always have higher priority than a subgroup of a lower QoS class.

e. Jain and Chawla (2017) present a novel buffer management scheme designed for supporting traffic differentiation (or prioritization) in DTNs. It supports a priority class model to provide a custom allocation of resources and employs a fuzzy message ranking mechanism for ordering messages in each traffic class queue. The proposed traffic differentiation scheme is inspired by the general guidelines of the CST traffic differentiation scheme (Sandulescu and Tehrani, 2008). It uses the same concept of dividing the available bandwidth and buffer space resources among the three traffic classes so as to ensure fairness and at the same time provide preferential treatment to expedited class messages. But two more message properties, namely the number of replicas (i.e., the number of times a message has been replicated to encountered nodes) and size have been used in addition to RTTL for taking dropping and scheduling decisions within each traffic class queue. The fuzzy-based message ranking algorithm is based on one of the author's previous works (Jain, Chawla, Soares, & Rodrigues, 2014).

3.4 Summary and Discussions

The scheduling of messages from a node buffer at any contact opportunity and message dropping on the arrival of a new message when buffer overflow condition arises, are two important decisions that have to be taken in the process of message delivery in DTNs. These decisions can have a significant impact on the performance of DTN routing protocols working in resource-constrained environments. A plethora of buffer management techniques have been proposed, which vary from the amount and type of information used for decision making. A comprehensive study of these techniques reveals the different message parameters that may be used for message ranking at the time of message transmission and buffer overflow (Jain, Chawla, Soares, & Rodrigues, 2014; Lakkakorpi, Pitkanen, & Ott, 2011):

- RTTL of a message, i.e. the remaining time for which the message will be active in the network, as after expiration of TTL the message will be discarded from the network.
- Age of the message, i.e. the time spent by the message in the network since its creation.
- Hop count or forward transmission count of a message, i.e. the number of hops a message has travelled from the source node to the current node.
- Number of replicas of a message, i.e. number of nodes in the network that have a message copy.
- Last forwarding time of the message, i.e. the last time stamp when a message was forwarded by a node.
- Size of the message.

- Distance to the destination.
- Delivery cost of the message such as the meeting probability of a node to meet the destination of the message.
- Contact statistics between nodes.

Most of the works have developed local-knowledge-based buffer management policies as they are easy to design and require minimal information, while global buffer management policies, although they guarantee optimal solutions, are complex due to the requirement of network-wide information about all the messages or their estimates for practical implementation. Certain buffer management techniques are based on the delivery predictability metric or distance metrics; hence, they are only applicable to routing protocols that use these kinds of metrics, such as in Davis, Fagg, & Levine (2001), Burgess, Gallagher, Jensen, & Levine (2006), and Phanse and Lindgren, (2006).

Hop count or forward transmission count, and replica count: these parameters denote the extent to which a message has been spread in the network and are the most important parameters for message prioritization; their accurate estimation is a critical task in a distributed manner.

Another distinguishing feature is the number of message parameters used by the buffer management policies for decision making. Some buffer management policies use one or the other message parameter individually (Tang, Chai, & Weng, 2012; Davis, Fagg, & Levine, 2001; Naves, Moraes, & Albuquerque, 2012; Ramanathan, Hansen, & Basu, 2007; Li, Zhao, Liu, & Liu 2009; Soares, Farahmand, & Rodrigues, 2010; Bjurefors, Gunningberg, Rohner, & Tavakoli, 2011; Rohner, Bjurefors, Gunningberg, McNamara, & Nordstrom, 2012; Ayub and Rashid, 2010; Rashid, Ayub, Soperi, Zahid, & Abdullah, 2011; Abdullah, Soperi, Zahid, & Ayub, 2012; Rashid, Ayub, Soperi, Zahid, & Abdullah, 2013; and Liu, Wang, Zhang, & Zhou, 2011) while others combine multiple message parameters for deciding message priority (Kim & Shin, 2011; Jain, Chawla, Soares, & Rodrigues, 2014; Prodhan, Das, Kabir, & Shoja, 2011; Rashid, Ayub, & Abdullah, 2015; Silva, Miniy, & Loureiro., 2017; Chen, Yao, Zong, & Wang, 2017; Pan, Ruan, Zhou, Liu, & Songet, 2013; Mathurapoj, Pornavalai, & Chakraborty, 2009; and Senttawatcharawanit, Yamada, & Haque, 2013). The latter has shown better performance. Message parameter aggregation is another area of research which needs special attention so that each message parameter may be given due weightage.

3.5 Scope for Future Work

Most of the works have tested their schemes for one or two routing protocols. A comprehensive analysis of the application of buffer management schemes on different classes of routing protocols such as flooding-based, replication-based, controlled-replication-based, history-based, social-based, etc. is required to analyze their effect on each and then to determine which type of buffer management scheme is most suitable for which kind of routing protocols. Buffer management and routing strategy are two complementary aspects of the DTN message delivery process which should work hand in hand and when combined properly may result in improved performance in terms of delivery ratio, delivery delay, and overhead ratio.

In order to get optimal solutions for global buffer management schemes as well as for some of the local buffer management schemes, different estimates of message replica counts have been used. The performance of these buffer management schemes is highly dependent on the accuracy of these estimates. This is difficult to accomplish in DTNs, where the node neighborhood is dynamic due to frequent network partitions. The reliable estimation of the number of copies of a message based on local knowledge is one of the key challenges in implementing buffer management policies based on message replica count.

The use of the congestion control mechanism can make the buffer scheduling and drooping policies more effective. Rather than simply dropping the messages when buffer overflow occurs, if the nodes can detect congestion in a network area then it may prevent wasting its resources during that time. Nodes can exchange their free buffer space, in addition to prediction of contact time, and prevent the unnecessary dropping of messages and save those messages which are already in the node buffer. Hence, buffer management should be an integral part of congestion control rather than a separate aspect.

A comprehensive buffer management technique is required which is adaptive to the challenging network environments for which DTNs have actually been designed.

References

Ayub, Q., & Rashid, S. (2010). Effective buffer management policy DLA for DTN routing protocols under congestion. *International Journal of Computer and Network Security*, Vol 2, Issue 9, 118–121.

Balasubramanian, A. (2007). DTN routing as a resource allocation problem. *Proceedings of the Conference on Applications, Technologies, Architectures, and Protocols for Computer Communications (SIGCOMM)*, Kyoto, Japan, 373–384.

Bjurefors, F., Gunningberg, P., Rohner, C., & Tavakoli, S. (2011). Congestion avoidance in a data-centric opportunistic network. *ACM ICN*, Toronto, Canada, 32–37.

Burgess, J., Gallagher, B., Jensen, D., & Levine, B. N. (2006). MaxProp: Routing for vehicle-based disruption-tolerant networks. *25th IEEE International Conference on Computer Communications (INFOCOM) Proceedings*, Barcelona, Spain, 1–11.

Cerf, V., Burleigh, S., Hooke, A., Torgerson, L., Durst, R., Scott, K., Fall, K., and Weiss, H. (2007). Delay-Tolerant Networking Architecture. *RFC 4838*, DOI 10.17487/RFC4838, https://www.rfc-editor.org/info/rfc4838.

Chen, Y., Yao, W., Zong, M., & Wang, D. (2017). An effective buffer management policy for opportunistic networks. *CollaborateCom 2016, LNICST 2016*, Beijing, China, 242–251.

Davis, J. A., Fagg, H. A., & Levine, B. N. (2001). Wearable computers as packet transport mechanisms in highly partitioned ad-hoc networks. *5th International Symposium on Wearable Computing*, Zurich, Switzerland, 141–148.

Elwhishi, A., Ho, P. H., Naik, K., & Shahida, B. (2013). Novel message scheduling framework for delay tolerant networks routing. *IEEE Transactions on Parallel & Distributed Systems*, Vol. 24, Issue 5, 871–880.

Gupta, A. (2013). Routing in delay tolerant networks (DTNs). Dept. of Computer Science & Engineering, IIT, Kharagpur. https://pdfs.semanticscholar.org/presentation/4c15/6e8a80058bfc8a94e13e077ec15e6f78fb14.pdf?_ga=2.95549596.1073448070.1535460729-812226833.1535460729, accessed August 8, 2018.

Iranmanesh, S. (2016). A novel queue management policy for delay-tolerant networks. *EURASIP Journal on Wireless Communications and Networking*, Vol. 88, 1–23.

Jain, S., & Chawla, M. (2017). A fuzzy logic based buffer management scheme with traffic differentiation support for delay tolerant networks. *Journal of Telecommunication Systems*, Vol. 68, Issue. 2, 319–335.

Jain, S., Chawla, M., Soares, V. N., & Rodrigues, J. P. (2014). Enhanced fuzzy logic based spray and wait routing protocol for delay tolerant networks. *International Journal of Communication Systems*, Vol. 29, Issue. 12, 1820–1843.

Kim, K., & Shin, S. (2011). Enhanced buffer management policy that utilises message properties for delay-tolerant networks. *IET Communications*, Vol. 5, Issue. 6, 753–759.

Krifa, A., Barakat, C., & Spyropoulos, T. (2012). Message drop and scheduling in DTNs: Theory and practice. *IEEE Transactions on Mobile Computing*, Vol. 11, Issue 9, 1470–1483.

Krifa, C. B. (2008). Optimal buffer management policies for delay tolerant networks. *5th Annual IEEE Communications Society Conference on Sensor, Mesh and Ad Hoc Communications and Networks*, San Francisco, CA, 260–268.

Lakkakorpi, J., Pitkanen, M., & Ott, J. (2011). Using buffer space advertisements to avoid congestion in mobile opportunistic DTNs. *Proceedings of the 9th IFIP TC 6 International Conference on Wired/wireless Internet Communications (WWIC)*, Vilanova i la Geltru, Spain, 386–397.

Le, T., Kalantarian, H., & Gerla, M. (2016). A joint relay selection and buffer management scheme for delivery rate optimization in DTNs. *Proceedings of 17th International Symposium on A World of Wireless, Mobile and Multimedia Networks (WoWMoM))*, Coimbra, Portugal, 1–9.

Li, Y., Zhao, L., Liu, Z., & Liu, Q. (2009). N-Drop: Congestion control strategy under epidemic routing in DTN. *Proceedings of the International Conference on Wireless Communications and Mobile Computing: Connecting the World Wirelessly*, Leipzig, Germany, 457–460.

Lindgren, A., Doria, A., & Schelen, O. (2003). Probabilistic routing in intermittently connected network. *ACM SIGMOBILE Mobile Computing and Communication Review*, Vol. 7, Issue 3, 19–20.

Liu, Y., Wang, J., Zhang, S., & Zhou, H. (2011). A buffer management scheme based on message transmission status in delay tolerant networks. *IEEE Global Telecommunications Conference (GLOBECOM)*, Kathmandu, Nepal, 1–5.

Liu, Y., Wang, K., Guo, H., Lu, Q., & Sun, Y. (2017). Social-aware computing based congestion control in delay tolerant networks. *Mobile Networks Application*, Vol. 22, Issue 2, 174–185.

Lo, S. C., Chiang, M. H., Liou, J. H., & Gao, J. S. (2011). Routing and buffering strategies in delay tolerant networks: Survey and evaluation. *40th International Conference on Parallel Processing Workshops*, Taipei City, Taiwan, 91–100.

Makhlouta, J., Harkous, H., Hutayt, F., & Artail, H. (2011). Adaptive fuzzy spray and wait: Efficient routing for opportunistic networks. *International Conference on Selected Topics in Mobile and Wireless Networking, iCOST,*, Shanghai, China, 64–69.

Marano, A., & Socievole, S. (2012). Evaluating the impact of energy consumption on routing performance in delay tolerant networks. *8th International Conference on Wireless Communications and Mobile Computing*, Limassol, Cyprus, 481–486.

Mathurapoj, G., Pornavalai, A., & Chakraborty, C. (2009). Fuzzy-spray: Efficient routing in delay tolerant ad-hoc network based on fuzzy decision mechanism. *IEEE International Conference on Fuzzy Systems*, Jeju Island, South Korea, 104–109.

Matzakos, P., Spyropoulos, T., & Bonnet, C. (2015). Buffer management policies for DTN applications with different QoS requirements. *IEEE Global Communications Conference: Wireless Communications*, San Diego, CA, 1–7.

Naves, J. F., Moraes, I. M., & Albuquerque, C. (2012). LPS and LRF: Efficient buffer management policies for delay and disruption tolerant networks. *37th Annual IEEE Conference on Local Computer Networks*, Clearwater, FL, 368–375.

Nelson, S. C., Bakht, M., & Kravets, R. (2009). Encounter based routing in delay tolerant networks. *Proceedings of IEEE INFOCOM*, Rio de Janeiro, Brazil, 338–341.

Pan, D., Ruan, Z., Zhou, N., Liu, X., & Song, Z. (2013). A comprehensive-integrated buffer management strategy for opportunistic networks. *EURASIP Journal on Wireless Communications and Networking*, 2013, Issue. 103, 1–10.

Phanse, A., & Lindgren, K. (2006). Evaluation of queueing policies and forwarding strategies for routing in intermittently connected networks. *1st International Conference on Communication Systems Software & Middleware*, New Dehli, India, 1–10.

Prodhan, A. T., Das, R., Kabir, H., & Shoja, G. C. (2011). TTL based routing in opportunistic networks. *Journal of Network and Computer Applications*, Vol. 34, Issue 5, 1660–1670.

Ramanathan, R., Hansen, R., & Basu, P. (2007). Prioritized epidemic routing for opportunistic networks. *ACM MobiOpp*, San Juan, Puerto Rico, 62–66.

Rashid, S., Abdullah, A. H., Soperi, M., Zahid, M., & Ayub, Q. (2012). Mean drop: An effectual buffer management policy for delay tolerant network. *European Journal of Scientific Research*, Vol. 70, Issue 3, 396–407.

Rashid, S., & Ayub, Q. (2010). T-Drop: An optimal buffer management policy to improve QOS in DTN routing protocols. *Journal of Computing*, Vol. 2, Issue 10, 46–50.

Rashid, S., Ayub, Q., & Abdullah, A. H. (2015). Reactive weight based buffer management policy for DTN routing protocols. *Wireless Personal Communications*, Issue 3, 993–1010.

Rashid, S., Ayub, Q., Soperi, M., Zahid, M., & Abdullah, A. H. (2011). E-drop: An effective buffer management policy for DTN routing protocol. *International Journal of Computer Applications*, Vol. 13, Issue 7, 8–13.

Rashid, S., Ayub, Q., Soperi, M., Zahid, M., & Abdullah, A. H. (2013). Message drop control buffer management policy for DTN routing protocols. *Journal of Wireless Personal Communications*, Vol. 72, Issue 1, 653–669.

Rohner, C., Bjurefors, F., Gunningberg, P., McNamara, L., & Nordstrom, E. (2012). Making the most of your contacts: Transfer ordering in data-centric opportunistic networks. *ACM MobiOpp*, Zurich, Switzerland, 53–59.

Sandulescu, G., & Tehrani, N. S. (2008). Opportunistic DTN routing with window-aware adaptive replication. *Proceedings of the 4th Asian Conference on Internet Engineering (AINTEC)*, Bangkok, Thailand, 103–112.

Senttawatcharawanit, T., Yamada, S., & Haque, M. E. (2013). Message dropping policy in congested social delay tolerant networks. *10th IEEE International Conference on Computer Science and Software Engineering*, Maha Sarakham, Thailand, 116–120.

Shen, Z. L. (2013). SEDUM: Exploiting social networks in utility based distributed routing for DTNs. *IEEE Transactions on Computers*, Vol. 62, Issue 1, 83–97.

Shin, K., & Kim, K. (2012). Traffic management strategy for delay-tolerant networks. *Journal of Network and Computer Applications*, Vol. 35, Issue 6, 1762–1770.

Silva, M. D., Nunes, I. O., Miniy, A. F., & Loureiro, A. A. (2017). ST-Drop: A novel buffer management strategy for D2D opportunistic networks. *IEEE Symposium on Computers and Communications (ISCC)*, Heraklion, Greece, 1300–1305.

Soares, V. N., Farahmand, F., & Rodrigues, J. J. (2010). Performance analysis of scheduling and dropping policies in vehicular delay-tolerant networks. *IARIA International Journal on Advances in Internet Technology*, Vol. 3, Issues 1 and 2, 137–145.

Soares, V. N., Farahmand, F., & Rodrigues, J. J. (2011). Traffic differentiation support in vehicular delay tolerant networks. *Journal of Telecommunication Systems*, Vol. 48, 151–162.

Souza, C., Mota, E., Galvao, L., Manzoni, P., & Cano, J. C. (2014). Drop less known strategy for buffer management in DTN nodes. *Proceedings of the Latin America Networking Conference (LANC)*, Montevideo, Uruguay.

Spyropoulos, T., Turletti, T., & Obraczka, K. (2009). Routing in delay-tolerant networks comprising heterogeneous node populations. *IEEE Transactions on Mobile Computing*, Vol. 8, Issue 8, 1132–1147.

Sun, Y., & Cao, Z. (2013). Routing in delay/disruption tolerant networks: A taxonomy, survey and challenges. *IEEE Communications Surveys and Tutorials*, Vol. 15, Issue 2, 654–677.

Tang, L., Chai, Y., & Weng, B. (2012). Buffer management policies in opportunistic networks. *Journal of Computational Information Systems*, Vol. 8, Issue 12, 5149–5159.

Wanga, E., Yanga, Y., & Wub, J. (2015). A Knapsack-based buffer management strategy for delay-tolerant networks. *Journal of Parallel and Distributed Computing*, Vol. 86, 1–15.

Wikipedia. (n.d.). *Differentiated services-Wikipedia, the free encyclopedia.* Retrieved from Wikipedia: https://en.wikipedia.org/wiki/Differentiated_services.

Zhang, X., Neglia, G., Kurose, J., & Towsley, D. (2007). Performance modelling of epidemic routing. *Computer Networks*, Vol. 51, Issue 10, 2867–2891.

Zhou, J., Lin, Y., Zhou, S., & Liu, Q. (2016). Community-based adaptive buffer management strategy in opportunistic network. *International Conference on Security, Privacy and Anonymity in Computation, Communication and Storage (SpaCCS)*, Zhangjiajie, China, Vol. 10067, 16–25.

4

Taxonomy of Mobility Models

Jyotsna Verma

CONTENTS

4.1 Introduction 59
 4.1.1 Background 60
 4.1.2 Mobility Model Characteristics 61
 4.1.3 Classification of Mobility Models 62
4.2 Mobility Models 63
 4.2.1 Trace-Based Models 64
  4.2.1.1 Trace-Based Analysis 64
 4.2.2 Stochastic Mobility Model 65
  4.2.2.1 Random-Based Mobility Models 65
 4.2.3 Synthetic Models 68
  4.2.3.1 Temporal-Dependency–Based Mobility Models 68
  4.2.3.2 Spatial-Dependency–Based Mobility Models 70
  4.2.3.3 Geographical-Restriction–Based Mobility Model 72
  4.2.3.4 Evaluation of Synthetic Mobility Models 74
 4.2.4 Map-Based Mobility Models 75
  4.2.4.1 Route-Based Map Mobility Model 75
  4.2.4.2 Rush Hour (Human) Traffic Model 75
  4.2.4.3 Working Day Movement Model 76
  4.2.4.4 Shortest Path Map-Based Movement 76
  4.2.4.5 Evaluation of Map-Based Mobility Models 77
 4.2.5 Social-Network–Based Mobility Models 77
  4.2.5.1 Community-Based Mobility Models 77
  4.2.5.2 Social Network Models 77
4.3 TRANSIMs 78
4.4 Testing Tools 78
4.5 Impact of Mobility Models on the Performance of Opportunistic Networks 79
4.6 Future Research Directions in Mobility Modeling 80
References 80

4.1 Introduction

With the proliferation of information technology, we have now entered into the networked world, in which large numbers of small handheld mobile communication devices like smart phones, PDAs, tablets, etc., communicate and share data with each other through infrastructure-less networks; this has led to the development of mobile ad hoc networks

(MANETs). MANETs are an autonomous system of mobile devices that have a self-configuring and distributed routable networking environment. Mobile devices in MANETs can form networks dynamically, without relying on any pre-existing infrastructure for easy access and effective communication within the network. Nevertheless, because of frequent topology changes and partition, the network must guarantee end-to-end communication (i.e., paths must be established between the transmitting ends before packets can be routed) and should have a dedicated path in prior. However, in reality, this is not the case; communication can be disturbed due to the very nature of various physical and environmental factors such as short-range communications, mobility patterns of nodes, physical obstruction like buildings, trees, etc. The mobility behavior and diverse mobility patterns of nodes in MANETs affects the performance of the wireless networks such as protocols, the communication traffic pattern, the performance of the network, the connectivity of nodes, packet dropping with an unreachable destination between the transmitting end, etc., and, hence, disrupt the communications.

Thus, there comes the concept of opportunistic networking; this is an extension of MANETs which is built entirely on users' devices and intermittently connects mobile devices to enable communication within the network based on their geographical proximity. Attracting significant attention from academia and the research community, opportunistic networking has emerged as a new communication paradigm in wireless networks which store and forward the packets instead of discarding them until the communication link is established. The opportunistic mobile wireless networks are able to be much more tolerant than MANETs, despite frequent path breaks, node mobility and reconfigurations, by exploiting the users' mobility in order to increase the network capacity. Mobility modeling in opportunistic networks is now the paramount topic in the research community in identifying realistic mobility models that can be deployed in complex real-world scenarios, like emergency operations, search and rescue operations, disaster recovery, etc.

4.1.1 Background

The history of opportunistic networks (OppNets) is not very long; it was started with the proposal for the Interplanetary Internet (IPN) by NASA, MITRE and others, which was funded by the Defense Advanced Research Projects Agency (DAR-PA). The initial IPN architecture was developed by Vint Cerf and others for the necessity of coping with the significant delays and packet corruption of deep-space communications. Later, Kevin Fall (2002) used the same concept and designed terrestrial networks that can resist the longer delays incurred by intermittent connectivity and invented the term *Delay tolerant network* (DTN), also known as *intermittently connected network* or *disruption tolerant network*, as the nodes in such networks experience brief link duration. Furthermore, the DTN research group (DTNRG) developed architecture and published RFC 4838 and RFC 5050 in 2007 to define a common abstraction to software running on disrupted networks. RFC 5050, known as the Bundle Protocol, defines the series of contiguous data blocks as a bundle, where each bundle contains semantic information to allow the application to make progress where an individual block may not. When the participating nodes are in communication range of each other, bundles are routed opportunistically between them in store-carry-forward manner and are often called mobile opportunistic network.

The communication in opportunistic networks is based on the proximity of mobile users using short-range mobile devices (such as smartphones and tablets) via wireless connections (such as Bluetooth or Wi-Fi technologies). The network topologies in opportunistic network scenarios are usually insufficient to maintain end-to-end connectivity and are assumed to be highly dynamic. The ability to store data on the devices and forward it to other devices carried by the mobile users makes the communication peer-to-peer in manner. However, taking the literature into consideration, it is proved that network performance is heavily reliant on the behavior of node movement. Hence, some part of the research should be dedicated toward the study of mobility traces and the formulation of realistic mobility models, as mobility is the key issue in OppNets. Many researchers are seeking ways to address different aspects of OppNets (Zhang, 2006; Martinez et al., 2011, Pereira et al., 2012, Grasic and Lindgren, 2012) and have analyzed factors such as network traffic, mobility and connectivity models, node characteristics and routing issues.

In conventional MANETs, networking protocols were generally evaluated using stochastic movement models such as the Random Walk Mobility model (RWM) (Einstein, 1956), the Random Waypoint mobility model (RWP) (Johnson and Maltz, 1996), the Random Direction Mobility model (RDM) (Royer et al., 2001) etc. However, users' mobility is dependent upon the users' movement behavior, their social and personal characteristics and environmental factors (Aschenbruck et al., 2011) because in reality users' movement is rarely random (Yoon et al., 2003); thus, the conventional stochastic random-based mobility models fail to evaluate the networking protocols accurately in OppNets. In the literature, there are various studies which demonstrate that users' social and personal behavior has a significant correlation with the users' movement patterns. For instance, in recent studies and experimental analysis, it has been proven that users frequently visit a few locations where they spend the most of their time (Song et al., 2010) and with which they have strong social relationships (Phithakkitnukoon et al., 2012). Furthermore, they rarely travel long distances and very often travel over short distances (Gonzalez et al., 2008). Hence, based on the mobility characteristics discussed in the following section, an appropriate selection of mobility traces and models is necessary in order to emulate the movement behavior of humans and vehicles.

4.1.2 Mobility Model Characteristics

Mobility models are used to represent the movement pattern of the nodes and how its location, acceleration and velocity changes over time in wireless ad hoc networks (WANETs); this in turn affects the performance of network protocols, applications and systems. Mobility models accurately exhibit the behavior of the mobile users' mobility in an OppNets, which is decisive for the evaluation of protocols for a specific type of mobility scenario. They are studied to predict the future state of network topology, control route reconstruction, minimize disruptions, reduce overheads, eliminate transmission of control packets and find routes in a timely manner. There are various characteristics/properties pertaining to the users' mobility. On the basis of behavioral patterns, human mobility can be categorized into three levels: strategic, tactical and operational (Hoogendoorn and Bovy, 2004). The *strategic* level describes the daily movement patterns of an individual, such as shopping, going to work or engaging in outdoor activities. The *tactical* level focuses on scheduling activities and route choice based on the set of activities (i.e., which is the shortest or fastest path to the destination) and the availability of time depending on the environmental factors (e.g., obstacles on the path or traffic congestion). The *operational* level describes the physical process of human movement (Hoogendoorn and Bovy, 2004). This level considers walking or driving speed, interaction with other nodes due to collision avoidance and queuing.

Each of the structural levels has a specific impact on the performance of opportunistic mobile systems. Inter-contact time is a crucial parameter for most routing protocols as it directly determines the message delay and the probability of successful message delivery. The strategic and tactical level decisions affect the inter-contact time, like the distribution of time between two consecutive meetings of specific nodes. Decisions taken on the operational level can affect the node connectivity and duration of contact, which determine the existence of the multi-hop path between distant users communicating over relay nodes.

Most of the research effort in mobility modeling is on the tactical and strategic level; operational-level mobility modeling, such as in (Jardosh et al., 2003; Helgason et al., 2010), has drawn less attention, but is thoroughly studied for planning emergency and evacuation strategies, where capturing the properties of the individual movements is of high importance. Currently, spatio-temporal properties of human mobility (Hsu et al., 2007; Mei and Stefa, 2009; Lee et al., 2009), and social aspects (Musolesi and Mascolo, 2007; Boldrini and Passarella, 2010) are characterized by popular mobility models.

Numerous models have been developed and they often provide a better approximation of human movement than the models commonly used in networking. There are numbers of parameters, like inter-contact time, contact time, remaining inter-contact time, return time, etc., which govern the nature of mobility. By studying these parameters, researchers try to model the real-life mobility traces. A number of mobility models have also been proposed based on the human social community and the behavior of social structures. Modeling mobility should investigate the actual scenario that determines which characteristics of mobility are necessary to capture and which characteristics can be abstracted away from the model to avoid unnecessary complexity. There are three properties of mobility models, which are discussed in the following (Sichitiu et al., 2009):

1. *Realism*: This describes the degree of accuracy to which the movement of mobile nodes works in a real scenario.
2. *Diversification*: This is the ability of mobility models to work in different scenarios, environments and with different types of mobile nodes.
3. *Complexity*: This measures the computational resources required to produce the simulation traces.

Mobility characteristics include the speed, predictability of movement pattern, and uniformity of mobile nodes in communication networking. Thus, the characteristics and properties of mobility models form the basis of classification of mobility models.

4.1.3 Classification of Mobility Models

Mobility models are broadly classified into three categories: trace-based models, synthetic models and stochastic models. The realistic mobility model is crucial for reliable performance evaluation and tracing mobile hosts in real-world scenarios. There is limited availability of real traces in the public domain due to the lack of a real working system in wireless ad hoc networks. They are pertinent to specific scenarios and it is very difficult to generalize their validity.

Synthetic models are the mathematical models that capture the movement of nodes by imposing constraints, like obstacles, pathways, etc., whereas the stochastic models are idealistic models that primarily rely on the random movements of the nodes (without imposing constraints) like random walk, random waypoint mobility models, etc. In general, synthetic models are largely preferred as they are not trace-driven, but they generally fail to evaluate the protocols accurately, so it is necessary to validate the mobility models by realistic traces.

Furthermore, many mobility models based on map, route and social networks are proposed in the literature. Map and route-based mobility models require traffic and user mobility patterns in order to evaluate and deal with networking issues in the opportunistic network, whereas a social-network-based mobility model follows social network theory. For the exponent, the social network mobility model captures the nature of moving individuals between inter and intra groups like disaster relief, battlefield, etc., and relies upon the structure of the relationship among the individuals. There is a significant amount of mobility models for the opportunistic networks. But no generalized mobility model is available so far which can be used for all the application scenarios; mostly they are limited to a few application scenarios. There are various mobility models which can help to form the taxonomy of mobility models (Figure 4.1).

4.2 Mobility Models

Mobility is the prominent feature of opportunistic networks, thus considering the mobility characteristics; the formulation of the realistic mobility model is very important, as the unrealistic model can give unrealistic results during simulation and cannot be deployed in the real world when evaluating the performance of the systems. Communication traffic patterns and mobility models are the key parameters of the protocol simulation. Thus, the formulation of mobility models that accurately mimic the expected real-world scenario is necessary. Many mobility models for the MANETs are reconsidered by exploiting the real users' movement behavior traces for the opportunistic networks.

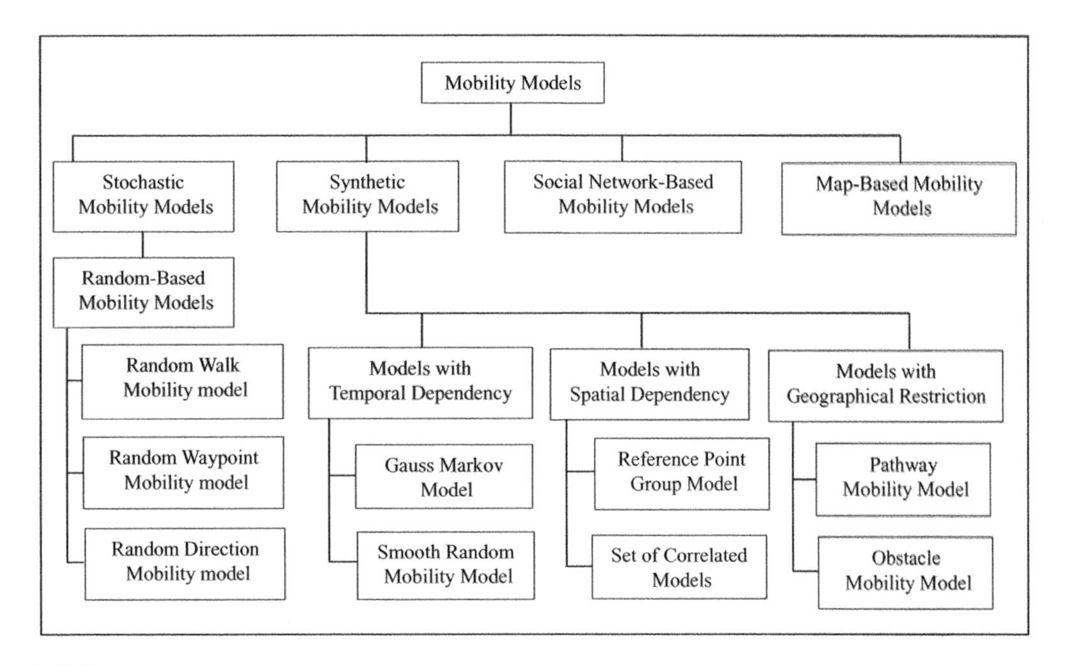

FIGURE 4.1
Taxonomy of mobility models.

4.2.1 Trace-Based Models

Trace-based models can have an insight into actual mobility patterns by tracing mobile hosts in real-world scenarios. For instance, experiments were conducted among students and researchers in Cambridge by the University of Cambridge and the Intel Research Laboratory (Chaintreau et al., 2005). They collected data about human movement by the Bluetooth devices and studied their co-location patterns. Similar projects were also carried out like the Wireless Topology Discovery project (McNett and Voelker, 2005), which studied the mobility characteristics and the access patterns of wireless PDA users on the University of California San Diego (UCSD) campus; the largest wireless local area network (WLAN) network mobility and geographic measurement project in the campus of Dartmouth College (Kotz and Henderson, 2005); and projects for collecting real traces for the mobile networking community, which were also initiated at Dartmouth (Kotz and Essien, 2005).

There are three methods for acquiring the traces (Aschenbruck et al., 2011). The first method is to trace the location of devices by monitoring with a particular tool. Currently Global Positioning System (GPS) is used for monitoring the localization system. The second method is to use the communication system for monitoring the communication devices. The accuracy of monitoring the communications is based upon the density of the access point to which communication devices communicate. This method may not be very precise, but it can be used to validate the mobility models. The third method is to acquire contact traces using Bluetooth or WLAN by monitoring the contacts among the mobile devices. For an opportunistic network, it is more interesting to have contacts between the devices than the actual location of devices, which helps to inspect the movement of devices. Currently, derivation of location-based traces from the traces of contacts is in focus, so that such characteristics may help in developing a new model and validating the existing one (Whitbeck et al., 2010). There are several repositories of real data traces like CRAWDAD (Kotz and Henderson, 2005),[1] UNC/FORTH,[2] MobiLib[3] which contain real traces.

4.2.1.1 Trace-Based Analysis

There are several problems involved with the trace-based models. The ping-pong effect can be observed between WLAN access points; for example, a building can have various access points and users, without moving into reality, recurrently switch between the access points (Yoon et al., 2006). This problem can be solved by accumulating the data over the access points. This approach definitely overcomes the ping-pong effect, but it comes with the cost of losing location accuracy. Another challenge with trace-based analysis is the variations in time, in which there is the existence of variation in cellular traces for different days of the week (Verkasalo et al., 2007). To derive mobility models by acquiring a sufficient number of samples is another challenge in the analysis of the traces. One method to deal with this problem is to accumulate traces by analyzing the parameter of traces from the several similar application scenarios. A similar nature of mobility of WLAN users in school campuses is observed, for example, "friendship" relationships between the students are asymmetric as they never encounter other students on the campus (Hsu and Helmy, 2005).

To generate new traces which resemble the real traces, WLAN traces are analyzed and are used for designing mobility models (Tuduce and Gross, 2005). The relevant statistics are extracted from the WLAN traces that record the session length of each user at each access point (AP) for processing. The mobility model creates a "realistic" number of

[1] http://crawdad.cs.dartmouth.edu/
[2] www.ist-mome.org/database/MeasurementData/index4475.html?cmd=datadetail&id=3873
[3] http://nile.cise.ufl.edu/MobiLib/

nodes at each access point by using the same distribution with a "realistic session length (Tuduce and Gross, 2005). In the past few years, acute research is being done by the researchers in the area of trace-based modeling for several application scenarios. Analysis of different scenarios of GPS traces and pre-processing of data is mandatory due to GPS trace error which causes a small change in direction (Lee et al., 2009; Rhee et al., 2011). There are several interesting characteristics of human mobility that should be tried to model the traces for different outdoor scenarios. The traces for modeling the campus scenarios are taken from the WLAN networks of the campus and proposed a WLAN mobility model (Tuduce and Gross, 2005). The traces of WLAN taken from the campus scenario show that it fails to capture all the user movement in the campus scenario and show various limitations while evaluating the performance of the communication systems.

There are various mobility models (Kim et al., 2006; Walsh et al., 2008; Hsu et al., 2009) that model campus scenarios and provide trace parameterized solutions, but the Campus Waypoint model (McNett and Voelker, 2005) and the statistical mobility model (Kotz and Essien, 2005) do not consider all the challenges discussed previously (i.e., ping-pong effect and variations in traces). The data traces match various other characteristics of the real-life environment like movement trace from a corporate office building, university campus and public networking (Kotz and Essien, 2005; Tang and Baker, 2000). For many scenarios, like battlefield scenarios, office scenarios (Minder et al., 2005; Rojas et al., 2005) and city scenarios (Rojas et al., 2005), data traces are not available; hence there is significant need of accurate mobility traces for various types of application scenarios.

4.2.2 Stochastic Mobility Model

Mobility models under stochastic mobility are very simple and relatively easy to study but show little or no resemblance to realistic scenarios. They generally rely on the random movement of nodes. In this section we discuss different stochastic mobility models.

4.2.2.1 Random-Based Mobility Models

The mobile nodes in random-based mobility models move independently in the network without any restrictions, with randomly chosen direction, speed and destination. The random-based mobility models are very easy to implement; hence, they are used in different types of simulation studies. Popular random-based mobility models are described in subsequent subsections and we briefly evaluate random-based mobility models in Section 4.2.2.1.4.

4.2.2.1.1 Random Walk Mobility Model

The Random Walk Mobility Model (Einstein, 1956), also called Brownian motion, is a very popular mobility model and was proposed by Einstein in 1926. In this, the nodes move independently from one location to another with randomly chosen direction and speed from predefined ranges between $[0, 2\pi]$ and $[\text{minspeed}, \text{maxspeed}]$ respectively at constant direction traveled d or at constant time interval t. If a node during movement reaches the simulation boundary, it bounces back with an angle of incoming direction and continues its movement along that direction and this effect is known as *border effect* (Bettstetter and Wagner, 2002). Many variations of random mobility model have been developed, including the 1D, 2D, 3D and d-D walks, and it is the most widely used mobility model because of its simplicity and wide availability (Bar-Noy et al., 1995; Decker, 1995; Rubin and Choi, 1997;

Zonoozi and Dassanayake, 1997; Garcia-Luna-Aceves and Madruga, 1999). The Random Walk mobility model helps to investigate the different set of system parameters by getting the mean cell sojourn time $E(S)$ of random movement of nodes (Rubin and Choi, 1997). It is used to track the random movement of nodes and calculates channel holding time and the handover number by partitioning the whole area into several regions according to previous, current and next motion directions of a mobile node (Zonoozi and Dassanayake, 1997). A pre-designed state transition matrix is used to give a movement pattern of mobile nodes by characterizing the mean duration of stay in the current position and the probability of choosing a moving path (Decker, 1995). The Random Walk model is a memoryless mobility model which is independent of its previous velocity and direction and thus is not considered suitable for wireless scenarios because it generates unrealistic mobility patterns (Liang and Haas, 1999).

4.2.2.1.2 Random Waypoint Mobility Model

The Random Waypoint mobility model is a very popular and widely used mobility model for evaluating the performance of routing protocols in wireless networks (Broch et al., 1998; Chiang and Gerla, 1998; Garcia-Luna-Aceves and Spohn, 1999; Johansson et al., 1999). It was proposed by Johnson and Maltz (1996). Random Waypoint extends on the Random Walk mobility model by incorporating a pause time between the change in direction and/or speed. In this mobility model, as shown in Figure 4.2, each mobile node moves independently and begins its movement by staying at one location for a specific period of pause time.

Once the time expires, it chooses a random destination and speed between the predefined ranges [minspeed, maxspeed] within the simulation area. The mobile node then moves to the randomly selected destination with the randomly selected speed and again stays for the specified period of pause time and repeats this process until the simulation ends. This model bears resemblance to the movement of the suspended particles in a fluid, so it is often called Brownian motion. RWP uses a bounding rectangle so, assuming the world is a torus, when nodes reach the destination they can bounce from the edge. The use of a bounded rectangle exponentially decomposes the time between the nodes which are in the wireless range of each other (Cai and Eun, 2007) and pose challenges toward RWP being realistic. However, by removing bounding rectangle, RWP becomes a power law and makes RWP more realistic.

It's very difficult to imagine realistic scenarios for the RWP mobility model. But, due to its simplicity, several scenarios can be modeled; for example, in campus scenarios taxies in a city are fairly easy to diversify by only changing the speed and pause time intervals, and, if needed, increasing the size of the bounding rectangle for modeling a large campus.

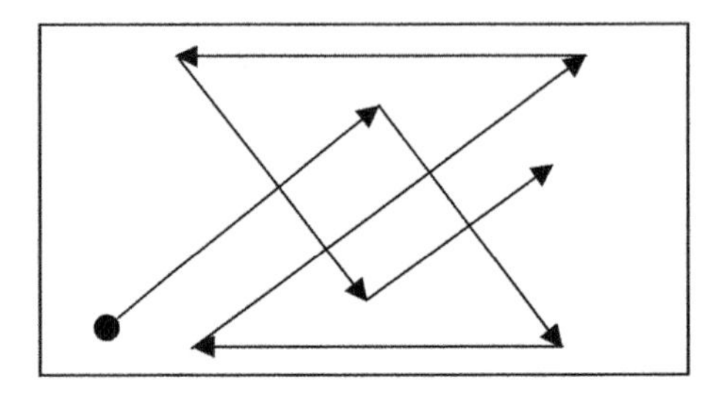

FIGURE 4.2
Node movement in the Random Waypoint Model.

4.2.2.1.3 Random Direction Mobility Model

The mobile nodes in the Random Direction Model, as shown in Figure 4.3 chooses a random direction and with a random speed it travels along that direction until it encounters an edge. Once it reaches the edge it again chooses a direction and repeats the process until the simulation ends. With a little modification to RDM, a Modified Random Direction mobility model is proposed by Royer et al. (2001).

In this model, there is no longer any need to travel to the simulation boundary to change direction. A mobile node can travel in a randomly chosen direction and can select a destination anywhere along that direction of travel. Like the RWM and RWP, the RDM model also exhibits sudden changes in speed and direction; hence, it is also an unrealistic mobility model. To avoid sudden changes, edge effect and unrealistic behavior in the mobility model, a Smooth Mobility model (SM), where the speed changes gradually, is proposed (Haas, 1997). In the Smooth Mobility model, each node is characterized by speed and direction. We will discuss this mobility model in detail in a later section. The position (x, y) of a node and its motion vector (v, θ), where v is the speed of a node and θ is the direction of a node are periodically updated at every Δt seconds as follows:

$$x(t+\Delta t) = x(t) + v(t)\cos(\theta(t)) \tag{4.1}$$

$$y(t+\Delta t) = y(t) + v(t)\sin(\theta(t)) \tag{4.2}$$

$$v(t+\Delta t) = \min[\max(v(t)+\Delta v, 0), V_{\max}] \tag{4.3}$$

$$\theta(t+\Delta t) = \theta(t) + \Delta\theta \tag{4.4}$$

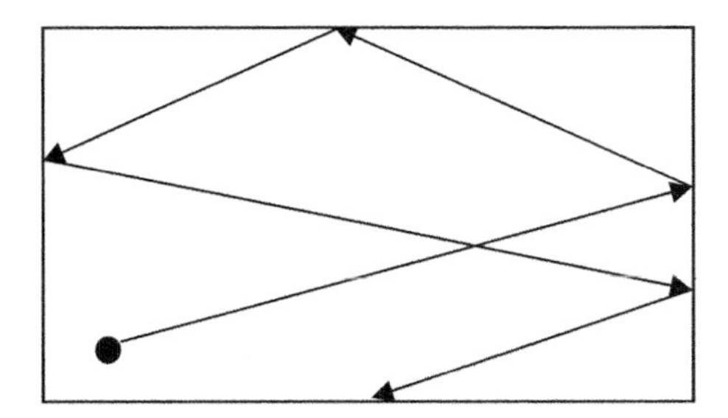

FIGURE 4.3
Node movement in Random Direction Model.

4.2.2.1.4 Evaluation of Stochastic Mobility Models

Random mobility models are formulated to model the movement of nodes in a very simplified way. They are widely accepted because of their simplicity but are not able to model most of the characteristics of realistic scenarios. Following are some characteristics of realistic scenarios that random mobility models are not able to capture:

1. *Temporal dependencies*: In many realistic scenarios, mobility behavior such as sudden changes in acceleration, sudden braking and sharp turns may not occur frequently; instead, they occur incrementally and there is a smooth change in

direction. In random mobility models, the velocity is a memoryless random process so they are not able to model realistic mobility behavior adequately.

2. *Spatial dependencies*: Nodes in most realistic scenarios do not move independently; rather, they move in a correlated manner, like on battlefields, during search and rescue operations and in museum touring. Node behavior may be influenced by neighborhood nodes and may work in a correlated manner which is not possible with random mobility models.
3. *Geographical dependencies*: Random mobility models can move freely within the simulation area without any restriction, but in a real scenario there may be a possibility of obstacles like trees, buildings, streets. Random mobility models do not take geographical obstruction into consideration and fail to mimic the real environment scenarios.

The Random Waypoint mobility model is very a flexible mobility model used to evaluate multicast protocols of the ad hoc networks, but it is restricted to cover a small portion of the simulation area as it does not move far from its initial position. In this model, initially all the mobile nodes are distributed randomly within the simulation area, and after the performance investigation of the Random Waypoint mobility model, it is found that the nodes' speed and pause time have a complex relationship with each other. For example, a mobile node with fast speed and shorter pause time provide more stability in the network than a node with slow speed and longer pause time. The Random Direction mobility model is an unrealistic mobility model, e.g., it is very unlikely that people would evenly distribute themselves within the simulation area and stay for a specific pause time at the edge of the given area. The modified random direction model, does not allow the nodes to remain at the edge; rather, they can change direction and can pause before reaching the simulation boundary. The movement pattern of the node in the Random Direction model is very similar to the movement pattern of the node in the Random Walk model with pause time.

Stochastic models fail to model real environments and many mobility behaviors of opportunistic networks. Thus, several mobility models were proposed to model the mobility behavior that exists in OppNets.

4.2.3 Synthetic Models

Synthetic models are the mathematical models that capture the movement of nodes by imposing constraints, like obstacles, pathways, etc. They do not rely on the random movement of nodes; rather, they move in a correlated manner to capture the movement of the nodes realistically. In this section we discuss various synthetic mobility models.

4.2.3.1 Temporal-Dependency–Based Mobility Models

The physical constraint of the mobile nodes affects the velocity of the mobile node which gradually or abruptly changes. The temporal dependency property describes the velocity of the nodes which are nearby to each other at two time instances i.e., the current velocity of the node is dependent upon the previous velocity. The temporal-dependency-based mobility model considers the property of temporal dependency and is proposed in the literature; here, two mobility models popular under this category are discussed.

4.2.3.1.1 Gauss–Markov Model

The Gauss–Markov mobility model was proposed by Liang and Haas (1999) for predicting the user mobility pattern in the wireless personal communication service (PCS) network. This model is created to circumvent the limitations of the randomness in the random-based mobility models by using velocity as a tuning parameter to vary the degree of randomness in the movement pattern.

In this mobility model each mobile node is initially assigned a current speed and direction, and the movement of each node occurs by updating its speed and direction at fixed intervals of time. The speed and direction of each mobile node are correlated over time and are calculated upon the previous speed and direction of the corresponding mobile node by using Equations 4.5 and 4.6 and the updated position of the node at each time interval is calculated based on the current position, speed and direction of the movement of the node by using Equation 4.7 and 4.8:

$$s_t = \alpha s_{t-1} + (1-\alpha)\bar{s} + \sqrt{(1-\alpha^2)} \cdot s_{x_{t-1}} \tag{4.5}$$

$$d_t = \alpha d_{t-1} + (1-\alpha)\bar{d} + \sqrt{(1-\alpha^2)} \cdot s_{d_{t-1}} \tag{4.6}$$

$$x_t = x_{t-1} + s_{t-1}\cos d_{t-1} \tag{4.7}$$

$$y_t = y_{t-1} + s_{t-1}\sin d_{t-1} \tag{4.8}$$

where, s_t and d_t are the updated speed and direction of the MN at time interval t.

$\alpha = 0 \le \alpha \le 1$, is the tuning parameter used to vary the randomness.

s and d are constants representing the mean value of speed and direction as $t \to \infty$ and $s_{x_{t-1}}$ and $s_{d_{t-1}}$ are random variables from a Gaussian distribution.

(x_t, y_t) and (x_{t-1}, y_{t-1}) are the x and y coordinates of the mobile node's position at the n_{th} and $(n-1)_{\text{st}}$ time intervals, respectively, and s_{t-1} and d_{t-1} are the speed and direction of the mobile node, respectively, at the $(n-1)_{\text{st}}$ time interval.

To prevent the nodes from exhibiting the undesired edge effect, they are forced to stay away from the simulation boundary by modifying the mean direction in the above equation when they are at a certain distance from the simulation boundary.

4.2.3.1.2 Smooth Random Mobility Model

Another temporal-dependency-based mobility model is a smooth random mobility model. This mobility model was proposed by Haas (1997), to avoid the unrealistic behavior of the nodes, the edge effect and the sudden changes of speed during the movement of the node, which were all present in the random-based mobility models (Figure 4.4).

In this mobility model the speed of the mobile node changes gradually rather than with sudden acceleration and sharp turns. Each node is defined by a motion vector (v, θ) where v and θ are the speed and direction respectively. At regular intervals of time, the position and the motion vector of the mobile node is updated based on the following equation:

$$v(t+\Delta t) = \min\left[\max\left(v(t) + \Delta v, 0\right), v_{\max}\right] \tag{4.9}$$

$$\theta(t+\Delta t) = \theta(t) + \Delta\theta \tag{4.10}$$

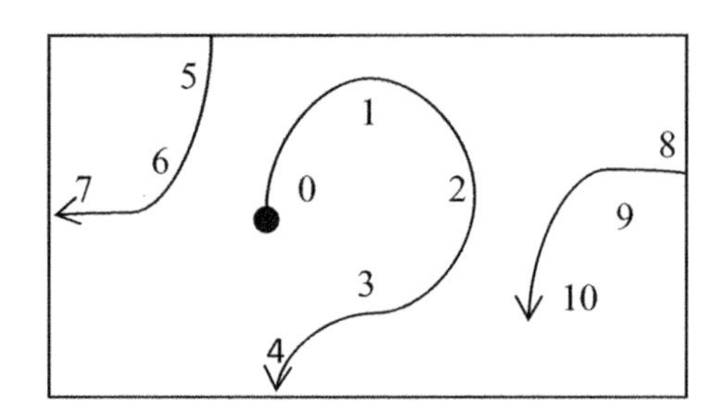

FIGURE 4.4
Node movement in Smooth Random Mobility Model.

$$x(t+\Delta t) = x(t) + v(t)\cos(\theta(t)) \quad (4.11)$$

$$y(t+\Delta t) = y(t) + v(t)\sin(\theta(t)) \quad (4.12)$$

where, Δv and $\Delta\theta$ is the change in speed and direction respectively and are random variables chosen to be small for a smooth trajectory of the node; x and y are the position of the mobile node and $v_{\max}$ is the maximum speed.

4.2.3.2 Spatial-Dependency–Based Mobility Models

The mobility models under this category are characterized by the spatial dependency property, in which the movement of the mobile node is correlated by the neighboring nodes. Various spatial-dependency-based mobility models, including the Reference Point Group Mobility model (RPGM), the Pursue mobility model, the Community mobility model and the Nomadic community model, are proposed in the literature which are briefly discussed below.

4.2.3.2.1 Reference Point Group Mobility Model

The Reference Point Group mobility model is one of the most popular mobility models, proposed by Hong et al. (1999). In the Reference Point Group mobility model, each mobile node in a group is randomly distributed around the predefined reference point $\overrightarrow{RP}$, which allows individual random movement of nodes along with the group motion behavior. Each group of nodes has a logical center that defines the entire group movement via the group motion vector, $\overrightarrow{GM}$. The location of the individual reference point is updated according to the group's logical center when the reference point moves from t to $t+1$. The motion vector $\overrightarrow{GM}$ can be designed on the basis of predefined paths or can be randomly chosen. Once the group motion vector reaches the checkpoint it again computes the next checkpoint and moves to that checkpoint. With proper selection of checkpoint the RPGM can model various realistic application scenarios. The RPGM model (Hong et al., 1999) has modeled three application scenarios:

1. *In-place mobility model*: In this model, different working groups in different adjacent locations within the simulation area are working with different movement patterns.
2. *Overlap mobility model*: In this model, different working groups work within the same simulation area with different movement pattern.

3. *Convention mobility model*: In this model, different working groups work in their respective adjacent locations with each other and, depending upon the scenario, some nodes from one working group can move to another working group working in the adjacent location.

RPGM can model various other realistic application scenarios such as military, disaster relief, search and rescue, etc.

4.2.3.2.2 *Column Mobility Model*

The Column mobility model was proposed by Sanchez (n.d.), and is a spatially dependent group mobility model useful for searching and scanning purposes. In this mobility model, a set of mobile nodes uniformly moves forward in a particular direction by forming a line. For example, a group of soldiers marching toward their enemy by forming a line. A minor modification of the mobility model is described by Sanchez (n.d.) where a set of mobile nodes are placed in a single-file line and are allowed to follow one another about their initial position. The node's reference position is updated by the following equation:

$$\text{next\_reference\_position} = \text{prev\_reference\_position} + \text{advance\_vector} \tag{4.13}$$

where, prev_reference_position is the initial location of the node, advance vector is the predefined offset for moving the reference grid, which is calculated by a random distance and a random angle between 0 to π.

$$\text{next\_position} = \text{next\_reference\_position} + \text{random\_vector} \tag{4.14}$$

The node's next position is updated by the sum of $\text{next\_reference\_position}$ and random_vector, which is a random offset given in the above equation.

4.2.3.2.3 *Pursue Mobility Model*

The Pursue mobility model is another spatially-dependent-based mobility model (Sanchez, n.d.; Bergamo et al., 1996), in which mobile nodes attempt to track a particular target. It can be used for law enforcement and signal source tracking, as the nodes in the model try to capture the single target node. This is also based on the Reference Point Group mobility model and the node being pursued follows the Random waypoint mobility model for the movement within the simulation area (Figure 4.5).

The position of the node in the Pursue mobility model is updated by the following equation:

$$\begin{aligned}\text{Next\_position} = {} & \text{Prev\_position} + \text{acceleration}\left(\text{target} - \text{prev\_position}\right) \\ & + \text{random vector}\end{aligned} \tag{4.15}$$

The next position of the node is calculated based on the current position of the node, a random vector which is a tuning parameter to maintain the randomness for the effective tracking of the mobile node being pursued. Target is the expected position of the node being pursued.

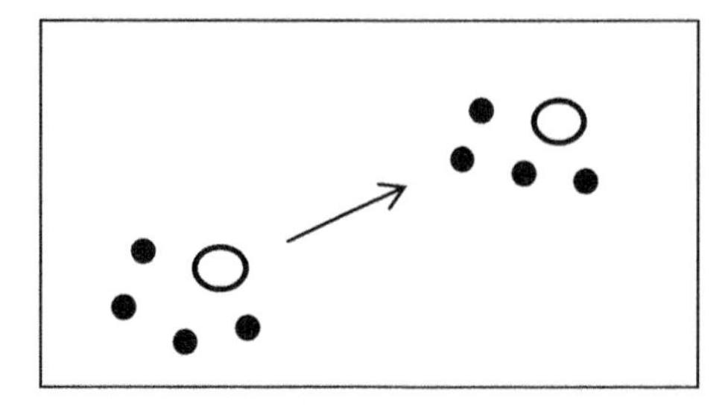

FIGURE 4.5
Movement of 4 nodes in Pursue Mobility Model.

4.2.3.2.4 Nomadic Community Mobility Model

The Nomadic Community mobility model (Sanchez, n.d.; Sanchez and Manzoni, 2001) is another type of correlated mobility model in which a group of mobile nodes randomly moves together from one location to another. Each node in the mobility model follows an individual mobility model, such as a random mobility model, to move around the reference point. When the reference point changes, the whole group of nodes moves to another location defined by the reference point and begin to move around the newly changed reference point. Compared to the Column mobility model, the Nomadic Community model shares the reference grid and the movement of nodes is sporadic (Figure 4.6), whereas the Column mobility model has its own reference point and the movement of nodes is more or less constant.

The general movement of the group determines the reference point of each node. The position of the node is updated by the following equation:

$$\text{next\_position} = \text{Predefined\_reference\_position} + \text{random\_vector} \tag{4.16}$$

Each node's predefined reference point can be offset by the random vector for the random motion of the node. The Nomadic Community mobility model can be used in various types of scenarios, such as military scenarios, agricultural scenarios, etc.

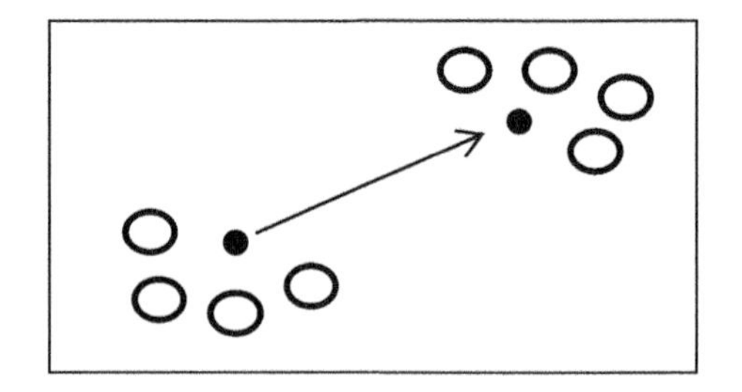

FIGURE 4.6
Movement of 4 nodes in Nomadic Community Mobility model.

4.2.3.3 Geographical-Restriction–Based Mobility Model

In random mobility models, mobile nodes are free to move anywhere within the simulation area without any geographical restrictions like environmental obstacles such as pathways, streets, buildings, etc. Geographical-restriction-based mobility models are models which are bounded by environmental restrictions. Various mobility models are proposed under this category, and are briefly discussed in the following.

4.2.3.3.1 *Pathway Mobility Model*

The Pathway mobility model is a geographical-restriction-based mobility model where the model is bounded by pathways. In this model, the map is predefined for the simulation area and it models the map as a random graph where vertices in the graph represent the buildings and edges represent pathways (Tian et al., 2002). The movement of the mobile node begins by placing the node onto the edges, i.e., pathways. The node begins its movement to a randomly chosen destination and moves toward the destination through the shortest path following the edges. On reaching the destination, it stays there for a specified pause time, then moves again toward a new randomly chosen destination. The Pathway mobility model also shows some form of randomness, but it is bounded by pathways, i.e., it can move anywhere within the simulation area but only on the pathways, unlike random-based mobility models. The Pathway mobility model can be applied in various scenarios, like on university campuses, streets etc.

4.2.3.3.2 *Obstacle Mobility Model*

The Obstacle mobility model is another geographical-restriction-based mobility model which tries to model the movement of the node considering the environmental obstacles. Environmental obstacles affect the movement pattern of the nodes as when the node encounters the obstacle it has to change its path. Therefore, obstacles play a significant part in modeling the mobility of the node. Various mobility models were proposed by the researchers to avoid the environmental obstacles. Johansson et al. (1999) developed three realistic scenarios to avoid the obstacles.

1. *Event coverage scenario*: In this scenario, a group of people are highly mobile and frequently changing location within the scenario.
2. *Conference scenario*: With 50 people in a conference room scenario; only a few people are moving with low mobility and most of them are static.
3. *Disaster relief scenario*: In the disaster relief scenario, some nodes are highly mobile and some are less mobile depending upon the application scenario.

In all of the three scenarios discussed above, the obstacles are placed within the simulation area as a rectangular box and the mobile node tries to avoid the obstacles by choosing the correct movement trajectory. Jardosh et al. (2003) tries to investigate the effect of the obstacles on the mobility of the nodes. They placed the obstacle in the simulation field to model the buildings of the UCSB campus. The authors found that the people in the campus follow a predefined pathway within the simulation area rather than moving randomly. So, a Voronoi graph is computed for the construction of the pathways based on the building's location. Once the pathway graph is constructed by joining the intersection of buildings or obstacles with Voronoi graph, nodes are restricted on the pathways, i.e., they are allowed to travel on the pathways and can enter and exit the buildings because of the campus scenario. Nodes move in a predefined pathway graph through the shortest distance to the randomly selected destination. The shortest distance in Voronoi graph is computed by the Dijikstras algorithm.

The restriction of predefined pathways limits the application scenarios and makes the application scenarios more or less static; also, in some scenarios, choosing the shortest path to reach the destination is not possible. To address these issues, Huang (2005) proposed a Delaunay model in which nodes can avoid the obstacles without passing through them. The detour pathways are computed with the shortest pathways by avoiding obstacles and also with the number of pathways. But the detour method does not seem to be feasible as it requires complex calculation for the computation of the pathways and fails to represent the group motion behavior. Tan et al. (2002) proposes a realistic human motion group behavioral model based on the individual complex behavioral interaction. The geographic restriction restricts the movement of the nodes and in turn affects the mobility of the nodes in the network. Hence, researchers are trying to mimic the movement pattern which realistically captures the movement of the nodes in the network.

4.2.3.3.3 Manhattan Mobility Model

The Manhattan mobility model (Bai et al., 2003) is also a geographical-restriction-based mobility model which uses predefined grid topology for the movement of the nodes. It models the movement pattern of the nodes in well-organized streets of urban areas. It moves in a horizontal and vertical direction by using a probabilistic approach to move straight, left and right on the street. For moving straight at each intersection of nodes it chooses a probability of 0.5 and a probability of 0.25 for a left or right turn. It is not suitable for highway scenarios, but is allowed to move left or right with certain probabilities, i.e., it is allowed to change direction.

4.2.3.3.4 City Section Model

The City Section mobility model (Davies et al., 2000) mimics the realistic movement of the node in a section of a city. The mobile nodes in the mobility model follow the predefined path and are restricted by obstacles and traffic regulation as in the real scenarios. Each mobile node begins its movement from a defined point on the street toward the randomly chosen destination through the shortest path. After reaching the destination, it stays at that location for a specific pause time and again it chooses a random destination for the movement and repeats the process until the simulation ends. The City Section mobility model presents the realistic movement of the node by restricting the movement pattern, i.e., following the predefined path like in real scenarios as people move on a predefined path following traffic regulations. Researchers are considering improvements to the City Section mobility model and have improved a lot, like in the pause time of the model, incorporation of changes in velocity, increasing number of streets, expanding the simulation area, high speed road, etc.

4.2.3.4 Evaluation of Synthetic Mobility Models

Synthetic mobility models are designed for realistically modeling the movement of nodes in a correlated manner with physical constraints. They are widely accepted because of their realistic modeling of the movement pattern and group motion behavior. Following are some characteristics of realistic scenarios that synthetic mobility models are able to capture:

1. *Temporal dependencies*: Synthetic mobility models are able to deal with realistic scenarios with mobility behavior where nodes exhibit incremental acceleration and smooth changes of direction within the simulation area.

2. *Spatial dependencies*: The nodes in most of the synthetic mobility models do not move independently; rather, they move in a correlated manner, like on the battlefield, search and rescue operations and in museum touring. They influence the neighborhood nodes and may work in a correlated manner.
3. *Geographical dependencies*: Synthetic mobility models deal with geographical restrictions as in a real scenario there may be a possibility of obstacles like trees, buildings, streets. Hence, geographical-restriction-based mobility models take geographical obstructions into consideration and mimic the real environment scenarios.

Synthetic models are able to model real environments and many mobility behaviors, including the group motion behavior of opportunistic networks. Thus, several mobility models were proposed to model the mobility behavior that exists in OppNets.

4.2.4 Map-Based Mobility Models

Map-based mobility models are based on the traffic and user mobility patterns in order to evaluate and deal with networking issues in the opportunistic network. Nodes in map-based mobility models, move randomly on a predetermined path defined by the map data in a well-known text format file. In map-based mobility models, a group of nodes can select a certain destination or point on the map and after traveling to a specified distance they pause for a certain pause time and then again begin their journey. There are various map-based mobility models which will discuss in subsequent Sections.

4.2.4.1 Route-Based Map Mobility Model

The Route-based Map Movement (RBMM) mobility model mimics the movement of nodes on the basis of predefined routes on the simulation maps. Nodes in this model have predetermined routes that they follow in the simulation map after getting the map data, but during the movement of nodes the model does not select the destination randomly; instead, they select the next destination in which they are currently moving. The routes in the model have many stop points which the nodes follow while traveling; they stop for a specified pause time on the stop points to reach the destination with shortest path for modeling the movement pattern (Ekman et al. 2008). Route-based Map Movement mobility model can be used to model the movement of various real-life application scenarios, like bus routes, train routes, etc. For modeling the vehicular motion behavior, a variant of the RBMM mobility model was proposed, namely STreet Random Waypoint (STRAW) mobility model, which uses real street scenarios, where the nodes travel on the predefined roads or routes defined by the street map data to model the realistic vehicular traffic pattern (Choffnes and Bustamante, 2005).

4.2.4.2 Rush Hour (Human) Traffic Model

The Rush Hour Traffic model (Seah et al., 2006) deals with the traffic condition of rush hour. The Rush Hour Traffic model focuses on high traffic area and choice of destination, as people tend to travel to nearby places rather than going further destinations. For deciding the node's destination, the distribution is done by a destination allocation algorithm based on the exponential distribution. Several improvements in the rush hour traffic model are required as it does not consider geographical restriction, temporal dependencies, etc.

4.2.4.3 Working Day Movement Model

The Working Day Movement model was proposed by Ekman et al. (2008) to capture movement patterns of real-life scenarios. This model was proposed for delay tolerant networks to deal with inter-contact time and contact time distributions that are found in the traces of real-world measurement experiments. The authors have validated their model with the ONE simulator. The model deals with the everyday work routine of people and tries to simulate the work routine and compare its statistical features to real-world traces. The Working Day model is developed with four submodels, depicting the work routine which most of the people follow every day and repeatedly.

1. *Home activity submodel*: This scenario is used in the evening and at night for modeling the home activities. In this, the node is assigned a point in the map as its home location and it walks for a short distance and remains there until the wakeup time. The model does not consider the movement inside the home (Ekman et al., 2008).
2. *Office activity submodel*: This model shows the movement of the employee in the office (e.g. from the employee's desk to the conference or meeting location) (Ekman et al., 2008).
3. *Evening activity submodel*: This models the activities which people can do in the evening in groups after work, like taking an evening walk, roaming around the streets, going into restaurants etc., (Ekman et al., 2008).
4. *Transport submodel*: This submodel models the movement of nodes between office, home and evening activities. This can be done by the walking submodel, the car submodel or the bus submodel (Ekman et al., 2008).

The nodes move on a map which contains all the locations of home, office and evening activities, and defines spaces and routes for the movement of the nodes. The Working Day Movement model tries to model each scenario which people follow in everyday life.

4.2.4.4 Shortest Path Map-Based Movement

The Shortest Path Map-based Movement (SPMBM) mobility model (Keränen and Ott, 2007) proposed for DTNs is based on the shortest path available within the simulation map scenario. In this model, rather than moving randomly around the simulation map, a shortest path is chosen over the available paths between two random nodes and Points of Interest (POIs) from the simulation map, as people mostly use the shortest path to travel. Point of Interests (POI) is the places on the simulation map that node within the group travel. In the Shortest Path Map-based mobility model, all the nodes are placed randomly on the simulation map and then nodes travel to a certain destination in the simulation map following the Dijikstra algorithm to discover the shortest path from the available paths on the simulation map. After reaching a specified destination, the nodes pause for a certain period of time and again move to a newly selected destination through shortest path.

4.2.4.5 Evaluation of Map-Based Mobility Models

Map-based mobility models generally model movement pattern based on the map, the routes people follow, the traffic pattern, the working activities pattern, etc. These mobility models are useful in modeling various real-life scenarios, but the map-based mobility models have certain limitations in that the paths in the model are predefined, which, in some cases, is not possible in real life. The nodes always follow the predefined route of the map data, which fails to accurately represent real human movement. Researchers are looking forward to model more real-life application scenarios which are related to route- and map-based movements.

4.2.5 Social-Network–Based Mobility Models

The Social-network-based mobility model captures the behavior of movement based on human decisions and the social nature of humans or the entities just like the social behavior of humans in a battlefield or during disaster relief, etc., and relies upon the structure of the relationships among the individuals.

4.2.5.1 Community-Based Mobility Models

The Community-based Mobility Model (CMM) (Musolesi and Mascolo, 2006) is a social-network-based mobility model which mimics the movement pattern based on social network. The CMM mobility model groups the node according to the community to which they belong. Nodes with the same community are grouped as friends, and nodes which belong to different communities are grouped as non-friends. Initially a cell is allotted to each community where they share a social link between the friends and non-friends community for the movement of the nodes in the network. The gregarious behavior of the nodes is the main drawback of this model where, when a node decides to exit the community, all other nodes follow the node belonging to the same community.

To resolve the limitations, a Home-cell Community-Based Mobility (HCM) model is proposed. According to this model, which is a variant of the CMM model, other the than home community some nodes can also have social links with the other community (Hsu et al., 2007). Hence, the HCM mobility model deals with the gregarious behavior, and the probability of the nodes leaving their home community for the destination community are based on the social links of the node with the other nodes belonging to the destination community (Veeramani et al., 2011).

4.2.5.2 Social Network Models

Herrmann (2003) proposed a social mobility model in which the movement of the users is influenced by the relationships among them. In this model, artificial users periodically move between a set of abstract locations. A set of abstract locations, which they have to visit at fixed time, is given to the users. Musolesi et al. (2004) proposed a social network model is based on social network theory as it depends upon the human relationships. The model allows a number of hosts to be grouped together based on the social relationships among them and mapped to a topographical space. It models the social network through a weighted graph of the relationships. Social network models are based on the social relationship between the nodes, which affects the movement pattern of the node in the network; hence, it is necessary to model the social behavior based on social network theory.

4.3 TRANSIMs

The TRansportation ANalysis SIMulation System (TRANSIMs) (Smith et al., 1995) is a US-DOT, Federal Transit Administration, Federal Highway Administration and the Environmental Protection Agency sponsored integrated system of travel forecasting and decision support tools developed to conduct analysis and modeling of the regional transportation system. Based on the cellular automata (a concept originally discovered in 1940s by Stanislaw Ulam and John Von Neuman) microsimulator, TRAINSIMs is capable of simulating the movements and activities of individual travelers and vehicles through the transportation network of a large urban population. With its ability to model the multimodal transportation conditions, TRANSIMS makes the transportation planning process accurate as well as flexible.

4.4 Testing Tools

Simulation is the most common method, and an invaluable tool, in analyzing the performance of ad hoc network protocols, as it quickly explores the large systems; they are an imitation of the real-world system, with key characteristics of selected systems. Among the different methods: analytical, simulation, emulation and testbed experiments for validating network protocols, simulation offers several advantages, including reproducibility, parameters isolation, scalability and metrics exploration. Among numerous available simulators, such as GloMoSim (Zeng et al., 1998), OPNET Modeler (2004), NetSim (Wong et al., 1990), etc., ns-2 (a monarch extension) (The ns-2 Project, 2008), is the most popular simulation tool for evaluating the performance of network protocols. It is a discrete event simulator written in C++ and OTcl. Its active development and maintenance stopped in 2010 but volunteers from the user community are still developing it. Ns-3 (The ns-3 Project, 2003) is the actively developed new version of ns-2; it is written in C++ and Python with scripting capability, and is not a backward compatible extension of ns-2. The class of discrete event simulators includes many simulators which simulate not only the complex computer system but can also simulate biological, industrial, economic and systems of various other disciplines. Generally, these types of simulators perform the operation of systems as a discrete sequence of events (i.e., each sequence of events occurs at a particular instant of time and no change of system occurs in between consecutive events). SimEvents (Clune et al., 2006), SIMUL 8 (Hauge and Paige, 2004), Simio (Kelton et al., 2011), GoldSim (Guide et al., 2007), FlexSim (Gelenbe and Guennouni, 1991), ExtendSim (Diamond et al., 2007), Enterprise Dynamics (Rabelo et al., 2005), Simcad Pro (Adra, 2005) etc., are examples of commercial discrete event simulators. Simula (Dahl and Nygaard, 1965), OMNET++ (Varga, 2005), Parsec (Bagrodia et al., 1998), CPN Tools (2005), JaamSim (King and Harrison, 2013) are examples of open source discrete event simulators.

Despite the several advantages offered by simulation, it still may not accurately model realistic scenarios and is not able to reflect the true performance of the network system. Results obtained by different simulators may show significant differences and divergent results, which is shown by many authors. There is a significant amount of difference in the results obtained by some popular simulators (OPNET Modeler,

GloMoSim and ns-2) (Cavin et al., 2002). Taking mobility modeling into consideration, the presence of few set of nodes and limited duration of trace results create critical issues in simulation. Therefore, there is thriving interest in testing systems and networking protocols with various tools that reproduce the movement traces accurately. STRAW (Choffnes and Bustamante, 2005) uses real vehicular US cities for modeling the vehicular traffic and provides more accurate simulation results by using a vehicular mobility model. Mobility simulator (Haerri et al., 2005) is another simulator that realistically models scenarios, funded by a Department of Transportation (DOT) smart city challenge. TRANSIM (Smith et al., 1995) is another tool that forecasts the transportation planning and emissions analysis, and was discussed in detail in Section 4.3. Tools for simulations and the validations of the results are increasing and are open for future work.

4.5 Impact of Mobility Models on the Performance of Opportunistic Networks

Opportunistic networks can work in a highly mobile environment, which in turn affects the performance of the systems. Not surprisingly, the use of different mobility models show significant differences in the performance of the networking protocols. The highly dynamic network, mobility characteristics and variation in the performance metrics has affected the performance of networking protocols, since messages in opportunistic networks are transferred in peer-to-peer fashion when two nodes are in contact with each other. The throughput of a node can increase with the highly dynamic network as compared to the fixed networks (Grossglauser and Tse, 2001). In general, opportunistic networks are characterized by a small diameter, a so called "small world" between the mobile nodes (Chaintreau et al., 2007). The results are validated by capturing the human mobility and computing all the paths that impact on the diameter of the growing network size of an opportunistic network (Chaintreau et al., 2007). For performance evaluation of the mobility model, protocol dependent and independent metrics are used (Bai et al., 2003); protocol independent metrics are directly extracted from the traces of the mobility models. They are independent of the networking protocols, whereas protocol dependent metrics are dependent on the network protocols running on the system. These metrics influence the performance evaluation of the system like link duration, link breakage etc., (Lenders et al., 2006). Throughput and delay performance are affected by the node correlation in the RPGM model (Ciullo et al., 2011). The analysis of the clustered movement of nodes in the routing algorithm for the opportunistic network shows a reduction in throughput and an increase in delay for sparse but highly clustered distributions of nodes. Flooding time has great impact on opportunistic networks with realistic mobility models (Shin et al., 2007). Moreover, the performance of protocols and contact trace-based metrics is strongly affected by the diffusive behavior of mobile nodes (Kim and Lee, 2010).

The social mobility impact on opportunistic networks was simulated and the results show that the exploitation of the node movement behavior always gives better results; with a greater degree of sociability, the packet delivery probability will be greater with minimal delay (Ciullo et al., 2011). There are several other interesting areas where the mobility model plays a significant role in opportunistic networks.

4.6 Future Research Directions in Mobility Modeling

The broad concept and challenges for mobility models in opportunistic networks described in the chapter have been widely understood and are a topic of active research in the research community. The concept of opportunistic networks is built over the foundation laid by the MANETs. The need of network protocol evaluation and the problems associated with the mobility opens the door for further research. Researchers are looking forward to the challenges involved in developing new mobility models that combine the best properties of other models for the accurate modeling of mobility patterns. An avenue of future work is to scrutinize the entire entities involved in the real world for the production of the realistic mobility model. Further, a minimum standard for the thorough evaluation of the performance of different mobility models should be allowed in conjunction with the study of mobility metrics. Finally, acquiring traces for realistic application scenarios is a very significant challenge in view of future research.

References

Adra, H. (2005, December). CreateASoft Inc.: Simcad Pro 7.1. In *Proceedings of the 37th Conference on Winter Simulation*, Orlando, Florida (p. 72). Winter Simulation Conference.

Aschenbruck, N., Munjal, A., and Camp, T. (2011). Trace-based mobility modeling for multi-hop wireless networks. *Computer Communications*, 34(6), 704–714.

Bagrodia, R., Meyer, R., Takai, M., Chen, Y. A., Zeng, X., Martin, J., and Song, H. Y. (1998). Parsec: A parallel simulation environment for complex systems. *Computer*, 31(10), 77–85.

Bai, F., Sadagopan, N., and Helmy, A. (2003, March). IMPORTANT: A framework to systematically analyze the Impact of Mobility on Performance of RouTing protocols for Adhoc NeTworks. In *INFOCOM 2003. Twenty-second Annual Joint Conference of the IEEE Computer and Communications. IEEE Societies*, San Francisco, CA, (Vol. 2, pp. 825–835).

Bar-Noy, A., Kessler, I., and Sidi, M. (1995). Mobile users: To update or not to update? *Wireless Networks*, 1(2), 175–185.

Bergamo, M., Hain, R. R., Kasera, K., Li, D., Ramanathan, R., and Steenstrup, M. (1996). System design specification for mobile multimedia wireless network (MMWN)(draft). DARPA project DAAB07-95-C-D156.

Bettstetter, C. and Wagner, C. (2002). The spatial node distribution of the random waypoint mobility model. *WMAN*, 11, 41–58.

Boldrini, C. and Passarella, A. (2010). HCMM: Modelling spatial and temporal properties of human mobility driven by users' social relationships. *Computer Communications*, 33(9), 1056–1074.

Broch, J., Maltz, D. A., Johnson, D. B., Hu, Y. C., and Jheva, J. (1998, October). A performance comparison of multi-hop wireless ad hoc network routing protocols. In *Proceedings of the 4th Annual ACM/IEEE International Conference on Mobile Computing and Networking*, Dallas, Texas (pp. 85–97).

Cai, H. and Eun, D. Y. (2007, September). Crossing over the bounded domain: From exponential to power-law inter-meeting time in manet. In *Proceedings of the 13th Annual ACM International Conference on Mobile Computing and Networking*, Montréal, Québec (pp. 159–170).

Cavin, D., Sasson, Y., and Schiper, A. (2002, October). On the accuracy of MANET simulators. In *Proceedings of the Second ACM International Workshop on Principles of Mobile Computing*, Toulouse, France (pp. 38–43).

Chaintreau, A., Hui, P., Crowcroft, J., Diot, C., Gass, R., and Scott, J. (2005). *Pocket Switched Networks: Real-world Mobility and its Consequences for Opportunistic Forwarding* (No. UCAM-CL-TR-617). University of Cambridge, Computer Laboratory.

Chaintreau, A., Mtibaa, A., Massoulie, L., and Diot, C. (2007, December). The diameter of opportunistic mobile networks. In *Proceedings of the 2007 ACM CoNEXT Conference*, New York, NY (p. 12). ACM.

Chiang, C. C. and Gerla, M. (1998, October). On-demand multicast in mobile wireless networks. In *Network Protocols, 1998. Proceedings. Sixth International Conference*, Austin, TX (pp. 262–270). IEEE.

Choffnes, D. R. and Bustamante, F. E. (2005, September). An integrated mobility and traffic model for vehicular wireless networks. In *Proceedings of the 2nd ACM International Workshop on Vehicular Ad Hoc Networks*, Cologne, Germany (pp. 69–78). ACM.

Ciullo, D., Martina, V., Garetto, M., and Leonardi, E. (2011). Impact of correlated mobility on delay-throughput performance in mobile ad hoc networks. *IEEE/ACM Transactions on Networking (TON)*, 19(6), 1745–1758.

Clune, M. I., Mosterman, P. J., and Cassandras, C. G. (2006, July). Discrete event and hybrid system simulation with simevents. In *Proceedings of the 8th International Workshop on Discrete Event Systems*, Ann Arbor, MI (pp. 386–387).

Dahl, O. J. and Nygaard, K. (1965). SIMULA-A language for programming and description of descrete event systems. Tech. rep., Norwegian Computing Center.

Davies, V. A. (2000). *Evaluating Mobility Models within an Ad Hoc Network* (Master's thesis, advisor: Tracy Camp, Dept. of Mathematical and Computer Sciences. Colorado School of Mines).

Decker, P. (1995). http://www.comnets.rwth-aachen.de/dpl/thesis/node1.html.

Diamond, B., Lamperti, J. S., Krahl, D., Nastasi, A., and Damiron, C. (2007). *ExtendSim User Guide*. San Jose: Imagine That Inc.

Einstein, A. (1956). *Investigations on the Theory of the Brownian Movement*. Courier Corporation.

Ekman, F., Keränen, A., Karvo, J., and Ott, J. (2008, May). Working day movement model. In *Proceedings of the 1st ACM SIGMOBILE Workshop on Mobility Models*, Hong Kong, China (pp. 33–40). ACM.

Fall, K. (2003, August). A delay-tolerant network architecture for challenged internets. In *Proceedings of the 2003 Conference on Applications, Technologies, Architectures, and Protocols for Computer Communications*, Karlsruhe, Germany (pp. 27–34). ACM.

Garcia-Luna-Aceves, J. J. and Madruga, E. L. (1999, March). A multicast routing protocol for ad-hoc networks. In *INFOCOM'99. Eighteenth Annual Joint Conference of the IEEE Computer and Communications Societies. Proceedings. IEEE*, New York, NY (Vol. 2, pp. 784–792). IEEE.

Garcia-Luna-Aceves, J. J. and Spohn, M. (1999, October). Source-tree routing in wireless networks. In *Seventh International Conference on Network Protocols, 1999. (ICNP'99) Proceedings*, Toronto, Ontario (pp. 273–282). IEEE.

Gelenbe, E. and Guennouni, H. (1991). FlexSim: A flexible manufacturing system simulator. *European Journal of Operational Research*, 53(2), 149–165.

Gonzalez, M. C., Hidalgo, C. A. and Barabasi, A. L. (2008). Understanding individual human mobility patterns. *Nature*, 453(7196), 779–782.

Grasic, S. and Lindgren, A. (2012, August). An analysis of evaluation practices for dtn routing protocols. In *Proceedings of the Seventh ACM International Workshop on Challenged Networks*, Istanbul, Turkey (pp. 57–64). ACM.

Grossglauser, M. and Tse, D. (2001). Mobility increases the capacity of ad-hoc wireless networks. In *INFOCOM 2001. Twentieth Annual Joint Conference of the IEEE Computer and Communications Societies. Proceedings. IEEE*, Achorage, AK (Vol. 3, pp. 1360–1369). IEEE.

Guide, G. U. S. (2007). GoldSim Technology Group.

Haas, Z. J. (1997, October). A new routing protocol for the reconfigurable wireless networks. In *1997 IEEE 6th International Conference on Universal Personal Communications Record, 1997. Conference Record*, San Diego, CA (Vol. 2, pp. 562–566). IEEE.

Haerri, J., Fiore, M., Filali, F., Bonnet, C., Chiasserini, C. F., and Casetti, C. (2005). *A Realistic Mobility Simulator for Vehicular Ad Hoc Networks*. Eurécom Technical Report, Institut Eurécom, France.

Hauge, J. W. and Paige, K. N. (2004). *Learning SIMUL8: The Complete Guide*. PlainVu.

Helgason, Ó., Kouyoumdjieva, S. T., and Karlsson, G. (2010, February). Does mobility matter? In *2010 Seventh International Conference on Wireless On-demand Network Systems and Services (WONS)* (pp. 9–16). IEEE.

Henderson, T., Kotz, D., and Abyzov, I. (2008). The changing usage of a mature campus-wide wireless network. *Computer Networks*, 52(14), 2690–2712.

Herrmann, K. (2003, September). Modeling the sociological aspects of mobility in ad hoc networks. In *Proceedings of the 6th ACM International Workshop on Modeling Analysis and Simulation of Wireless and Mobile Systems*, San Diego, CA (pp. 128–129). ACM.

Hong, X., Gerla, M., Pei, G., and Chiang, C. C. (1999, August). A group mobility model for ad hoc wireless networks. In *Proceedings of the 2nd ACM International Workshop on Modeling, Analysis and Simulation of Wireless and Mobile Systems*, Seattle, WA (pp. 53–60). ACM.

Hoogendoorn, S. P. and Bovy, P. H. (2004). Pedestrian route-choice and activity scheduling theory and models. *Transportation Research Part B: Methodological*, 38(2), 169–190.

Hsu, W. J. and Helmy, A. (2005). Impact: Investigation of mobile-user patterns across university campuses using wlan trace analysis. *arXiv preprint cs/0508009*.

Hsu, W. J., Spyropoulos, T., Psounis, K., and Helmy, A. (2007, May). Modeling time-variant user mobility in wireless mobile networks. In *INFOCOM 2007. 26th IEEE International Conference on Computer Communications. IEEE*, Barcelona, Spain (pp. 758–766). IEEE.

Hsu, W. J., Spyropoulos, T., Psounis, K., and Helmy, A. (2009). Modeling spatial and temporal dependencies of user mobility in wireless mobile networks. *IEEE/ACM Transactions on Networking (ToN)*, 17(5), 1564–1577.

Huang, D. (2005, March). Using delaunay triangulation to construct obstacle detour mobility model. In *Wireless Communications and Networking Conference, 2005, IEEE*, New Orleans, LA (Vol. 3, pp. 1644–1649). IEEE.

Jardosh, A., Belding-Royer, E. M., Almeroth, K. C., and Suri, S. (2003, September). Towards realistic mobility models for mobile ad hoc networks. In *Proceedings of the 9th Annual International Conference on Mobile Computing and Networking*, San Diego, CA (pp. 217–229). ACM.

Johansson, P., Larsson, T., Hedman, N., Mielczarek, B., and Degermark, M. (1999a, August). Scenario-based performance analysis of routing protocols for mobile ad-hoc networks. In *Proceedings of the 5th Annual ACM/IEEE International Conference on Mobile Computing and Networking*, Seattle, WA (pp. 195–206). ACM.

Johansson, P., Larsson, T., Hedman, N., Mielczarek, B., and Degermark, M. (1999b, August). Routing protocols for mobile ad-hoc networks-a comparative performance analysis. In *Proceedings of the 5th International Conference on Mobile Computing and Networking (ACM MOBICOM'99)*, Seattle, WA (pp. 195–206).

Johnson, D. B. and Maltz, D. A. (1996). Dynamic source routing in ad hoc wireless networks. In *Mobile Computing* (pp.153–181). Springer, Boston, MA.

Kelton, W. D., Smith, J. S., and Sturrock, D. T. (2011). *Simio & Simulation: Modeling, Analysis, Applications*. Learning Solutions. McGraw-Hill, New York, NY.

Keränen, A. and Ott, J. (2007). *Increasing Reality for dtn Protocol Simulations*. Helsinki University of Technology, Tech. Rep.

Kim, M., Kotz, D., and Kim, S. (2006, April). Extracting a mobility model from real user traces. In *INFOCOM 2006. 25th IEEE International Conference on Computer Communications. Proceedings*, Barcelona, Spain (pp. 1–13). IEEE.

Kim, S. and Lee, C. H. (2010). Superdiffusive behavior of mobile nodes and its impact on routing protocol performance. *IEEE Transactions on Mobile Computing*, 9(2), 288–304.

King, D. H. and Harrison, H. S. (2013, July). JaamSim open-source simulation software. In *Proceedings of the 2013 Grand Challenges on Modeling and Simulation Conference* (p. 1). Society for Modeling & Simulation International.

Kotz, D. and Essien, K. (2005). Analysis of a campus-wide wireless network. *Wireless Networks*, 11(1–2), 115–133.

Kotz, D. and Henderson, T. (2005). Crawdad: A community resource for archiving wireless data at dartmouth. *IEEE Pervasive Computing*, 4(4), 12–14.

Lee, K., Hong, S., Kim, S. J., Rhee, I., and Chong, S. (2009, April). Slaw: A new mobility model for human walks. In *INFOCOM 2009, IEEE*, Rio de Janeiro, Brazil (pp. 855–863). IEEE.

Lenders, V., Wagner, J., and May, M. (2006, May). Analyzing the impact of mobility in ad hoc networks. In *Proceedings of the 2nd International Workshop on Multi-hop Ad Hoc Networks: From Theory to Reality*, Florence, Italy (pp. 39–46). ACM.

Liang, B. and Haas, Z. J. (1999, March). Predictive distance-based mobility management for PCS networks. In *INFOCOM'99. Eighteenth Annual Joint Conference of the IEEE Computer and Communications Societies. Proceedings. IEEE*, New York, NY (Vol. 3, pp. 1377–1384). IEEE.

Martinez, F. J., Toh, C. K., Cano, J. C., Calafate, C. T., and Manzoni, P. (2011). A survey and comparative study of simulators for vehicular ad hoc networks (VANETs). *Wireless Communications and Mobile Computing*, 11(7), 813–828.

McNett, M. and Voelker, G. M. (2005). Access and mobility of wireless PDA users. *ACM SIGMOBILE Mobile Computing and Communications Review*, 9(2), 40–55.

Mei, A. and Stefa, J. (2009, April). SWIM: A simple model to generate small mobile worlds. In *INFOCOM 2009, IEEE*, Rio de Janeiro, Brazil (pp. 2106–2113). IEEE.

Minder, D., Marrón, P. J., Lachenmann, A., and Rothermel, K. (2005, June). Experimental construction of a meeting model for smart office environments. In *Proc. of the Workshop on Real-World Wireless Sensor Networks, SICS Technical Report* (p. 09).

Musolesi, M., Hailes, S., and Mascolo, C. (2004, October). An ad hoc mobility model founded on social network theory. In *Proceedings of the 7th ACM International Symposium on Modeling, Analysis and Simulation of Wireless and Mobile Systems*, Venice, Italy (pp. 20–24). ACM.

Musolesi, M. and Mascolo, C. (2006, May). A community based mobility model for ad hoc network research. In *Proceedings of the 2nd International Workshop on Multi-hop Ad Hoc Networks: From Theory to Reality*, Florence, Italy (pp. 31–38). ACM.

Musolesi, M. and Mascolo, C. (2007). Designing mobility models based on social network theory. *ACM SIGMOBILE Mobile Computing and Communications Review*, 11(3), 59–70.

OPNET Technologies Inc. (2004). Opnet modeller. http://www.opnet.com/products/modeler/home.html.

Pereira, P. R., Casaca, A., Rodrigues, J. J., Soares, V. N., Triay, J., and Cervelló-Pastor, C. (2012). From delay-tolerant networks to vehicular delay-tolerant networks. *IEEE Communications Surveys & Tutorials*, 14(4), 1166–1182.

Phithakkitnukoon, S., Smoreda, Z., and Olivier, P. (2012). Socio-geography of human mobility: A study using longitudinal mobile phone data. *PloS One*, 7(6), e39253.

Rabelo, L., Helal, M., Jones, A., and Min, H. S. (2005). Enterprise simulation: A hybrid system approach. *International Journal of Computer Integrated Manufacturing*, 18(6), 498–508.

Rhee, I., Shin, M., Hong, S., Lee, K., Kim, S. J., and Chong, S. (2011). On the levy-walk nature of human mobility. *IEEE/ACM Transactions on Networking (TON)*, 19(3), 630–643.

Rojas, A., Branch, P., and Armitage, G. (2005, October). Experimental validation of the random waypoint mobility model through a real world mobility trace for large geographical areas. In *Proceedings of the 8th ACM International Symposium on Modeling, Analysis and Simulation of Wireless and Mobile Systems*, Montréal, Quebec (pp. 174–177). ACM.

Royer, E. M., Melliar-Smith, P. M., and Moser, L. E. (2001). An analysis of the optimum node density for ad hoc mobile networks. In *IEEE International Conference on Communications, 2001. ICC 2001*, Helsinki, Finland (Vol. 3, pp. 857–861). IEEE.

Rubin, I. and Choi, C. W. (1997). Impact of the location area structure on the performance of signaling channels in wireless cellular networks. *IEEE Communications Magazine*, 35(2), 108–115.

Sánchez, M. and Manzoni, P. (2001). ANEJOS: A java based simulator for ad hoc networks. *Future Generation Computer Systems*, 17(5), 573–583.

Sanchez, M. (1998). Mobility models. http://www.disca.upv.es/misan/mobmodel.htm, page accessed on October 15th 2017.

Seah, W. K., Lee, F. W., Mock, K. W., Ng, E. K., and Kwek, M. Q. (2006, September). Mobility modeling of rush hour traffic for multihop routing in mobile wireless networks. In *Vehicular Technology Conference, 2006. VTC-2006 Fall. 2006 IEEE 64th*, Montréal, Quebec (pp. 1–5). IEEE.

Shin, I. R. N. M., Lee, S. H. N. K., and Chong, S. (2007). Human mobility patterns and their impact on routing in human-driven mobile networks. In *HotNets-VI*.
Sichitiu, M. L. (2009). Mobility models for ad hoc networks. In *Guide to Wireless Ad Hoc Networks* (pp. 237–254). Springer, London.
Smith, L., Beckman, R., Anson, D., Nagel, K., and Williams, M. (1995). *TRANSIMS: Transportation Analysis and Simulation System* (No. LA-UR-95-1664; CONF-9504197-1). Los Alamos National Lab., NM (United States).
Song, C., Koren, T., Wang, P., and Barabási, A. L. (2010). Modelling the scaling properties of human mobility. *Nature Physics*, 6(10), 818–823.
Tan, D. S., Zhou, S., Ho, J., Mehta, J. S., and Tanabe, H. (2002, January). Design and evaluation of an individually simulated mobility model in wireless ad hoc networks. In *Communication Networks and Distributed Systems Modeling and Simulation Conference* (pp. 61–68).
Tang, D. and Baker, M. (2000, August). Analysis of a local-area wireless network. In *Proceedings of the 6th Annual International Conference on Mobile Computing and Networking*, Boston, MA (pp. 1–10). ACM.
The ns-2 Project. (2008). http://www.isi.edu/nsnam/ns/.
The ns-3 Project. (2008). http://www.nsnam.org/.
Tian, J., Hahner, J., Becker, C., Stepanov, I., and Rothermel, K. (2002, April). Graph-based mobility model for mobile ad hoc network simulation. In *Simulation Symposium, 2002. Proceedings. 35th Annual*, San Diego, CA (pp. 337–344). IEEE.
Tools, C. P. N. (2005). Computer tool for coloured petri nets. http://wiki.daimi.au.dk/cpntools/cpntools.wiki.
Tuduce, C. and Gross, T. (2005, March). A mobility model based on WLAN traces and its validation. In *INFOCOM 2005. 24th Annual Joint Conference of the IEEE Computer and Communications Societies. Proceedings IEEE*, Miami, FL (Vol. 1, pp. 664–674). IEEE.
Varga, A. (2005). OMNeT++ discrete event simulation system version 3.2 user manual. http://www.omnetpp.org/doc/manual/usman.html.
Veeramani Mahendran, S. K. A., and Murthy, C. S. R. (2011, January). A realistic framework for delay-tolerant network routing in open terrains with continuous churn. In *Distributed Computing and Networking: 12th International Conference, ICDCN 2011, Bangalore, India, January 2–5, 2011, Proceedings* (p. 407). Springer Science & Business Media.
Verkasalo, H. (2007, August). Contextual usage-level analysis of mobile services. In *Fourth Annual International Conference on Mobile and Ubiquitous Systems: Networking & Services, 2007. MobiQuitous 2007*, Philadelphia, PA (pp. 1–8). IEEE.
Walsh, C., Doci, A., and Camp, T. (2008). A call to arms: It's time for REAL mobility models. *ACM SIGMOBILE Mobile Computing and Communications Review*, 12(1), 34–36.
Whitbeck, J., de Amorim, M. D., and Conan, V. (2010, February). Plausible mobility: Inferring movement from contacts. In *Proceedings of the Second International Workshop on Mobile Opportunistic Networking*, Pisa, Italy (pp. 110–117). ACM.
Wong, S. Y. (1990). TRAF-NETSIM: How it works, what it does. *ITE Journal*, 60(4), 22–27.
Yoon, J., Liu, M., and Noble, B. (2003, March). Random waypoint considered harmful. In *INFOCOM 2003. Twenty-second Annual Joint Conference of the IEEE Computer and Communications. IEEE Societies*, San Francisco, CA (Vol. 2, pp. 1312–1321). IEEE.
Yoon, J., Noble, B. D., Liu, M., and Kim, M. (2006, June). Building realistic mobility models from coarse-grained traces. In *Proceedings of the 4th International Conference on Mobile Systems, Applications and Services*, Uppsala, Sweden (pp. 177–190). ACM.
Zeng, X., Bagrodia, R., and Gerla, M. (1998, May). GloMoSim: A library for parallel simulation of large-scale wireless networks. In *Twelfth Workshop on Parallel and Distributed Simulation, 1998. PADS 98. Proceedings*, Banff, Alberta (pp. 154–161). IEEE.
Zhang, Z. (2006). Routing in intermittently connected mobile ad hoc networks and delay tolerant networks: Overview and challenges. *IEEE Communications Surveys & Tutorials*, 8(1), 24–37.
Zonoozi, M. M., and Dassanayake, P. (1997). User mobility modeling and characterization of mobility patterns. *IEEE Journal on Selected Areas in Communications*, 15(7), 1239–1252.

5

Taxonomy of Routing Protocols for Opportunistic Networks

Khaleel Ahmad, Muneera Fathima, and Khairol Amali bin Ahmad

CONTENTS

5.1 Introduction 87
5.2 Geographic-Based 88
5.2.1 Contention-Based Forwarding (CBF) 89
5.2.2 Geographic Random Forwarding (GeRaF) 89
5.2.3 Geographic Opportunistic Routing (GOR) 89
5.2.4 Multi-Rate Geographic Opportunistic Routing (MGOR) 90
5.2.5 Location Aided Opportunistic Routing (LAOR) 90
5.2.6 Position-Based Opportunistic Routing (POR) 91
5.2.7 Topology- and Link-Quality-Aware Geographical Opportunistic Routing (TLG OR) 91
5.2.8 Link Quality and Geographical Aware Opportunistic Routing (LinGo) 91
5.2.9 Resilient Opportunistic Mesh Routing (ROMER) 91
5.2.10 Directed Transmission Routing Protocol (DTRP) 92
5.2.11 Opportunistic Routing in Dynamic Ad Hoc Networks (OPRAH) 92
5.2.12 Cooperative Opportunistic Routing in Mobile Ad Hoc Networks (CORMAN) 93
5.2.13 Cross-Layer Link Quality and Geographical-Aware Beaconless Opportunistic Routing (XLinGo) 93
5.2.14 Topology-Assisted Geographic Opportunistic Routing (To Go) 93
5.3 Link State Aware 94
5.3.1 Opportunistic Multi-Hop Routing for Wireless Networks (ExOR) 94
5.3.2 Economy: A Duplicate Free Opportunistic Routing 94
5.3.3 MAC Independent Opportunistic Routing (MORE) 95
5.3.4 Code OR: Opportunistic Routing in Wireless Mesh Networks with Segmented Network Coding 95
5.3.5 Slide OR: Online Opportunistic Network Coding in Wireless Mesh Networks 95
5.3.6 XCOR (Synergistic Interflow Network Coding and Opportunistic Routing) 95
5.3.7 Cumulative Coded Acknowledgment (CCACK) 96
5.3.8 O3: Optimized Overlay-Based Opportunistic Routing 96
5.3.9 Simple Opportunistic Adaptive Routing Protocol for Wireless Mesh Networks (SOAR) 96
5.3.10 Opportunistic AnyPath Forwarding (OAPF) 97
5.4 Context-Aware 97
5.4.1 Partially Context-Aware 97
5.4.1.1 Probabilistic Routing Protocol Using History of Encounters and Transitivity (PRoPHET) 97

5.4.1.2 MaxProp ... 98
5.4.1.3 MobySpace Routing ... 98
5.4.1.4 Bubble Rap ... 98
5.4.1.5 PRoPHET+: An Adaptive PRoPHET-Based Routing ... 99
5.4.1.6 Resource Allocation Protocol for Intentional DTN Routing (RAPID) ... 99
5.4.2 Fully Context-Aware ... 99
5.4.2.1 Context-Aware Routing (COR) ... 99
5.4.2.2 History-Based Opportunistic Routing Protocol (HiBOp) ... 101
5.4.2.3 SimBet ... 101
5.5 Probabilistic ... 101
5.5.1 Epidemic Routing ... 101
5.5.2 Spray and Wait: An Efficient Routing Scheme for Intermittently Connected Mobile Networks ... 102
5.5.3 Fixed Point Opportunistic Routing (FPOR) ... 102
5.5.4 Opportunistic Flooding (OR Flood) ... 102
5.5.5 Delegation Forwarding ... 102
5.5.6 Encounter-Based Routing (EBR) ... 103
5.5.7 MaxOPP: A Novel Opportunistic Routing for Wireless Mesh Networks ... 103
5.5.8 Optimal Probabilistic Forwarding (OPF) ... 103
5.5.9 Optimal Opportunistic Forwarding (OOF) ... 104
5.6 Optimization-Based ... 104
5.6.1 Utility-Based ... 104
5.6.1.1 Node-Constrained Opportunistic Routing (Consort) ... 104
5.6.1.2 Optimized Multi-Path Network Coding (OMNC) ... 104
5.6.1.3 Opportunistic Residual Expected Network Utilities (OpRENU) ... 104
5.6.1.4 Time Sensitive Utility-Based Opportunistic Routing (TOUR) ... 105
5.6.1.5 Dice ... 105
5.6.1.6 Auto-Adjustable Opportunistic Acknowledgment/Timer-Based Routing (JOKER) ... 105
5.6.2 Learning-Based ... 105
5.6.2.1 Adapt Opportunistic Routing ... 106
5.6.2.2 Opportunistic Routing with Learning Algorithm (ORL) ... 106
5.6.2.3 Selfish Aware ... 106
5.6.3 Graph-Based ... 106
5.6.3.1 Shortest Multi-Rate Anypath First (SMAF) ... 106
5.6.3.2 Least Cost Anypath Routing (LCAR) ... 107
5.6.3.3 Polynomial Time Algorithm for Multi-Rate Anypath Routing (PTAS MRA) ... 107
5.6.3.4 Multi-Constrained Anypath Routing (MAP) ... 107
5.6.3.5 PLASMA: A New Routing Paradigm for Wireless Multi-Hop Networks ... 107
5.6.3.6 Localized Opportunistic Routing (LOR) ... 108
5.7 Cross-Layer ... 108
5.7.1 Physical-Layer-Aware (PHY-Aware) ... 108
5.7.1.1 Interference Limited Opportunistic Relaying (ILOR) ... 108
5.7.1.2 Parallel Opportunistic Routing (Parallel OR) ... 108
5.7.1.3 Simple and Practical Opportunistic Routing (SPOR) ... 109
5.7.1.4 High-Speed Opportunistic Routing (HS OR) ... 109
5.7.1.5 Energy Efficient Opportunistic Routing (EEOR) ... 109

5.7.2 MAC Aware 109
5.7.2.1 QoS Oriented Opportunistic Routing (QOR) 110
5.7.2.2 Opportunistic Routing for Low Power and Lossy Networks (ORPL) 110
5.7.2.3 Opportunistic Routing with Congestion Diversity (ORCD) 110
5.7.2.4 Opportunistic Routing for Wireless Sensor Networks (ORW) 110
5.7.3 PHY and MAC Aware 111
5.7.3.1 Protocol for Retransmitting Opportunistically (PRO) 111
5.7.3.2 Maximizing Transmission Opportunities in Wireless Multi-Hop Network (MTOP) 111
5.7.3.3 Cross-Layer Aided Energy-Efficient Opportunistic Routing (CL EE) 111
5.7.3.4 Sensor Context-Aware Adaptive Duty-Cycled Beaconless Opportunistic Routing (SCAD) 112
5.8 Conclusion 112
References 112

5.1 Introduction

An opportunistic network (OppNet) is a kind of mobile ad hoc network (MANET) which does not require any predefined end-to-end path to deliver a message to a pair of nodes. OppNets are an infrastructure-less network, i.e. it does not require any repeaters or routers to be placed in the network. Here, nodes are carried by any wireless devices which are nearer to the destination; they may be movable or immovable. MANETs and OppNets do not have any knowledge about the nodes wishing to communicate (Verma & Anurag, 2011).

The two main parameters to improve route optimization are mobility models and routing protocols. Mobility models are those which depict the path or motion of mobile nodes (MNs) while routing protocols help in forwarding messages as per the directions. To optimize any given route, these two parameters must be selected depending on the nature of the network, the traffic, etc. Hence, proper selection of these parameters will ensure optimization.

Due to its highly dynamic nature, i.e. frequent changes in topology, routing in the opportunistic network is a challenge. Since routing helps us in finding the next best hop, here comes a challenge to design an efficient routing protocol by which throughput can be increased. In some cases, a piece of information called *context* is given to the network in order to assist the routing decision process, but this may not be available every time. Context may be in the form of the previous node's history, the social activities of people and probabilities depending on when the node last got connected (aging). It has been shown that nodes encounter depends on the user's social behavior since in most cases mobile devices are carried by users (Xiaohua & Song, 2014).

Based on these considerations, the taxonomy of routing protocols is characterized into six categories:

1. Geographic-Based
2. Link-State-Aware
3. Context-Aware

4. Probabilistic
5. Optimization-Based
6. Cross-Layer

5.2 Geographic-Based

These routing protocols were introduced to provide geographic locations to ad hoc networks. The nodes within a specified geographic range are allowed to form a network and exchange messages (Figure 5.1).

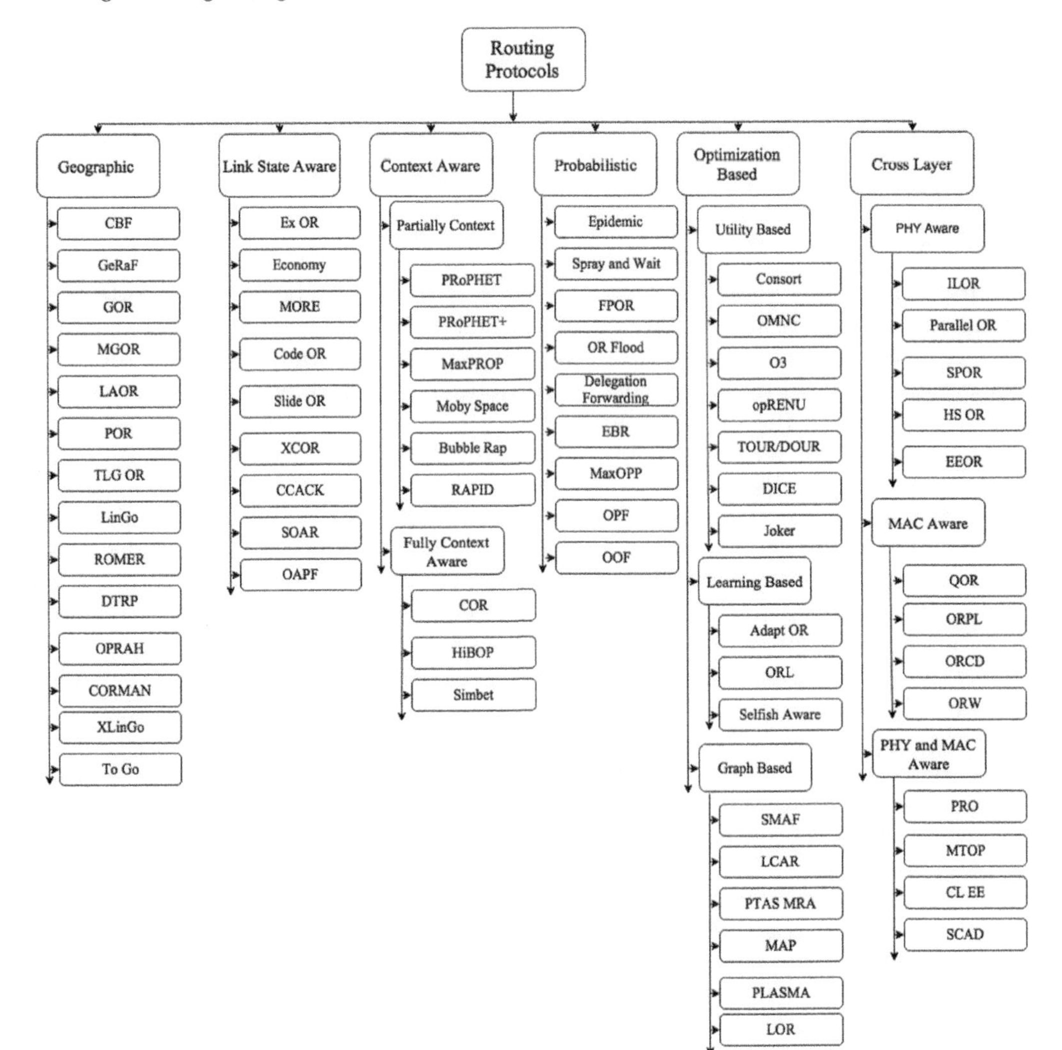

FIGURE 5.1
Taxonomy of routing protocols.

5.2.1 Contention-Based Forwarding (CBF)

CBF is a geographical routing protocol designed for vehicular ad hoc networks (VANETs) (Füßler, Widmer, Käsemann, Mauve & Hartenstein., 2003). Each node contains its previous hop's ID, its destination ID and a packet ID. If the node received is not its final destination, then the packet shall be forwarded again.

The packet progress for node *n* is calculated as:

$$P_n = \text{dist}(1,d) - (n,d)$$

where dist is its Euclidean distance; l and *d* are its last hop and destination respectively.

The timer value is calculated as:

$$t = \begin{cases} \tau(1-(R_n / R_{\max})) & 0 \le R_n < R_{\max} \\ \infty & \text{Otherwise} \end{cases}$$

R_{max} is the maximum radio range; R_n is the range of node *n*; τ is a maximum forwarding delay; based on the value of *t*, the number of participants are known. If it is infinite, then the packet is dropped, otherwise it is forwarded after *t* seconds unless it is overheard by the same ID. In such a case the timer is canceled. By simulation, it is proved that the throughput in CBF is about 100% if connections are available; overhearing is also less due to there being no duplicate messages. At the destination, a final acknowledgment (ACK) is sent to all direct nodes informing about the successful reception.

5.2.2 Geographic Random Forwarding (GeRaF)

A routing protocol proposed by B.S. Raj et al. is a completely unique forwarding technique that is supported by geographic location and the random choices of relaying nodes. A greedy geographic algorithm is proposed. It is attractive in wireless device networks due to its potency and quantifiability. However, greedy geographic routing causes long routing models and may not work well with random network topologies. The best theme is selected to maximize the efficiency of the network (Raj, Naveenraj & Gopinath, 2015).

5.2.3 Geographic Opportunistic Routing (GOR)

GOR is the combination of geographical and opportunistic routing. Geographic routing has an advantage in that it is not required to maintain any tables. It just makes decisions based on neighboring nodes. Opportunistic routing is always better than traditional routing (TR). GOR analyzes the tradeoff between the reliability, the packet advancement and the medium access control (MAC) coordination time cost. A local parameter is being introduced called the *expected one-hop throughput* (EOT) to balance all the factors. Maximum EOT must be achieved (Zeng, Lou, Yang & Brown, 2007). The idea of achieving maximum EOT comes from the following:

1. Since the whole-path achievable throughput is less than the per-hop throughput, if we maximize EOT there is every chance that the throughput may increase.

For a given candidate set

$$R(M_j) = L_p * \frac{Y * X}{P_M t_r + Z * X}$$

Here $X = \pi_{w=0}{}^{i-1}\bar{P}_{jw}, Y = \sum_{i=1}^{r} a_{ji} * p_{ji}$ and $Z = \sum_{i=1}^{r} t_i * p_{ji}$

$$P_M = \pi_{i=1}{}^{r}\left(1 - P_i\right), \bar{P}_{jw} = \left(1 - P_{jw}\right) \text{and} \, t_i = T_s + T_f\left(i\right)$$

$M_j = \{S_{ji}, \ldots S_{jr}\}$ which is an ordered set of nodes in M with priority.

2. The path delay is a summation of per-hop delays, which is actual delay caused by transmitting packets and candidate coordination. As the per-hop delay factors (T_s&$T_f(i)$) form together, t_i is in the denominator, so that as delay reduces, throughput increases.
3. EOT also considers the packet advancement factor; so by increasing potential, we can reduce hop counts, by which the message interference and delay are reduced.

We calculate an upper bound by which we got to know its concavity, which shows that even if we involve a large number of forwarding candidates and hence raise the chances for the packet to come closer to the destination the gained benefit is just marginal.

5.2.4 Multi-Rate Geographic Opportunistic Routing (MGOR)

K. Zeng et al. conducted a comprehensive study on the impact of interference, multiple rates, candidate selection and prioritization on maximize throughput. Considering wireless interference and the unique properties of opportunistic routing (OR), they proposed a concept of concurrent transmitter sets to show the constraints imposed by the transmission conflicts of OR and to formulate the maximum end-to-end throughput problem as maximum flow linear programming. Two multi-rate OR parameters are introduced: i. expected medium time (EMT) and ii. expected advancement rate (EAR), and their corresponding distribution of local rate. As for the candidate selection schemes, one of them are least medium time OR (LMT OR) and the other is multi-rate geographic OR (MGOR). When simulated, MGOR outperforms GOR (Zeng, Lou & Zhai, 2008).

5.2.5 Location Aided Opportunistic Routing (LAOR)

Routing in a wireless channel is a challenge. GOR was designed to work well with unreliable transmissions by exploiting the broadcast nature and using a spatial diversity of the network topology. Since GOR cannot support multi-channel, LAOR was designed (Zeng, Yang & Lou, 2009). LAOR considers candidate selection, prioritization and coordination with respect to GOR. LAOR proposed a local parameter called *one-hop throughput* (OEOT), to define the tradeoff between the one-hop's packet advancement and packet forwarding time. Different algorithms are proposed based on local rate adaptation and candidate selection algorithms to optimize the network. After rigorous simulation, results have shown that multi-rate GOR, incorporating rate adaptation and candidate selection, outperforms GOR. MGOR shows high performance with a high delivery rate and a lower delay than single-rate GOR.

5.2.6 Position-Based Opportunistic Routing (POR)

S. Yang et al. proposed a novel protocol called POR. It considers the complete benefits of the broadcast nature of wireless channel as well as opportunistic forwarding. Data packets are forwarded to multicast using multiple forwarders. A predefined list is inserted in the header of the packet using local information, and the packet moves in the given direction. The redundancy and randomness make it more efficient and robust. POR has negligible overheard and better scalability. The main cost incurred here is due to increased computational resources since many packets are forwarded to upper layers. It requires much buffer space, but if the time slot is properly used buffer usage can be reduced. POR performs well not only in normal conditions but also works well in hostile environments (Yang, Zhong, Yeo, Lee & Boleng, 2009).

5.2.7 Topology- and Link-Quality-Aware Geographical Opportunistic Routing (TLG OR)

OR takes advantage of both the broadcast nature and the spatial diversity of wireless transmissions to improve the performance of ad hoc networks (Zhao, Rosario, Braun, Cerqueira, Xu & Huang 2013). In most of the existing protocols, a predefined path is set and the nodes are allowed to follow that. Instead, OR lets us choose the next hop by the receiver side, by letting multiple receivers coordinate and decide which one can be a better forwarder. TLG OR uses network metrics like network topology, link quality and geographic location to perform coordination. TLG OR performs well both in terms of quality of service (QoS) and quality of experience (QoE).

5.2.8 Link Quality and Geographical Aware Opportunistic Routing (LinGo)

The Internet of things (IoT) has a huge impact on safety and environmental monitoring multimedia applications, which aim to minimize emergency response times and are also used to predict hazardous events. In this scenario, OR allows routing in a distributed manner, i.e. in hop-by-hop fashion, rather than based on the predefined end-to-end path, which is not a reliable solution. This enables video dissemination of a monitored object with QoE support to users, in the IoT platform. The existing model has considered only a single metric for candidate selection, including link quality or geographic information which may cause high packet loss and minimizes video perception from the user's point. The proposed LinGo is a cross-layer link-quality-aware and geographic-aware opportunistic protocol (Rosario, Zhao, Braun, Cerqueira, Santos & Li, 2013). It is designed for video dissemination in the mobile multimedia IoT platform. LinGo improves routing by considering multiple metrics, such as link quality, geographic location and energy. LinGo gives better results in terms of QoE when compared to existing models.

5.2.9 Resilient Opportunistic Mesh Routing (ROMER)

Wireless mesh networks (WMNs) provide robust and high throughput for wireless users, with the help of high-speed access points (HAPs) installed with advanced antennas communicated over a wireless channel to produce an indoor/outdoor broadband backhaul. This backbone is used to effectively send traffic to gateways (APs) which have high-speed connections to the wired Internet. ROMER is a resilient opportunistic routing protocol that balances long-term route stability and short-term opportunistic performance (Yuan, Yang, Wong, Lu & Arbaugh, 2005). It builds a runtime and forwarding mesh for every packet that offers candidate routes to it. ROMER delivers redundant data copies both in a controlled and randomized manner over the candidate forwarding mesh.

5.2.10 Directed Transmission Routing Protocol (DTRP)

In sensor networks, stable and reliable routing is a crucial issue. To address this, mica2 motes running Tiny OS are implemented. In DTRP, motes are tested to achieve better results (Nassr, Jun, Eidenbenz, Hansson & Mielke, 2007). DTRP is a multipath proactive routing protocol specifically designed to improve the scalability and reliability of the network. DTRP is based on beacon messages, which periodically originate at the sink node and are flooded throughout the network by sensor nodes. Unlike other single path distance-vector routing protocols, beacon messages are not used to resolve the next hop node for the destination. Instead, they are used to determine the hop count distance value between the sink and the sensor nodes. Some assumptions are made, such as when any data arrives at the node, it knows the number of hops from the originating source to the destination (sink) and it also knows the hop distance between itself and the destination, which is obtained from the beacon. The third piece of information, the hop distance between the originating source and the reached node, is easily available in the packet header.

Three values d_1, d_2, and d_3 are described as:

D_1: The shortest path's hop count distance between the originating source nodes to the destination (obtained from beacon packets)

D_2: The hop count distance a packet travelled before reaching a node (obtained from data packet header)

D_3: The shortest path's hop count distance between a relaying node and the destination.

Using these three pieces of information, a node makes a decision whether or not to reflood.

$$P^{tr} = e^{m\alpha}$$

$\alpha = d_1 - (d_2 + d_3)$ and m is a tunable parameter. Typically it is assumed that $m \geq 0$, if $m = 1$, the scenario is same as a full flood.

5.2.11 Opportunistic Routing in Dynamic Ad Hoc Networks (OPRAH)

The overhearing of any other participant's traffic is called *interference*. This is a major issue while routing, but most of the time it is ignored. There are two kinds of routing protocols:

1. *Reactive*: It attempts to identify the route only at the time of connection.
2. *Proactive*: It tries to identify the connections between nodes and store the information route table prior to connection.

Both reactive and proactive protocols work similarly in OPRAH; they attempt to identify the route that lasts longer. After extensive simulation, results show that OPRAH consistently finds a shorter path with an average of 77% of the length of the ad hoc on-demand distance vector (AODV) path (Westphal, 2006).

5.2.12 Cooperative Opportunistic Routing in Mobile Ad Hoc Networks (CORMAN)

Z. Wang et al. proposed a novel routing protocol called CORMAN. The link quality of wireless channels was the major issue in data dissemination until research into utilizing its characteristics was undertaken. Link quality is a factor which changes drastically over different geographical areas. The combination of link quality variation and the broadcasting nature of wireless channels has given a new scope for research, and it is named *cooperative communication*. CORMAN is an extension of Extremely Opportunistic Routing (ExOR) in order to satisfy node mobility (Wang, Chen & Li, 2012). These are basic components of CORMAN:

1. *PSR (proactive source routing)*: It provides each node with complete knowledge about the other nodes in the network.
2. *Large-scale live update*: As the packets progress, the nodes which are listed as forwarders may vary if any change in topology occurs.
3. *Small-scale retransmission*: Some nodes that are not in the listing are also used for the retransmission of data if this seems to be helpful.

All these 3 parameters are used to achieve an efficient cooperative communication.

5.2.13 Cross-Layer Link Quality and Geographical-Aware Beaconless Opportunistic Routing (XLinGo)

Due to its nature, flying ad hoc networks (FANETs) must be prepared for any kind of topology changes. Even the user experience should be better except in cases of buffer overflow and high packet loss ratio. To this end, XLinGo is proposed (Rosario, Zhao, Braun, Cerqueira, Santos & Alyafawi, 2014), which improves the transmission of concurrent multiple videos flows on FANETs by generating and maintaining reliable persistent multi-hop routes. XLinGo relies on information such as a set of cross-layer, human-related for routing decisions, such as performance metrics and QoE. Results show that XLinGo achieves better multimedia dissemination with QoE support and robustness in multi-flow, multi-hop and mobile network environments.

5.2.14 Topology-Assisted Geographic Opportunistic Routing (To Go)

Road topology information is being used recently to improve geographic routing in wireless vehicular networks. Due to its unreliable nature, road-topology-assisted geographic routing is a challenge. Here, a review of conventional and topology-assisted geographical routing is done to investigate the robust routing protocols which can address unreliable networks (Lee, Lee & Gerla, 2010). To Go is introduced, which integrates topology-assisted geographic routing with opportunistic forwarding algorithms such as a next hop prediction algorithm (NPA), which determines a packet target node; a forwarding set selection (FSS) algorithm to get a set of candidate forwarding nodes; and a priority scheduler, which reduces redundant packets based on distance-based timer.

The set of candidates is selected using a simple junction prediction algorithm with topology knowledge and enhanced beaconing. The forwarding set is then adjusted to reduce duplication and collisions. After simulation, results show that To Go performs better than existing geographical routing protocols which do not use road topology.

5.3 Link State Aware

Next, we consider the routing protocols which address the issue of network reliability by taking their delivery probabilities into account. Data is analyzed at each node to ensure a high transmission rate.

5.3.1 Opportunistic Multi-Hop Routing for Wireless Networks (ExOR)

ExOR is an integrated routing and MAC protocol (Biswas and Morris, 2004). Unlike other protocols, ExOR checks each hop of the path by which a message has been sent. It checks whether intermediate nodes have received a message or not.

In ExOR, not all nodes are chosen to transmit data packets. Only a few best nodes are chosen and they are prioritized; depending on the priority, the node with the highest priority communicates first. ExOR faces four major issues:

1. The nodes should agree on which subset of them has received each packet. Since agreement includes communication, one should take care that the agreement protocol does not cause any overhead.
2. Among all the nodes which have received a packet, the one 'closest' to the destination must forward the packet.
3. From the large, dense network it should choose only potential forwarders since using too many participants increases the costs of the agreement. Only the best forwarders are chosen.
4. To avoid collisions, simultaneous transmissions should be avoided.

ExOR operates in the batch formats. The best nodes are listed and they are prioritized based on the estimated cost to destination. In real time, the results are quite impressive with a factor of 2%–4% improvement.

5.3.2 Economy: A Duplicate Free Opportunistic Routing

OR outperforms traditional routing but the duplication of messages causes issues in OR. Economy is a duplication-free protocol (Hsu, Liu & Seah, 2009). It assigns a token to all data packets. In Economy, each node has a packet buffer and an ACK manager. The ACK manager records a list of the data packets received or sent, to avoid duplication. As the data packet is received at the node, the node checks for the presence of the token. If the token is present, the data packet is dropped or else it is considered.

The token is actually a control packet that includes source address and acknowledgments. Here we have two types of acknowledgments:

1. *Cumulative ACK*: All sequence numbers (SN) till this number must have already been acknowledged.
2. *Discrete ACK*: The one which is on the list of SN but not in range.

Token passing is done in such a way as to acknowledge the sender and inform it that the data packet has reached the destination. Therefore, token passing is done from the destination and passed to the source. Economy outperforms ExOR.

5.3.3 MAC Independent Opportunistic Routing (MORE)

ExOR uses MAC with routing to impose strictness on routers' access to medium, although ExOR does not use complete features of 802.11 MAC. It prevents spatial reuse and thus may not utilize the resource properly. MORE combines both opportunistic routing and network coding (NC) in a natural way to provide opportunistic routing without node coordination (Chachulski, Jennings, Katti & Katabi, 2007).

MORE, a MAC independent protocol, mixes all the packets randomly. Randomness ensures that the packet once heard will not be heard again, in order to avoid duplication. MORE does not require any special schedule to communicate with routers and runs directly on 802.11. MORE achieved higher throughput than traditional prior opportunistic routing protocols.

5.3.4 Code OR: Opportunistic Routing in Wireless Mesh Networks with Segmented Network Coding

Y. Lin et al. proposed Code OR. With the help of NC, OR can be done with less complexity. The NC performs segmented NC, by partitioning data into several segments and sending them one by one. Until the acknowledgment of one data set is required, the other is not sent. Due to which, performance gets affected. Because of this, Code OR, which transmits multiple segmented data sets in parallel by properly utilizing the bandwidth, is designed. Code OR performs well in realistic scenarios if a data packet is small and the window size is large (Lin, Li & Liang, 2008).

5.3.5 Slide OR: Online Opportunistic Network Coding in Wireless Mesh Networks

Using NC, opportunistic routing can be performed in a simple and practical way. Due to some issues, the protocol using NC needs to divide into multiple segments and encode only packets within the same segment. However, it is challenging to know the exact time the next segment should be transmitted. To overcome this, Slide OR is designed (Lin, Liang & Li, 2010); it encodes source packets and places them in sliding fashion so that coded packets of one window may be used to decode source packets in another window. Slide OR outperforms existing models with multiple segments.

5.3.6 XCOR (Synergistic Interflow Network Coding and Opportunistic Routing)

OR and NC have their individual benefits. Interflow NC is expected to have a poor performance gain against highly lossy environments, whereas OR is expected to have a high performance gain. Conversely, as the number of flows in the network increases, the gains in OR are reduced whereas interflow NC has high gain.

For this reason, and since NC is performing better even if the number of flows increases, there comes an idea to combine OR and NC together called XCOR (Koutsonikolas, Hu & Wang, 2008). XCOR integration may have 2 issues:

1. Care should be taken to avoid duplicate transmissions
2. Interflow NC assumes a fixed path, i.e. its intermediate nodes have a predefined path, but that's not the case with OR, it does not have any information about the next hop. Because of this, the integration seems infeasible.

There is no clear answer as to whether OR or NC offers better throughput. The final answer depends on the specified network topology and traffic.

5.3.7 Cumulative Coded Acknowledgment (CCACK)

D. Koutsonikolas et al. proposed CCACK. It works based on network coding, which has significantly simplified the designs of routing protocols. However, NC-based protocols face many problems, like how many coded packets to send at once etc. To avoid overhead due to messages, the number of transmissions were heuristically calculated offline. To avoid this, CCACK was designed. CCACK allows nodes to acknowledge, allow the coded traffic in the upstream in a simple way, oblivious to less rate and negligible overhead. This was tested on 22 nodes in the 802.11 WMN test bed. After rigorous simulation, it was shown that CCACK outperforms MORE (Koutsonikolas, Wang & Hu, 2010).

5.3.8 O3: Optimized Overlay-Based Opportunistic Routing

A novel approach that uses interflow NC in opportunistic routing is explained (Han, Bhartia, Qiu & Rozner, 2011). A unique feature is that it reduces end-to-end delay. But opportunistic routing uses intraflow NC, which reduces the spread of information to nodes, which in turn minimizes interflow NC. To address the challenge, initially the opportunistic routing and interflow NC are decoupled and a novel solution is proposed where an overlay network performs overlay routing and interflow coding without any concern about packet loss. Secondly, an underlying network uses optimized opportunistic routing as well as rate limiting to provide efficient and reliable overlay links. Based on this model, an optimization algorithm is designed which jointly optimizes opportunistic routes, rate limits, interflow and intraflow NC. The QualNet simulation tool was used and results show that:

1. Rate limiting improvises remarkable performance of routing.
2. OR benefits under high loss rates whereas interflow at low loss rates.
3. O3 notably outperforms state-of-the-art routing protocols.

5.3.9 Simple Opportunistic Adaptive Routing Protocol for Wireless Mesh Networks (SOAR)

Traditional routing does not give better performance due to unreliable and unpredictable networks. SOAR, which supports simultaneous flows in WMNs, is proposed (Rozner, Seshadri, Mehta & Qiu, 2009). It works on four major components to gain high throughput and fairness:

1. Adaptive path selection while leveraging path diversity should be such that it reduces duplicate transmissions.
2. Priority timer-based forwarding is used to select the best forward node to forward a packet.
3. Local loss recovery should efficiently detect and retransmit lost packets.
4. The adaptive rate control is determined so that it should be appropriate to network conditions.

SOAR significantly outperforms traditional routing, ExOR and seminal opportunistic routing protocol under a wide range of scenarios.

5.3.10 Opportunistic AnyPath Forwarding (OAPF)

Unlike wired networks, a wireless network cannot have point-to-point links due to fluctuations in the network. Fluctuations in the quality of link cause retransmission of packet data at the link layer. OAPF selects a set of candidate next hops in advance and any one of them is used to forward a packet (Zhong, Wang, Nelakuditi & Lu, 2006).

The routing candidate selection and the prioritization of data packets are the main issues to be addressed. Similar to ExOR, candidates are selected and prioritized based on best path expected transmission count (ETX) from candidate to destination. Not all nodes are considered; only a few best nodes are selected as forwarding candidates. This avoids redundancy, congestion and interference. The candidate that is selected and prioritized does not ensure to employ anypath forwarding. To resolve this situation a new metric called *expected anypath transmission* (EAX), is introduced for a pair of nodes which can capture an expected number of hops between them until they reach the destination. Candidate selection and prioritization are done with respect to EAX for better results.

5.4 Context-Aware

Routing in opportunistic networks becomes a challenge due to its dynamic nature and absence of knowledge about the nodes. In some cases, a piece of information such as the history of encounters or predefined path may be fed to the network. This is called *context-aware*.

5.4.1 Partially Context-Aware

The network only has partial knowledge about the nodes in the network.

5.4.1.1 Probabilistic Routing Protocol Using History of Encounters and Transitivity (PRoPHET)

A. Lindgren et al. proposed a probabilistically-based routing protocol that works on the history of transmissions. A metric called *delivery predictability* is proposed. Based on some threshold value, the nodes which have a delivery predictability more than the threshold are considered as the best nodes. Only these nodes are used to forward packets (Lindgren, Doria & Schelén, 2003).

$$P_{(a,b)} = P_{(a,b)old} + \left(1 - P_{(a,b)old}\right) * P_{init} \quad \text{where } P \varepsilon [0,1]$$

This is used to calculate delivery predictability. In PRoPHET, the next hop is chosen depending on delivery predictability. The higher the delivery predictability, the more chance the node has to be next forwarder. PRoPHET also considers aging and transitivity. If any node has not forwarded for a long time, then it is aged and given low priority. Transitivity is when node *a* frequently encounters node *b*, and node *b* frequently encounters node *c*, then the node *c* is probably a good forwarder destined for node *a*.

5.4.1.2 MaxProp

Delay-tolerant network (DTN) routing is quite a challenge due to breakage of networks (dynamic nature). MaxProp is a routing protocol which is designed to effectively route DTN messages (Burgess, Gallagher, Jensen & Levine, 2006). MaxProp works on prioritization; it prioritizes both the packets to be transmitted and the packets to be dropped. These priorities are done based on path likelihood and historical encounters. When two nodes discover each other, these are steps to exchange packets:

1. All the messages are transferred to the neighbor node.
2. Routing information such as the probability of meetings and vector listing is shared with the neighbor node.
3. Acknowledgment is given by delivered data
4. Nodes which follow the shortest path or have not travelled far in the network are given priority.

After a 60 day trial of 30 buses, it has been shown that MaxProp performs better than other protocols that schedule their meetings between peers (nodes).

5.4.1.3 MobySpace Routing

Routing in a DTN is a challenge due to its dynamic nature. However, routing can benefit if we consider the advantage of knowledge regarding node mobility. Leguay, Friedman and Conan (2006) have stated this problem with a generic algorithm, which makes use of high-dimensional Euclidean space called MobySpace. MobySpace evaluates the nodes based on how frequently they visited each possible location. MobySpace performs well when compared to other algorithms, especially when the routing of nodes has a high connection time. Hence, it is determined that the degree of homogeneity has a huge effect on routing (Burgess, Gallagher, Jensen & Levine, 2006).

5.4.1.4 Bubble Rap

To gain an understanding of human mobility in terms of social patterns, a forwarding algorithm for pocket switch networks (PSNs) is proposed, called Bubble Rap (Hui, Crowcroft & Yoneki, 2011). It combines both the information about community structure and the knowledge of node centrality to make forwarding decisions.

There are two intuitions behind the algorithm:

1. People must have varying roles, i.e. forwarding is done with popular nodes first then the rest.
2. To know social lives of people, i.e. to identify the member of the destination and use them as relays.

Together, it is called *bubble forwarding.*

A couple of assumptions are made to the algorithm:

1. Each node must and should be a part of at least one community. Single node communities may even exist.
2. Each node has its global rank in the whole network and a local rank within the community. If a node belongs to many communities, it may have multiple local ranks.

Forwarding in Bubble Rap is as follows. When a node has a message to be forwarded to another node, the node initially bubbles up the message according to hierarchy using its global rank, unless and until it reaches a node with the same community as that of the destination node. Now the local ranking system is used, the message bubbles up in the tree until it reaches the destination or the message expires. Every node need not have to know the ranks of all nodes, but must be able to compare ranking if encountered. In order to reduce costs, as the message is delivered the original carrier (source) must be able to delete the memory from the buffer.

5.4.1.5 PRoPHET+: An Adaptive PRoPHET-Based Routing

PRoPHET+ is an extension of PRoPHET (Huang, Lee & Chen, 2010). PRoPHET+ is designed to maximize message delivery rate and minimize the delay of nodes. It evaluates a deliverability value to decide routing paths for packets. Deliverability is evaluated using weighted functions which consist of nodes' buffer size, location, power, popularity and predictability value from PRoPHET. PRoPHET+'s weights are selected based on qualitative consideration. PRoPHET+ can outperform PRoPHET if logical weights are properly considered.

5.4.1.6 Resource Allocation Protocol for Intentional DTN Routing (RAPID)

Existing routing protocols primarily focus on increasing the likelihood of finding paths with limited knowledge, but these approaches may have only an incidental effect on routing metrics such as maximum or average delivery delay. RAPID is an intentional routing protocol, which intentionally boosts the performance of the particular routing metric (Balasubramanian, Levine & Venkataramani, 2007). RAPID considers DTN routing as a resource allocation problem and by using an in-band control channel metadata is propagated. Moreover, DTN routing protocols lack sufficient information; even if sufficient information is present, optimality is an NP-hard problem. Over a rigorous simulation on the test bed, results show that RAPID yields better gains than existing protocols.

5.4.2 Fully Context-Aware

Fully Context-Aware not only helps us in exploiting information, but they also help by using context information.

5.4.2.1 Context-Aware Routing (COR)

Z. Zhao et al. proposed the COR protocol. Regular list OR follows many restrictions. To avoid this, COR is designed. COR works in 3 steps:

1. Context information such as geographic progress, link quality and residual energy are considered for routing decisions.
2. All the eligible nodes are used for data forwarding.
3. The relative mobility of nodes is used for further connections.

In COR, it is assumed that every node knows its geographic position with the help of devices like GPS. Each node calculates its dynamic forwarding delay (DFD) based on available context information (Zhao, Rosario, Braun & Cerqueira, 2014). To portray the importance of context, weights are added.

$$\text{DFD} = \left(\alpha * L_Q + \beta * \text{Progress} + \gamma * R_E + \delta * \text{Live}\right) * \text{DFD}_{\max}$$

α, β, γ, δ are weight coefficients, together the sum is 1 i.e. $\alpha+\beta+\gamma+\delta=1$
$\text{DFD}_{\max}$ is the maximum delay allowed to each node
L_Q = link quality, R_E = Residual energy
COR outperforms other protocols which use only single metric.

$$L_Q = \begin{cases} 0 & \text{if } L_Q I_t > L_Q I_{\text{Good}} \\ \dfrac{L_Q I_{\max} - L_Q I_2}{L_Q I_{\max}} & \text{if } L_Q I_{\text{Bad}} < L_Q I_t < L_Q I_{\text{Good}} \\ 1 & \text{if } L_Q I_t < L_Q I_{\text{Bad}} \end{cases}$$

L_Q: Packet delivery ratio is used to classify links in 3 categories

1. Connected (PDR > 90%)
2. Transitional (10% < PDR < 90%)
3. Disconnected (PDR < 10%)

Depending on this, we define bonds for good links and bad links in threshold values. $L_Q I_{\text{Good}}$ and $L_Q I_{\text{Bad}}$. $L_Q\ I_t$ is measured in $L_Q I_{\text{value}}$ of the link. $L_Q I_{\max}$ is the maximum value of $L_Q I_t$.

Progress: The node with large geographical progress toward the destination has a small value.

$$\text{Progress} = \begin{cases} 2R - P_i / 2R & \text{if } \text{Dist}_{\text{S-D}} > R \\ 0 & \text{if } \text{Dist}_{\text{S-D}} < R \end{cases}$$

P_i is the progress of node *i*, *R* is radio range and $\text{Dist}_{\text{S-D}}$ is the distance between source and destination.

$$R_E = \begin{cases} E_0 - E_r / E_0 & \text{if } E_r > E_{\min} \\ 1 & \text{if } E_r < E_{\min} \end{cases}$$

A node with high residual energy (E_r) generates only a small value of 'residual energy'. E_0 is initial and E_r is residual energy.

$$\text{LIVE} = \frac{1}{(180 - \alpha / 180)^2 * T_{\text{LV}}}$$

α is an angle with which a node is moving. If $\alpha = 0$, the node is moving toward the destination and if $\alpha = 180$, the node is moving in the opposite direction. T_{LV} is the time the link can hold.

All four parameters are used to forward the data in COR.

5.4.2.2 History-Based Opportunistic Routing Protocol (HiBOp)

Opportunistic networks allow content sharing even in the absence of an end-to-end path. A context-aware framework called HiBOp is proposed, which learns and represents itself by context information; users' behavior and their social relations drive the forwarding process (Boldrini, Conti & Passarella, 2008). Context creation in HiBOp can be done in two ways:

1. People meeting and exchanging information about themselves
2. Remembering information about people who met in the past.

HiBOp exploits social-inspired relations which automatically learn how to behave and relate to each other. HiBOp performs better than epidemic and PRoPHET in both unicast communication and group communication in terms of resource consumption, network traffic and buffer space occupancy.

5.4.2.3 SimBet

Routing in DTN is a challenge due to its dynamic nature. DTN uses the store-carry-forward approach for routing, in which packet is kept at the node until it encounters another node. In the past few years, social-based routing protocol has been given much interest. In social-based routing protocol, we are exploiting the social behavior of the nodes to make better decisions. Social network analysis is a study which mainly focuses on the relation between social entities, patterns and the implications of their relationships. SimBet, a social-based routing protocol, is proposed, which works on similarity and betweenness centrality for routing packets (Patel & Gondaliya, 2015). The SimBet multi-copy routing scheme is introduced, which sends multiple copies of messages during encounter opportunity depend on the proportion of the SimBet utility value of the nodes which utilize the consent of EBR. Simulation results show that SimBet multi-copy outperforms both SimBet and epidemic in terms of delivery ratio, with lower overhead and latency for large buffer space and time to live (TTL).

5.5 Probabilistic

A major issue in routing is to find the next hop. Here, the next hop is determined by means of probabilities.

5.5.1 Epidemic Routing

Mobile ad hoc networks can communicate even in the absence of any predefined destination existing in the network. Networks are quite dynamic in nature due to the absence of any knowledge. Existing protocols are not sufficient to handle dynamic networks. Epidemic routing protocol is designed to allow message delivery in such cases where a connected path between source and sink may not exist (Vahadat & Becker, 2000).

The main goal of the epidemic protocol is to maximize delivery rate and to minimize latency while also minimizing all the resources used. In epidemic routing, messages are flooded in the network, so that they can reach their destination early and throughput can be increased. Many message copies flooded at once can cause much wastage of scarce resources. To avoid this, only a few are selected as the best nodes, which tend to travel toward the destination and only to these nodes are message copies sent. The receiving node has every chance to accept/reject the data packet. By using the best nodes, the redundancy of a network is reduced and resource wastage is minimized. The simulation was done on Monarch simulator, which shows 100% message delivery with reasonable resource consumption.

5.5.2 Spray and Wait: An Efficient Routing Scheme for Intermittently Connected Mobile Networks

Flooding is mostly used in DTNs to improve delivery rate. Although they have a high probability of delivery, they misuse a lot of energy and are affected by severe contention. To avoid this, a protocol called Spray and Wait is proposed (Spyropoulos, Psounis & Raghavendra, 2005). It sprays, i.e. sends the message, and waits until the nodes have reached the destination.

Initially, the message is flooded similar to epidemic routing. Once there are enough copies which can reach the destination, flooding is stopped. Despite its simplicity, Spray and Wait outperforms flooding-based protocols in terms of the number of transmissions and delivery delays.

5.5.3 Fixed Point Opportunistic Routing (FPOR)

A single copy and a multi-hop OR scheme for DTN is proposed in V. Conan et al. The estimates of inter-contact times are used as inputs. FPOR aims to minimize delivery time in case of independent exponential pairwise inter-contacts. We can see a loop-free forwarding and polynomial convergence that makes the scheme workable for DTNs. By replaying real connectivity traces, it will show impressive results in terms of delay and delivery ratio while minimizing overhead (Conan, Leguay & Friedman, 2008).

5.5.4 Opportunistic Flooding (OR Flood)

Flooding has been greatly used in wireless networks. But low duty cycle wireless sensor networks (WSNs) were ignored since the nodes in that network are mostly asleep and wake up asynchronously. So, this issue is to be considered.

Opportunistic flooding is being proposed for low duty cycled networks with unstable wireless links and preset working schedules (Guo, He, Gu, Jiang & He, 2009). Packets are sent opportunistically using links outside the energy optimal tree to reduce redundancy and flooding delay in transmissions. A forwarder selection method is proposed to reduce the hidden terminal problem and a link-quality-based backoff method to rectify simultaneous forwarding operations. After extensive simulation, results show that opportunistic flooding is close to optimal performance.

5.5.5 Delegation Forwarding

OppNets are characterized by unpredictable mobility, heterogeneity of contacts and lack of global knowledge. With all these conditions, successfully delivering a message with low cost is usually not possible. In many cases, to avoid costs associated with the network, only a few nodes are chosen to forward the data. However, this may not be a good situation since the nodes which we select as better nodes may not relay properly at the destination.

Delegation forwarding mainly focuses on reducing cost while improving its success rate and average delay (Erramilli, Chaintreau, Crovella & Diot, 2008). Two variants of delegation forwarding are analyzed, where quality is independent of the underlying contact rate and quality is identical to the nodes' contact rate. In both cases, delegation forwarding reduces the expected cost from $O(N)$ to $O(\sqrt{N})$, while ensuring high performance. The high performance here describes high success rate and low average delay. Delegation forwarding performs well in the real world too.

5.5.6 Encounter-Based Routing (EBR)

In many protocols, flooding is done to increase delivery rate. But instead, this increases overhead and wastes buffer resources due to multiple copies. EBR is a quota-based DTN routing protocol (Nelson, Bakht & Kravets, 2009). It works on the principle that 'the future rate of node encounters can be roughly predicted by past data'. It means that the nodes which are encountered frequently are likely to be more successful than others.

EBR limits the number of replicas of a message, reducing network usage and buffer resources. Routing in EBR is done based on the decisions of nodes, the number of encounters, showing preference to message exchange. Due to these factors, the delivery rate increases; this constantly reduces overhead in the network. In EBR, information about nodes encounter is purely a local parameter and can be tracked easily. Therefore, EBR can maintain low overhead; it only requires $O(n)$ routing message exchanges during every connection, and $O(n^2)$ is stored locally. EBR is also secured from black hole attacks.

5.5.7 MaxOPP: A Novel Opportunistic Routing for Wireless Mesh Networks

Opportunistic networks have a degrading factor due to lossy links and varying channel conditions. Opportunistic routing protocols use the broadcast nature of the wireless medium to allow hop-by-hop transmission. But most of the existing protocols define the hops previous to transmission. MaxOPP, a novel protocol, is an adaptive and flexible opportunistic routing algorithm able to select the candidate forwarders that can maximize throughput gain at each hop and at the runtime (Bruno, Conti & Nurchis, 2010). The selection of candidate forwarders is done dynamically based on network conditions. MaxOPP achieves high throughput even for bulk data transfer when compared to traditional shortest path protocols.

5.5.8 Optimal Probabilistic Forwarding (OPF)

Opportunistic routing outperforms traditional routing in many ways, such as efficient utilization of broadcasting and better usage of spatial diversity of the wireless medium. Many algorithms have been proposed using various techniques and in various environments. But there is no algorithm which designed an algebra-based protocol. So, here the OR algebra-based routing protocol has been proposed for inter-domain routing, which identifies the essential features of OR in the mathematical language of OR algebra (Lu & Wu, 2009). Here, OR chooses any path routing protocol. A set of candidate paths is already selected for any node pair. Only a single node is selected dynamically depending on the conditions and availability of the network. The set of candidate paths is regarded as a tree called *the OR tree*.

The OR algebra algorithm can also be applied using Dijkstra's shortest path, the Bellman-Ford algorithm. This helps us in designing a proper routing metric for opportunistic routing.

5.5.9 Optimal Opportunistic Forwarding (OOF)

Multi-copy forwarding is mostly chosen to reduce costs (Liu & Wu, 2012). Here, two types of multi-copy forwarding protocols are described: i. optimized opportunistic forwarding (OOF) and ii. OOF-. These protocols maximize delivery rate while minimizing delay, but the number of forwarding messages must not exceed the threshold. TTL is given in the packet header; once the time is collapsed, the packet is dropped. OOF and OOF- performs better when compared to the epidemic, Spray and Wait, etc.; But these protocols are affected by few limitations.

5.6 Optimization-Based

Some routing protocols mainly concentrate on how to optimize a network.

5.6.1 Utility-Based

It optimizes source utilization.

5.6.1.1 Node-Constrained Opportunistic Routing (Consort)

X. Fang et al. proposed the protocol Consort. It mainly focuses on the problem of how to choose an opportunistic route for every user so that the total utility in a WMN is increased with respect to node constraints. By combining primal-dual and subgradient methods, a distributed algorithm is designed, i.e. Consort. At every iteration, Consort updates the Lagrange multipliers in a distributed manner according to their user and node behavior, which is received from its previous iteration. Depending on the received Lagrange multipliers, the nodes adjust their own behavior and then update the Lagrange multipliers. Consort gives results close to optimal solution (Fang, Yang & Xue, 2011).

5.6.1.2 Optimized Multi-Path Network Coding (OMNC)

NC always plays a prominent role in unreliable networks. OMNC applies multiple paths to transfer the data packets to the destination and broadcast MAC is used to deliver packets between neighboring nodes (Zhang & Li, 2009). The coding and broadcast rates are allotted to the transmitter by a distributed algorithm to increase the advantage of NC and to reduce congestion. OMNC is a rate control protocol that improves throughput dramatically for long wireless networks. OMNC achieves remarkable throughput advantage over traditional routing and existing NC protocols for both single and multi-unicast situations.

5.6.1.3 Opportunistic Residual Expected Network Utilities (OpRENU)

OR is mostly used to compensate low packet delivery ratio. Here, OR is applied to utility-based routing where a successful packet delivery of data gives benefit (Wu, Lu & Li, 2008).

Its main function is to maximize utility. Here, maximum utility is defined as a function of the benefit and cost of transmission, since the link reliability of each delay determines the eventual packet delivery and, hence, utility. OR offers the capacity to increase reliability through opportunistic delays. Evaluating the optimality of the utility-based routing is done through OR without permitting any retransmissions. They have observed that the optimal scheme requires exhaustive searching of all paths between a node pair. Both optimal and heuristic solutions were proposed to select relays and determine prioritization between them.

5.6.1.4 Time Sensitive Utility-Based Opportunistic Routing (TOUR)

Cyclic-based mobile social networks (MSNs) are a new type of DTN. Mobile users periodically move around and get connected to their short-distance communication devices. A utility-based routing is designed for MSNs. In a network, if a message is sent successfully to the destination, it gets the positive benefit as reward, while otherwise there is zero benefit. Each delivery incurs some forwarding cost, be it a success or failure. So, the utility is the benefit minus the forwarding cost. Using this model, TOUR or Deadline-Sensitive Opportunistic Utility-Based Routing (DOUR) is designed (Xiao, Wu, Liu & Huang, 2013). Initially, every node determines an optimal forwarding sequence, which possesses a series of forwarding opportunities in a distributed and greedy manner. Then the messages are forwarded as per the model. TOUR is also extended for multi-copy and inherently makes a good tradeoff between benefit, delay and cost per message delivery.

5.6.1.5 Dice

NC has emerged as a promising technique in wireless mesh networks. However, inheritance resource competition in wireless networks is not considered. Dice, a game theoretic approach to optimize resource allocation for NC-based protocols is proposed by X. Zhang and B. Li. Dice considers the problem as a network game; participants are players. All the players share the bandwidth resource with the competition. In case the players want to cooperate, the hash bargaining algorithm is used. In case the players are selfish, optimum equilibrium can be achieved by pricing them. In both the cases, players should undergo localized optimization for subproblems: broadcast/coding rate allocation and multipath opportunistic routing. After extensive work, results show that Dice outperforms other heuristic models like MORE (Zhang & Li, 2008).

5.6.1.6 Auto-Adjustable Opportunistic Acknowledgment/Timer-Based Routing (JOKER)

Ramon Sanchez-Iborra and Maria-Dolores Cano proposed a protocol called JOKER.JOKER is an opportunistic routing protocol designed by considering some features of Better Approach To Mobile Ad hoc Networking (BATMAN). It is a novel protocol used in candidate selection and coordination phases, which helps us to increase energy efficiency in supporting multimedia traffic. JOKER is proposed for supporting video streaming with energy efficiency. When compared to BATMAN, JOKER outperforms in real time (Sanchez-Iborra & Cano, 2016).

5.6.2 Learning-Based

Learning can be helpful for routing, since we cannot always have context information.

5.6.2.1 Adapt Opportunistic Routing

A.A. Bhorkar et al. proposed a protocol called d-Adapt OR, i.e. a distributed Opportunistic Routing algorithm. It works precisely even with zero knowledge regarding network topology and channel statistics. It is being simulated in the QualNet simulator with other adaptive routing protocols and d-Adapt OR shows better results in practical settings. Here, long-term reward criteria are being considered, which ignores short-term performance. To perform for short-term, 'regrets' are considered. Regret is a function of horizon which estimates the loss of performance under any given adaptive algorithm relative to the performance of topology-aware (optimal) protocol. The number of regrets should be as much as possible. Congestion control must be done properly (Bhorkar, Naghshvar, Javidi & Rao, 2012).

5.6.2.2 Opportunistic Routing with Learning Algorithm (ORL)

An unknown local probabilistic broadcast model is considered in P. Tehrani et al. The main objective is to design an online learning algorithm that can manage the sequential selection of relaying nodes based on realizations of the probabilistic links. The performance merit which we are interested in is 'regret'. Regret is measured as expected additional cost accumulated over time with regard to a known model having optimal centralized routing. The online learning algorithm is designed for both centralized and distributed networks (Tehrani, Zhao & Javidi, 2013).

5.6.2.3 Selfish Aware

The performance of the network may be affected due to selfish nodes and congestion, so before selecting next relay we should consider the selfish nodes and the congestion present in the network. Here, a technique is proposed to improve the performance of the distributed adaptive OR protocol for multi-hop wireless ad hoc networks (Febeena & Vinitha, 2013). The algorithm follows a three-way handshake to find the next hop:

1. The message is forwarded to all neighbors; nodes that receive the data packet will send an acknowledgment. After waiting for some time, it checks whether every node has sent an ACK. If any node has not sent an ACK, then it will be considered as a selfish node. Selection of relative variable is done by $Sd\ (j,s) = 1$ where j belongs to S and $i \neq j$.
2. A request is sent to that node to participate in routing after dropping its memory.
3. If an ACK is received after FO broadcast, then $Sd\ (j,s) = 0$ and $Cd\ (j,s) = 1$, it means that the node is congested.

This helps us in finding whether the node is selfish or congested so that necessary action can be taken for forwarding them.

5.6.3 Graph-Based

Graph-theoretic tools are used for optimization.

5.6.3.1 Shortest Multi-Rate Anypath First (SMAF)

Nodes in multi-rate anypath first use both sets of next hops and the selected transmission rate to reach the destination. Transmission rate can be used to broadcast a node in the set;

from the set any node forwards the packet to the destination. Existing protocols do not jointly optimize both the next hop and transmission rate.

In a given network topology and a destination, we will be finding out both transmission rate and forwarding set such that the destination path is minimized. This is known as shortest multi-rate anypath problem. This is an open problem to find the transmission rate and the forwarding set that jointly optimizes distance. To solve this, the expected anypath transmission time (EATT) routing metric and the SMAF algorithm were designed (Laufer, Dubois-Ferriere & Kleinrock, 2009). It has similar complexity to Dijkstra's algorithm for multi-rate single path routing, but is easy to execute in link-state routing protocols.

5.6.3.2 Least Cost Anypath Routing (LCAR)

Many networks today use opportunistic routing with a single path routing to minimize the cost of the path. The LCAR problem has been discussed (Dubois-Ferriere, Grossglauser & Vetterli, 2011), i.e. in which way to assign a set of candidate nodes so that we can achieve minimum cost. There is a trade-off that as candidate relays increase, forwarding costs decrease, but on the other hand, the likeliness of 'veiling' away from the shortest path can be seen.

The LCAR algorithm can be applied to any network to achieve optimum results. LCAR shows the best results when incorporated into underlying coordination protocols: a link layer protocol, which randomly selects any node to forward. When LCAR is applied, candidate selection is done in such a way as to optimize the network.

LCAR is applied to low power, low rate wireless communication and presents a new wireless link layer technique to reduce energy consumption and conjunction of the network. LCAR routes are much more stable and robust than other single path protocols.

5.6.3.3 Polynomial Time Algorithm for Multi-Rate Anypath Routing (PTAS MRA)

As in multi-rate anypath routing, every node operates both the onset of next hop and the selected transmission rate to reach the destination as soon as possible. Earlier protocols haven't considered the issue to optimize both the parameters at once. To solve this PTAS MRA routing algorithm is introduced (Laufer, Dubois-Ferriere & Kleinrock, 2012). The results show that it performs the same execution time as regular shortest path algorithm and is therefore suitable to employ in any protocol.

5.6.3.4 Multi-Constrained Anypath Routing (MAP)

Anypath routing has been proposed to work well with unreliable networks utilizing the broadcast nature of medium and spatial diversity. Here, we consider anypath routing, which is subject to K constraints and show a polynomial time K-approximation algorithm. When $K=1$, the algorithm shows optimal polynomial time for the problem. If $K \geq 2$, this problem is known as an NP-hard problem. This is quite a simple algorithm and can be applied to any wireless routing protocol (Fang, Yang & Xue 2013).

5.6.3.5 PLASMA: A New Routing Paradigm for Wireless Multi-Hop Networks

R. Laufer et al. proposed a new routing protocol called PLASMA. Each packet is delivered over the best available path to any of the gateways. Neither the path nor the gateways are selected prior; as the packet travels through the network they are selected on the fly by the mesh routers. A distributed routing algorithm is proposed to optimize both sets of gateways and the transmission rate. A load balancing technique is applied to disperse traffic to the gateways. After simulation, results show that PLASMA outperforms the state-of-the-art multi-rate anypath routing algorithm (Laufer, Velloso, Vieira & Kleinrock, 2012).

5.6.3.6 Localized Opportunistic Routing (LOR)

Existing protocols such as ExOR and MORE cannot be applied to large-scale networks. So to overcome this, LOR is designed (Li, Mohaisen & Zhang, 2013). It uses the distributed minimum transmission selection (MTS-B) algorithm and divides a large-scale network into many nested close-node-sets (CNS's) using local information. LOR locally realizes the network and optimizes them. Since it does not need global topology, overhead is minimized. LOR portrayed better tradeoff between the universal optimality of the used forwarders list and the scalability assumed by causing overhead. It outperforms MORE & ExOR in terms of overhead, end-to-end delay and throughput.

5.7 Cross-Layer

The interaction between the network layer and other lower-lying layers is done for better performance results.

5.7.1 Physical-Layer-Aware (PHY-Aware)

Network throughput is improved by considering channel state information dynamics.

5.7.1.1 Interference Limited Opportunistic Relaying (ILOR)

ILOR is designed to work well with the interface of a network. In a wireless network, interference is the main issue. So, ILOR helps to avoid it. ILOR works under slow fading wireless environments where the time is usually hundreds of milliseconds (Bletsas, Dimitriou & Sahalos, 2010).

Transmission of messages takes place in two phases:

Phase I: assisted by opportunistic routing and Phase II: assisted by conventional regenerative relaying.

During Phase I, the source transmits toward a new destination while relay I forwards a number of messages to a different destination without any delay. Phase II: an opportunistically selected array $b \neq I$ forwards to the appropriate destination while the source forwards a message to a new destination. The participating relays should have a strong path between source and destination, while at the same time they must be quiet isolated from each other. Here, 'usefulness' is well defined using both noise and interference limited. This works well for slow fading, dense, interference limited networks.

5.7.1.2 Parallel Opportunistic Routing (Parallel OR)

In W.Y. Shin et al. OR is examined on power, delay and total throughput as the number of pairs increase to the operating maximum. They proposed a protocol called Parallel

OR; it is performed by many nodes simultaneously to increase opportunistic gain while controlling the inter-user interface. It is observed that Parallel OR portrays a net improvement in the overall power–delay trade-off over conventional routing. Such a gain is possible due to the maximized received signal power given by the multi-user diversity gain (Shin, Chung & Lee, 2013).

5.7.1.3 Simple and Practical Opportunistic Routing (SPOR)

G.Y. Lee and Z. J. Haas proposed a simple and practical OR algorithm which analyzes its performance along with a multi-hop wireless network path by considering link level interface within the network nodes. SPOR algorithm shows better results by significant improvement in throughput, mostly for short haul paths. Due to its easy implementation, it can be easily integrated into most of the routing protocols with just minor changes. SPOR algorithm provides a practical and effective approach for the implementation of any opportunistic routing in wireless networks (Lee & Haas, 2011).

5.7.1.4 High-Speed Opportunistic Routing (HS OR)

Since the existing OR protocols cannot meet all the requirements for high speed, multi-rate wireless mesh network, a new routing protocol, Practical OR (POR) is designed to overcome this. In POR, the packet is forwarded is along a path, which is an ordered list. Every poor path should satisfy feedback constraint, i.e. any node on the path must be able to receive from its next hop to ensure correct reception from its downstream nodes. The complete path of the node is specified in its header of the packet. Once the packet is received, it will not forward it. The node acknowledges the receiving status of the packet in a feedback frame, to avoid duplication. After receiving a successful acknowledgment, then the packets at the receiver are forwarded. The paths are selected based on a greedy algorithm (Hu, Xie & Zhang 2013).

POR mainly concentrates on high throughput of the network. The packets are divided into frames and calculate the checksum of nodes. The congestion control bit is used to avoid congestion in the network. POR performs high speed, multi-rate wireless mesh network that runs on Wi-Fi interface, has low complexity, supports TCP, supports multilink layer data rates and is capable of exploiting high efficacy.

5.7.1.5 Energy Efficient Opportunistic Routing (EEOR)

The basic idea to improve throughput is to allow any node in the forwarders list if it overhears and is near to the destination. The nodes in the forwarders list are prioritized and if the receiver has higher prioritized packet than the sender, then the packet is discarded. The key issue here is how to prioritize the forwarder list efficiently.

Here (Mao, Tang, Xu, Li & Ma, 2011), an assumption is made that every node has fixed transmission power (non-adjustable) as well as where a node is enabled to adjust its transmission power (adjustable). Energy-efficient algorithms are proposed for both the cases which are used to select and prioritize a forwarder list. Results show that EEOR outperforms ExOR.

5.7.2 MAC Aware

These are proposed for WSNs, since the lifetime and performance are dependent on MAC.

5.7.2.1 QoS Oriented Opportunistic Routing (QOR)

TR cannot be applied to WSNs since they face the reality of radio physics, including asymmetric and long-range transient links. QOR a QoS-based opportunistic routing is proposed for data collection in WSNs (Lampin, Barthel, Auge-Blum & Valois, 2012). Unlike traditional routing, QOR gains the advantage of OR to provide fast and reliable transmissions.

QOR has a threefold process:

1. A joint routing structure and an addressing scheme which allows us to identify a limited set of the node which may become opportunistic relays between a pair of nodes.
2. Cascaded acknowledgment is designed, which gives a reliable ACK and replication free forwarding.
3. Performance evaluation concludes that QOR effectively uses opportunistic links.

QOR reduces end-to-end delay and offers high delivery rates. QOR works well for dense networks as well as sparse networks.

5.7.2.2 Opportunistic Routing for Low Power and Lossy Networks (ORPL)

In B. Pavkovic et al. a problem is considered, i.e. RPL over IEEE 802.15.4 MAC layer. The two-layer operates in a different fashion; a directed acyclic graph can be seen in RPL, whereas a cluster-tree in IEEE 802.15.4. The coupling of them both is an issue to be considered. A modified cluster-tree is designed which can associate with several parent nodes, taking advantage of sufficient organization of super frame at MAC layer. This is a modified MAC layer which can opportunistically forward with RPL, with the possibility of forwarding packets to multiple nodes. Instead of using same parent node every time, we use different parent nodes as long as they are close to a sink node. We take advantage of opportunistic forwarding to support higher-priority, delay sensitive alarms, which should come to the sink node before the deadline along with low-intensity monitoring data. ORPL outperforms RPL in delivery ratio, delay and overhead (Pavković, Theoleyre & Duda, 2011).

5.7.2.3 Opportunistic Routing with Congestion Diversity (ORCD)

The routing protocols with minimum delay are mostly preferred. The major issue with the minimum delay routing protocol is in heavy traffic scenarios, where the network results in severe congestion and unbounded delay. To avoid this, ORCD is proposed (Naghshvar and Javidi, 2010). It is the combination of shortest path routing with back pressure routing. Based on the congestion control measure, packets are ranked accordingly and expected to have low overall congestion. A novel function called Lyapunov is used to ensure a bounded expected delay for all networks under allowable traffic. Two variations of ORCD are given

1. *Infreq-ORCD*: It allows outdated backlog data in its computation to rank the nodes.
2. *AD-ORCD*: It allows an asynchronous distribution.

5.7.2.4 Opportunistic Routing for Wireless Sensor Networks (ORW)

Opportunistic routing is significantly better than WSNs. WSNs are duty-cycled, i.e. they enter sleep mode frequently to extend the network's lifetime. However, OR assumes that nodes are awake. To avoid this scenario, an opportunistic routing scheme for WSNs is being proposed (ORW) (Ghadimi, Landsiedel, Soldati, Duquennoy & Johansson, 2014). ORW uses a metric called the *expected number of duty-cycled wakeups* (EDC) that reflects the number of duty-cycled wakeups that are required to successfully reach the destination. An algorithm is designed to optimize EDC, such that it significantly reduces delay and improves energy efficiency when compared to TR. The performance of ORW is evaluated both on simulator and test bed. ORW reduces radio duty cycles on average by 50%.

5.7.3 PHY and MAC Aware

They provide practical and efficient routing protocols.

5.7.3.1 Protocol for Retransmitting Opportunistically (PRO)

M.H. Lu et al. proposed an efficient retransmission protocol called PRO. PRO is a link layer protocol that allows overhearing nodes to act as relays that retransmit on behalf of the source after they gain knowledge about the failed transmission (Lu, Steenkiste & Chen, 2009). Relays with high connectivity to the destination have a higher chance of delivering a packet than the source.

PRO has 4 main features:

1. Channel reciprocity coupled with a runtime calibration process is used to determine the instantaneous link quality to the destination.
2. Local process filters out the poor relays.
3. A distributed relay selection algorithm selects the best set of eligible relays from the qualified relays and prioritizes them.
4. 802.11e enhanced distributed channel access (EDCA) is applied to make sure that high priority relays transmit high probability. PRO works well both in test beds and the real world.

5.7.3.2 Maximizing Transmission Opportunities in Wireless Multi-Hop Network (MTOP)

As WiFi networks are becoming more crowded these days, where 802.11 radios are used MTOPs would be helpful in dense networks where the hop distance is short enough and data rates are high. MTOP is taken from the opportunistic transmission protocol (TXOP), which forwards the multiple frames back-to-back (Lee, Yu, Shin & Suh, 2013). MTOP transmits the frame to be forwarded a number of hops consecutively to minimize MAC overhead. Collision-prone non-stop forwarding is safe via analysis and USRP/GINU radio-based experiments. The simulation was done on the broad extent on the OPNET simulator. Results show that MTOP works under a wide range of scenarios.

5.7.3.3 Cross-Layer Aided Energy-Efficient Opportunistic Routing (CL EE)

J. Zuo et al. proposed CL EE, which takes advantage of cross-layer information exchange. Information may be a kind of frame error rate (FER) in the physical layer, the maximum number of retransmissions in MAC layer and number of relays in the network layer. Energy-dissipation-based objective functions were used for calculating the end-to-end energy consumption of each potentially available route for both traditional and opportunistic routing. After simulation, a result shows that energy-efficient OR outperforms TR in terms of end-to-end delay, and throughput is also increased (Zuo, Dong, Nguyen, Ng, Yang & Hanzo, 2014).

5.7.3.4 Sensor Context-Aware Adaptive Duty-Cycled Beaconless Opportunistic Routing (SCAD)

The primary concern in WSNs is energy. If low power transmissions occur, they make wireless links unreliable, which leads to frequent topology changes resulting in packet retransmission which wastes energy. SCAD is proposed to avoid this (Zhao & Braun, 2014). It makes use of beaconless OR for WSNs. SCAD is a cross-layer routing protocol which uses the concept of beaconless OR in WSNs. The selection of hops is made based on a few factors in the network context. To gain a balance between performance and energy, duty cycles of sensor based on real-time traffic load and energy drain rates are used. SCAD outperforms other protocols both in terms of throughput and lifetime.

5.8 Conclusion

Routing in opportunistic networks is a challenge, due to its dynamic nature. Routing helps in optimizing a network, so selecting an appropriate routing protocol gives us better results. The taxonomy of routing protocols in the opportunistic network have been discussed to give a clear idea about OppNets to the reader. In the future, we will try to discuss some other new protocols. Until now, no routing protocol has been designed which works for all situations.

References

Balasubramanian, A., Levine, B., & Venkataramani, A. (2007). DTN routing as a resource allocation problem. *ACM SIGCOMM Computer Communication Review*, 37(4), pp.373–384.

Bhorkar, A., Naghshvar, M., Javidi, T., & Rao, B. (2012). An adaptive opportunistic routing scheme for wireless ad-hoc networks. *IEEE/ACM Transactions on Networking*, 20(1), pp.243–256.

Biswas, S. & Morris, R. (2004). Opportunistic routing in multi-hop wireless networks. *ACM SIGCOMM Computer Communication Review*, 34(1), pp.69–74.

Bletsas, A., Dimitriou, A., & Sahalos, J. (2010). Interference-limited opportunistic relaying with reactive sensing. *IEEE Transactions on Wireless Communications*, 9(1), pp.14–20.

Boldrini, C., Conti, M., & Passarella, A. (2008). Exploiting users' social relations to forward data in opportunistic networks: The HiBOp solution. *Pervasive and Mobile Computing*, 4(5), pp.633–657.

Bruno, R., Conti, M., & Nurchis, M. (2010). MaxOPP: A novel opportunistic routing for wireless mesh networks. *The IEEE Symposium on Computers and Communications*, Riccione, Italy, pp.255–260.

Burgess, J., Gallagher, B., Jensen, D., & Levine, B. N. (2006). MaxProp: Routing for vehicle-based disruption-tolerant networks. *Proceedings IEEE INFOCOM 2006*, Barcelona, Spain.

Chachulski, S., Jennings, M., Katti, S., & Katabi, D. (2007). Trading structure for randomness in wireless opportunistic routing. *ACM SIGCOMM Computer Communication Review*, 37(4), pp.169–180.

Conan, V., Leguay, J., & Friedman, T. (2008). Fixed point opportunistic routing in delay tolerant networks. *IEEE Journal on Selected Areas in Communications*, 26(5), pp.773–782.

Dubois-Ferriere, H., Grossglauser, M., & Vetterli, M. (2011). Valuable detours: Least-cost anypath routing. *IEEE/ACM Transactions on Networking*, 19(2), pp.333–346.

Erramilli, V., Chaintreau, A., Crovella, M., & Diot, C. (2008). Delegation forwarding. *Proceedings of the 9th ACM International Symposium on Mobile Ad Hoc Networking and Computing – MobiHoc*, Hong Kong, China, pp.251–259.

Fang, X., Yang, D., & Xue, G. (2011). Consort: Node-constrained opportunistic routing in wireless mesh networks. *Proceedings IEEE INFOCOM*, Shanghai, China, pp.1907–1915.

Fang, X., Yang, D., & Xue, G. (2013). MAP: Multi constrained any path routing in wireless mesh networks. *IEEE Transactions on Mobile Computing*, 12(10), pp.1893–1906.

Febeena, M. & Vinitha, V. (2013). Selfishness aware adaptive opportunistic routing for wireless ad-hoc network. *IOSR Journal of Computer Engineering*, from: National Conference in Emerging Technologies' 14 (NCET), 1, pp.86–88.

Füßler, H., Widmer, J., Käsemann, M., Mauve, M., & Hartenstein, H. (2003). Contention-based forwarding for mobile ad hoc networks. *Ad Hoc Networks*, 1(4), pp.351–369.

Ghadimi, E., Landsiedel, O., Soldati, P., Duquennoy, S., & Johansson, M. (2014). Opportunistic routing in low duty cycled wireless sensor networks. *ACM Transactions on Sensor Networks*, 10(4), pp. 1–39.

Guo, S., He, L., Gu, Y., Jiang, B., & He, T. (2009). Opportunistic flooding in low-duty-cycle wireless sensor networks with unreliable links. *IEEE Transactions on Computers*, 63(11), pp.133–144.

Han, M. K., Bhartia, A., Qiu, L., & Rozner, E. (2011). O3: Optimized overlay based opportunistic routing. *Proceedings of the Twelfth ACM International Symposium on Mobile Ad Hoc Networking and Computing – MobiHoc*, Paris, France, pp.1–11.

Hsu, C.J., Liu, H.I., & Seah, W. (2009). Economy: A duplicate free opportunistic routing. *Proceedings of the 6th International Conference on Mobile Technology, Application & Systems*, Nice, France, pp.1–6.

Hu, W., Xie, J., & Zhang, Z. (2013). Practical opportunistic routing in high-speed multi-rate wireless mesh networks. *Proceedings of the Fourteenth ACM International Symposium on Mobile Ad Hoc Networking and Computing – MobiHoc*, Bangalore, India, pp.127–136.

Huang, T.K., Lee, C.-K., & Chen, L.-J. (2010). PRoPHET+: An adaptive PRoPHET-based routing protocol for opportunistic network. *24th IEEE International Conference on Advanced Information Networking and Applications*, Perth, Australia, pp.112–119.

Hui, P., Crowcroft, J., & Yoneki, E. (2011). BUBBLE rap: Social-based forwarding in delay-tolerant networks. *IEEE Transactions on Mobile Computing*, 10(11), pp.1576–1589.

Koutsonikolas, D., Hu, Y., & Wang, C. (2008). XCOR: Synergistic interflow network coding and opportunistic routing. *ACM Annual International Conference MobiCom*, San Francisco, CA, pp.1–3.

Koutsonikolas, D., Wang, C., & Hu, Y. (2010). CCACK: Efficient network-coding-based opportunistic routing through cumulative coded acknowledgments. *Proceedings IEEE INFOCOM*, San Diego, CA, pp.1–9.

Lampin, Q., Barthel, D., Auge-Blum, I., & Valois, F. (2012). QoS oriented opportunistic routing protocol for wireless sensor networks. *IFIP Wireless Days*, Dublin, Ireland, pp.1–6.

Laufer, R., Dubois-Ferriere, H., & Kleinrock, L. (2009). Multirate anypath routing in wireless mesh networks. *IEEE INFOCOM 2009 – The 28th Conference on Computer Communications*, Rio de Janeiro, Brazil, pp.37–45.

Laufer, R., Dubois-Ferriere, H., & Kleinrock, L. (2012). Polynomial-time algorithms for multirate anypath routing in wireless multihop networks. *IEEE/ACM Transactions on Networking*, 20(3), pp.742–755.

Laufer, R., Velloso, P. B., Vieira, L. F. M., & Kleinrock, L. (2012). PLASMA: A new routing paradigm for wireless multihop networks. *Proceedings IEEE INFOCOM*, Orlando, FL, pp.2706–2710.

Lee, G. & Haas, Z. (2011). Simple, practical, and effective opportunistic routing for short-haul multihop wireless networks. *IEEE Transactions on Wireless Communications*, 10(11), pp.3583–3588.

Lee, J., Yu, C., Shin, K. G., & Suh, Y. (2013). Maximizing transmission opportunities in wireless multi hop networks. *IEEE Transactions on Mobile Computing*, 12(9), pp.1879–1892.

Lee, K., Lee, U., & Gerla, M. (2010). Geo-opportunistic routing for vehicular networks [Topics in automotive networking]. *IEEE Communications Magazine*, 48(5), pp.164–170.
Leguay, J., Friedman, T., & Conan, V. (2006). Evaluating mobility pattern space routing for DTNs. *Proceedings IEEE INFOCOM*, Barcelona, Spain, pp. 1–10.
Li, Y., Mohaisen, A., & Zhang, Z. (2013). Trading optimality for scalability in large-scale opportunistic routing. *IEEE Transactions on Vehicular Technology*, 62(5), pp.2253–2263.
Lin, Y., Li, B., & Liang, B. (2008). CodeOR: Opportunistic routing in wireless mesh networks with segmented network coding. *IEEE International Conference on Network Protocols*, Orlando, FL, pp.13–22.
Lin, Y., Liang, B., & Li, B. (2010). SlideOR: Online opportunistic network coding in wireless mesh networks. *Proceedings IEEE INFOCOM*, San Diego, CA, pp.1–5.
Lindgren, A., Doria, A., & Schelén, O. (2003). Probabilistic routing in intermittently connected networks. *ACM SIGMOBILE Mobile Computing and Communications Review*, 7(3), pp.19–20.
Liu, C. & Wu, J. (2012). On multicopy opportunistic forwarding protocols in nondeterministic delay tolerant networks. *IEEE Transactions on Parallel and Distributed Systems*, 23(6), pp.1121–1128.
Lu, M. & Wu, J. (2009). Opportunistic routing algebra and its applications. *The 28th Conference on Computer Communications*, Rio de Janeiro, Brazil, pp.2374–2382.
Lu, M.-H., Steenkiste, P., & Chen, T. (2009). Design, implementation and evaluation of an efficient opportunistic retransmission protocol. *Proceedings of the 15th Annual International Conference on Mobile Computing and Networking – MobiCom*, Beijing, China, pp.73–84.
Mao, X., Tang, S., Xu, X., Li, X., & Ma, H. (2011). Energy-efficient opportunistic routing in wireless sensor networks. *IEEE Transactions on Parallel and Distributed Systems*, 22(11), pp.1934–1942.
Naghshvar, M. & Javidi, T. (2010). Opportunistic routing with congestion diversity in wireless multi-hop networks. *Proceedings IEEE INFOCOM*, San Diego, CA, pp.496–500.
Nassr, M. S., Jun, J., Eidenbenz, S. J., Hansson, A. A., & Mielke, A. M. (2007). Scalable and reliable sensor network routing: Performance study from field deployment. *26th IEEE International Conference on Computer Communications*, Barcelona, Spain, pp.670–678.
Nelson, S., Bakht, M., & Kravets, R. (2009). Encounter-based routing in DTNs. *ACM SIGMOBILE Mobile Computing and Communications Review*, 13(1), pp.846–854.
Patel, C. M. & Gondaliya, N. (2015). Enhancement of social based routing protocol in delay tolerant networks. *International Journal of Computer Applications*, 122(4), pp.19–25.
Pavković, B., Theoleyre, F., & Duda, A. (2011). Multipath opportunistic RPL routing over IEEE 802.15.4. *Proceedings of the 14th ACM International Conference on Modeling, Analysis and Simulation of Wireless and Mobile Systems – MSWiM*, Miami, FL, pp.179–186.
Raj, B., Naveenraj, A., & Gopinath, A. (2015). Geographic random forwarding for ad-hoc and sensor networks multihop performance. *International Journal of Innovative Research in Computer and Communication Engineering*, 3(3), pp.2504–2509.
Rosario, D., Zhao, Z., Braun, T., Cerqueira, E., Santos, A., & Alyafawi, I. (2014). Opportunistic routing for multi-flow video dissemination over Flying Ad-Hoc Networks. *Proceeding of IEEE International Symposium on a World of Wireless, Mobile and Multimedia Networks*, Sydney, Australia, pp.1–6.
Rosario, D., Zhao, Z., Braun, T., Cerqueira, E., Santos, A., & Li, Z. (2013). A link quality and geographical-aware routing protocol for video transmission in mobile IoT. Technical Report IAM-13-001, March 28, 2013, University of Bern.
Rozner, E., Seshadri, J., Mehta, Y., & Lili Qiu (2009). SOAR: Simple opportunistic adaptive routing protocol for wireless mesh networks. *IEEE Transactions on Mobile Computing*, 8(12), pp.1622–1635.
Sanchez-Iborra, R. & Cano, M. (2016). JOKER: A novel opportunistic routing protocol. *IEEE Journal on Selected Areas in Communications*, 34(5), pp.1690–1703.
Shin, W., Chung, S., & Lee, Y. (2013). Parallel opportunistic routing in wireless networks. *IEEE Transactions on Information Theory*, 59(10), pp.6290–6300.
Spyropoulos, T., Psounis, K., & Raghavendra, C. S. (2005). Spray and wait: An efficient routing scheme for intermittently connected mobile networks. *Proceeding of the ACM SIGCOMM Workshop on Delay-tolerant Networking – WDTN*, Philadelphia, PA, pp.252–259.

Tehrani, P., Zhao, Q., & Javidi, T. (2013). Opportunistic routing under unknown stochastic models. *5th IEEE International Workshop on Computational Advances in Multi-sensor Adaptive Processing (CAMSAP)*, San Martin, France, pp.145–148.

Vahadat, A. & Becker, D. (2000). Epidemic routing for partially connected ad hoc networks. Technical Report CS-200006, Duke University, Durham, NC.

Verma, A. & Anurag, D. (2011). Integrated routing protocol for opportunistic networks. *International Journal of Advanced Computer Science and Applications*, 2(3), pp.85–92.

Wang, Z., Chen, Y., & Li, C. (2012). CORMAN: A novel cooperative opportunistic routing scheme in mobile ad hoc networks. *IEEE Journal on Selected Areas in Communications*, 30(2), pp.289–296.

Westphal, C. (2006). Opportunistic routing in dynamic ad hoc networks: The OPRAH protocol. *IEEE International Conference on Mobile Ad Hoc and Sensor Systems*, Vancouver, Canada, pp.570–573.

Wu, J., Lu, M., & Li, F. (2008). Utility-based opportunistic routing in multi-hop wireless networks. *The 28th International Conference on Distributed Computing Systems*, Beijing, China, pp.470–477.

Xiao, M., Wu, J., Liu, C., & Huang, L. (2013). TOUR: Time-sensitive opportunistic utility-based routing in delay tolerant networks. *Proceedings IEEE INFOCOM*, Turin, Italy, pp.2085–2091.

Xiaohua, W. & Song, C. (2014). Identify & measure social relations: Routing algorithm based on social relations in opportunistic networks. *IEEE 17th International Conference on Computational Science and Engineering*, Chengdu, China, 407–412.

Yang, S., Zhong, F., Yeo, C. K., Lee, B. S., & Boleng, J. (2009). Position based opportunistic routing for robust data delivery in MANETs. *IEEE Global Telecommunications Conference*, Honolulu, HI, pp.1–6.

Yuan, Y., Yang, H., Wong, S., Lu, S., & Arbaugh, W. (2005). ROMER: Resilient opportunistic mesh routing for wireless mesh networks. *IEEE workshop Wimesh*, Santa Clara, CA, pp.1–9.

Zeng, K., Lou, W., Yang, J., & Brown, D. (2007). On throughput efficiency of geographic opportunistic routing in multihop wireless networks. *Mobile Networks and Applications*, 12(5–6), pp.347–357.

Zeng, K., Lou, W., & Zhai, H. (2008). Capacity of opportunistic routing in multi-rate and multi-hop wireless networks. *IEEE Transactions on Wireless Communications*, 7(12), pp.5118–5128.

Zeng, K., Yang, Z., and Lou, W. (2009). Location-aided opportunistic forwarding in multirate and multihop wireless networks. *IEEE Transactions on Vehicular Technology*, 58(6), pp.3032–3040.

Zhang, X. & Li, B. (2008). Dice: A game theoretic framework for wireless multipath network coding. *Proceedings of the 9th ACM International Symposium on Mobile Ad Hoc Networking and Computing – MobiHoc*, Hong Kong, China, pp.293–302.

Zhang, X. & Li, B. (2009). Optimized multipath network coding in lossy wireless networks. *IEEE Journal on Selected Areas in Communications*, 27(5), pp.622–634.

Zhao, Z. & Braun, T. (2014). Real-world evaluation of sensor context-aware adaptive duty-cycled opportunistic routing. *39th Annual IEEE Conference on Local Computer Networks*, Edmonton, Canada, pp.124–132.

Zhao, Z., Rosario, D., Braun, T., & Cerqueira, E. (2014). Context-aware opportunistic routing in mobile ad-hoc networks incorporating node mobility. *IEEE Wireless Communications and Networking Conference (WCNC)*, Istanbul, Turkey, pp.2138–2143.

Zhao, Z., Rosario, D., Braun, T., Cerqueira, E., Xu, H., & Huang, L. (2013). Topology and link quality-aware geographical opportunistic routing in wireless ad-hoc networks. *9th International Wireless Communications and Mobile Computing Conference (IWCMC)*, Sardinia, Italy, pp.1522–1527.

Zhong, Z., Wang, J., Nelakuditi, S., & Lu, G. (2006). On selection of candidates for opportunistic any path forwarding. *ACM SIGMOBILE Mobile Computing and Communications Review*, 10(4), pp. 1–2.

Zuo, J., Dong, C., Nguyen, H., Ng, S., Yang, L., & Hanzo, L. (2014). Cross-layer aided energy-efficient opportunistic routing in ad hoc networks. *IEEE Transactions on Communications*, 62(2), pp.522–535.

6

Congestion-Aware Adaptive Routing for Opportunistic Networks

Thabotharan Kathiravelu and Nalin Ranasinghe

CONTENTS

6.1 Introduction to Opportunistic Networks 117
6.1.1 Technical Background in Opportunistic Networks 117
6.1.2 Content Distribution in Opportunistic Networks 118
6.1.3 Research Initiatives in Opportunistic Networks 118
6.2 Adaptive Routing in Opportunistic Networks 119
6.2.1 Neighborhood Determination and Message Forwarding 120
6.2.2 Performance Indicators 123
6.2.3 Performance Evaluation of the Adaptive Routing Protocol 123
6.3 Adaptive Routing Protocol with Congestion Avoidance 125
6.3.1 Congestion-Aware Adaptive Routing Protocol 125
6.3.2 Results and Discussion 128
References 130

6.1 Introduction to Opportunistic Networks

6.1.1 Technical Background in Opportunistic Networks

Technological advances in communications and systems architectures have provided users with small handheld mobile wireless devices with multiple communication interfaces, giving rise to the emergent field of ubiquitous and pervasive computing (Chaintreau et al., 2005; Jung et al., 2007; Kathiravelu and Pears, 2006; Pareschi et al., 2008). The portability of these mobile devices with multiple communication interfaces and with various data rates has encouraged the invention of many new networking architectures (Feeney and Nilsson, 2001; Camp et al., 2002; Bruno et al., 2005). Ad hoc and mobile ad hoc networks (MANETs) are two such networking architectures where portable devices are able to establish pair-wise network connectivity among peer devices using short-range wireless connectivity. These ad hoc networks have been found to be useful in scenarios where connectivity to the Internet is not readily available or where the communications infrastructure has been damaged due to natural and other disasters (Bruno et al., 2005). Opportunistic networks have emerged as a new paradigm in communication networking during the past few years and are considered a subclass of MANETs and delay-tolerant networking (Fall, 2003; Cerf et al., 2004). The proliferation of small-sized mobile handheld devices with multiple communication interfaces has paved the way for this newer communication paradigm where mobile wireless devices, when they are within communication range of each other, can exchange their content of interest opportunistically or can forward data packets opportunistically for other mobile devices using their local connectivity with other devices. These opportunistic exchanges set the stage for opportunistic networking and for the emerging paradigm of store, carry, and forward (alternatively known as pocket switched networks [PSNs] [Chaintreau et al., 2005]) with minimum support from the infrastructure. The intermittent connectivity between the mobile devices in opportunistic networking forces them to store the content in their own buffer, carry it while they are mobile, and do so until they meet a potential device to forward the content. It is the device mobility and the cooperation among the mobile devices that have enabled opportunistic networking to become a success in information sharing (Kathiravelu and Pears, 2006; Chaintreau et al., 2005; Leguay et al., 2006; Conti and Kumar, 2010; Pelusi et al., 2006). Since opportunistic networks are assumed to operate without the support of typical networking infrastructure, they are also suitable for being employed in developing regions of the world where network infrastructure is not available or is very much of limited range for political and economic reasons. Even in places where network infrastructure does exist, natural disasters could demolish the infrastructure, making opportunistic networking beneficial to people to exchange vital and critical information among themselves, especially for people who are involved in rescue and relief operations (Conti and Kumar, 2010; Pelusi et al., 2006). In developed regions of the world, opportunistic networks can play a major role in supplementing service quality in high-density social interaction scenarios (Hui et al., 2008; Leguay et al., 2006; LeBrun and Chuah, 2006; Pietilinen et al., 2009). The interesting inherent properties of opportunistic networks have attracted researchers around the world to explore their research problems from different perspectives. Some of the early work on opportunistic networking since its inception has looked for opportunistic connectivity information that was available in the form of connectivity traces (Chaintreau et al., 2005, 2007; Leguay et al., 2006; Kathiravelu and Pears, 2006). There have been a few more studies in characterizing the network behavior, identification of possible use case scenarios, and the development of applications that can adapt to inherent properties of opportunistic networks (Kathiravelu and Pears, 2006; Calegari et al., 2007). Further developments on the theoretical and experimental studies have paved the way for the design and implementation of routing protocols and forwarding algorithms for opportunistic networks (Wang et al., 2005; Song and Kotz, 2007; Chaintreau et al., 2007).

6.1.2 Content Distribution in Opportunistic Networks

The potential and unique characteristics of opportunistic networks have attracted researchers in finding ways of distributing content of interest using opportunistic contacts. Early work in content distribution using opportunistic networking began with PSNs (Hui et al., 2005a; Chaintreau et al., 2005). PSNs aim to convey messages for mobile human scenarios taking advantage of the local and global connectivity (Hui et al., 2005a, 2005b).

6.1.3 Research Initiatives in Opportunistic Networks

The unique properties of opportunistic networks pose new problems when popular mobility models are used in simulation-based studies. Using popular mobility models to model device mobility and device connectivity in these studies would also complicate the design process (Hsu et al., 2005; Kathiravelu and Pears, 2006; Konishi et al., 2005).

As new network features, such as resilience and community structure (Girvan and Newman, 2002; Hui et al., 2007), emerge, newer models that can represent graphs with these specific characteristics and their specific inherent properties are required. Detailed investigations on time-aggregated field contact trace sets reveal the presence of recurring interactions among nodes when the whole set of time-aggregated contacts is viewed; we call these recurring interactions clusters of contacts (CoCs) in the network. In simulation-based studies of opportunistic networks using mobility models, the ultimate aim is to extract the contact information that includes parameters, such as how long each pair of node is in contact and long each pair is not in contact. Therefore, researchers have started to look at generating connectivity traces directly using the measured probabilistic distribution information available from real field traces, instead of relying on any formal mobility models (Bonn, 2005; Calegari et al., 2007). The argument is that even when one uses such formal mobility models, what one does is model the node behavior to measure the device connectivity, and therefore the question that naturally arises is, why should not the connectivity be modeled directly based on estimated parameters? In addition, generating connectivity information directly relieves the designer from the hassle of identifying the appropriate formal mobility model and setting its parameters. Other advantages, such as the ability to conduct simulation-based studies with a varying user population and device connectivity patterns, could also be easily integrated in such connectivity generators, and traces could be produced, as opposed to conducting complex real field trace collection experiments.

6.2 Adaptive Routing in Opportunistic Networks

In one of our previous studies (Kathiravelu et al., 2009), we were able to model the behavior of an opportunistic network using its two high-level properties of predictability and connectedness. The probabilistically estimated information from these two can be utilized by nodes to make probabilistic estimations in making future opportunistic contacts.

In order to facilitate the forwarding of messages toward their intended destinations, the routing algorithm first requires the opportunistic exchange of the neighborhood information. As given in Kathiravelu et al. (2009), nodes utilize their past history of *contacts* and *inter-contacts* with their neighbors to probabilistically determine their future contact opportunities in the form of predictability and connectedness. When nodes meet each other opportunistically, they exchange their estimated predictability and connectedness information with their neighbors. Receiving nodes of such information shall then determine the best forwarder node based on the received information to forward the messages stored in their buffers.

The proposal for this new protocol is founded on the following principle: based on the self-similarity property, nodes can probabilistically estimate their predictability and connectedness with their neighboring nodes, and therefore to forward messages, they need not have to utilize any random probabilistic estimation. Once the nodes estimate their connectivity information about their neighbors, they can store such information in a well-defined data structure and can exchange the data structure during opportunistic contacts. Nodes that receive the connectivity information from their neighbors may update their connectivity table information based on the received connectivity information; therefore, their connectivity tables contain the most updated information about the potential forwarders at any given time. Each node may also maintain information about more than one forwarder for a given destination. This enables each node to choose the best forwarder among the possible multiple forwarders toward a given destination. A snapshot of an example case of the connectivity table maintained by a given node 6 in a scenario is shown in Table 6.1.

TABLE 6.1

Connectivity Table Containing Information about the Potential Forwarders toward Given Destinations and Their Estimated Contact Predictions

Destination Node	Forwarder Nodes with Highest Predictability	Estimated Connection Establishment Times	Estimated Contact Duration Times
1	4, 5, 7	$T_{1s}, T'_{1s}, T''_{1s}$	$(T_{1e} - T_{1s}), (T'_{1e} - T'_{1s}), (T''_{1e} - T''_{1s})$
2	3	T_{2s}	$(T_{2e} - T_{2s})$
4	4	T_{1s}	$(T_{1e} - T_{1s})$
9	9	T_{4s}	$(T_{4e} - T_{4s})$
10	7	T_{5s}	$(T_{5e} - T_{5s})$

Each row of the connectivity table may contain up to three best nodes arranged in increasing order of destination node identifiers that have the highest predictability of meeting or forwarding toward an intended destination with which the node had made contact in the past, the next expected contact establishment time, and the expected contact duration. If a node has had no chances of meeting a particular node, then it maintains the information of the node with the highest predictability, which might meet that particular node. By doing so, each node maintains a global view of the network connectivity irrespective of whether it will meet a given node in the future.

6.2.1 Neighborhood Determination and Message Forwarding

When a node receives the serialized connectivity table from its neighbors, it updates its own connectivity table with highest-predictability forwarder node details, and its associated connection establishment time and contact duration. The adaptive message-forwarding algorithms executed by each node are given in Algorithms 6.1 through 6.3.

Algorithm 6.1: Send (Connectivity Summary Table)

begin
  arrange predictability information as a compact data structure;
  upon meeting with a neighbor ;
  exchange connectivity summary table with the neighbor that has just met;
  wait for the next predetermined time interval;
  if *the neighbor is still in contact* **then**
    if *there is an update to the connectivity summary table* **then**
      exchange the connectivity summary table with the neighbor;

Algorithm 6.2: Receive (Connectivity Summary Table)

```
begin
    compare connectivity summary table that has just been received,
    with the node's own connectivity table;
    for each entry in the received connectivity summary table do
        if the best forwarder node's connection establishment occurs
        before that of the existing table entry then
            update the contact prediction information with the details
            of the received contact predictability information in the
            node's table;
            prepare packets that will be affected by this update to be
            forwarded towards this node;
```

We use real field connectivity experiment traces (Leguay et al., 2006) to determine the predictability and connectedness information, as opposed to applying random probabilistic approaches to determine the delivery predictability of contacts (Lindgren et al., 2003). In each node, its history of contacts with its neighboring nodes is maintained in the form of a list of contact durations and intercontact durations. Based on this contact history information, a node can make contact predictions about future encounters with its peer nodes, and these predictions drive the opportunistic forwarding mechanism.

A node maintains packets in its message buffer until it meets the intended best forwarder node. When a node receives a packet from another node, and if it is destined for the node itself, then it will pass the packet to the upper layers for processing. If the received packet is destined for some other node, then the node will determine whether it is in direct contact with the intended destination. If it is in direct contact with the destination, then it will forward the packet to the destination.

Algorithm 6.3: Receive (Data Packets)

```
begin
    if the current node is the intended destination of the packet then
        forward it to the upper layers;
    else
        look up the connectivity table for a matching destination entry;
        choose the entry whose connection establishment time occurs
        first as the forwarder;
        store message in the outgoing buffer;
    if the current node itself is the best node to forward then
        store the packet in the message buffer;
        wait till a contact occurs with the destination;
    else
        if the chosen forwarder is currently not in contact with the
        node then
            wait for a contact occurrence to occur;
        forward the packet towards the chosen node;
```

Otherwise, it will choose a node with the highest predictability from its connectivity table and then forward the packet to such a node when it meets that node, or if the node finds itself as the best node to forward, it will keep the packet in its buffer and wait until it meets the destination node.

Three well-known routing protocols, namely, epidemic (Vahdad and Becker, 2000), PRoPHET (Lindgren et al., 2003), and HiBOP (Boldrini et al., 2007), are simulated and compared with the proposed congestion-aware adaptive routing protocol (CAARP). The discrete event–based simulator JiST/SWANS (Barr et al., 2005) was used to simulate the generation and dissemination of traffic as described in Kathiravelu et al. (2009).

It is assumed that in the network, a hop-by-hop path can be dynamically established eventually between any pair of nodes in the simulation area, and that nodes do not fail but can be either within or without the radio range. As the simulation progresses, the connectivity between nodes becomes intermittent.

Figure 6.1 shows the simulation scenario where there are three clusters of mobile nodes with a high bandwidth link connecting clusters. Altogether, there are 12 nodes that constitute the three clusters, with each cluster having 4 nodes. The radio range of a mobile node within the cluster is 200 m, and the intracluster and intercluster bandwidth is 2 Mbps. Nodes in clusters generate traffic that has different traffic quality of service requirements.

In the simulation, the initial 30 minutes is considered a transient period. Then the simulation is run for a continuous period of 6 hours. At the end of the simulation, 1-hour stabilization time is allowed before statistics are collected.

In the simulation, the dynamic connectivity pattern between a pair of nodes defines the connectivity model. It is the connectivity model that determines the link-level connectivity between pairs of nodes, that is, establishment of newer communication links and the disconnection of existing communication links. We have made modifications to the native 802.11 link layer implementation in the JiST/SWANS simulator (Barr et al., 2005) to

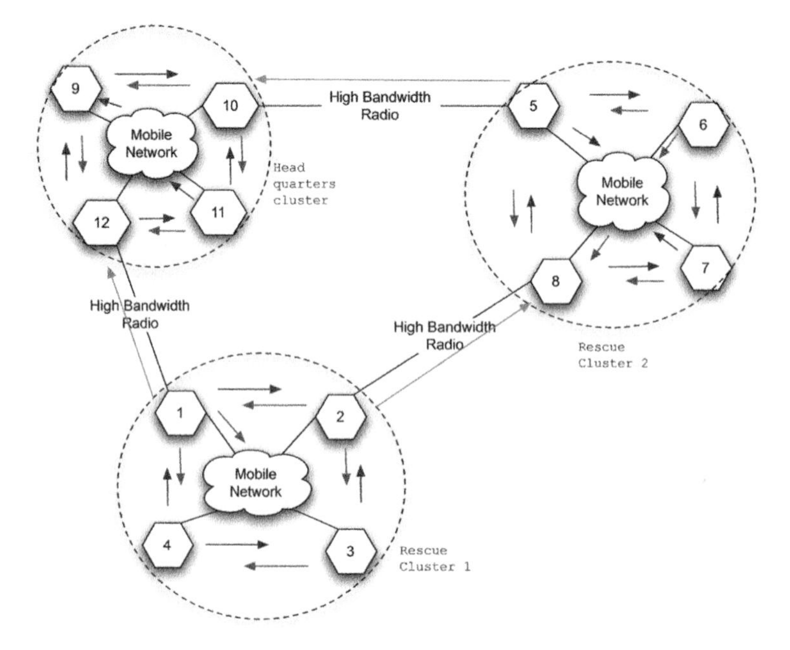

FIGURE 6.1
The simulation evaluation system set up with clusters of nodes.

simulate the contacts and intercontacts of the opportunistic network according to the connectivity model. As discussed in Kathiravelu et al. (2009), the members of the simulated search and rescue scenario make contacts among themselves within a time interval of 3–5 minutes. In our simulation, we have varied the connectedness and predictability of the system from 50% to 100%, in a total of 36 experimental runs.

6.2.2 Performance Indicators

As used by Lindgren and Phanse (2006) for a similar experimental study, we use the three buffer management policies, namely, the just drop (NOPO), MOFO (drop the most forwarded message first) and SHLI (evict the shortest lifetime first), in our simulation-based experiments to measure the performance.

We use message delivery ratio (MDR) and the average amount of energy spent by each node to transmit and receive messages, as well as in the sleep and idle states, as our performance indicators to measure the system performance under different combinations of predictability and connectedness in order to exactly identify the system performance while these two are varied.

In order to measure the energy spent in the transmitting, receiving, sleeping, and idling states, we use the energy consumption values given in Table 6.2 as proposed by Chen et al. (2002) for a 2 Mbps communicating radio and have adapted the method described by Friedman and Kogan (2009) for calculating the same. We monitor the state changes in each node while it switches from one activity to another and then measure the energy consumed.

We use the following equation to calculate the average energy expended by each node:

$$\text{Average Energy Expended per node} = \frac{\begin{array}{c}\text{Total energy spent for all the}\\ \text{activities by all nodes}\end{array}}{\text{Total number of nodes}} \tag{6.1}$$

6.2.3 Performance Evaluation of the Adaptive Routing Protocol

In Figures 6.2 through 6.4, we present the MDR for each of the buffer management policies separately for traffic type 1. First, a general observation can be made from all three plots with different buffer management policies. It is easy to observe that the adaptive protocol outperforms both the epidemic and PRoPHET routing protocols in MDR in all cases for any given buffer size. This confirms our hypothesis that by using the past history information, we can select the best future forwarder to achieve a higher MDR.

We can clearly observe that as expected, all three plot test cases with a predictability of 90% and connectedness of 50% performed better than the case with a predictability of 50% and connectedness of 50%. It is obvious that a 90% confidence in predictability will indeed achieve a higher MDR. It is also significant to observe that under the same operating

TABLE 6.2

Power Consumption in the Tx (Transmit), Rx (Receive), Idle, and Sleep Modes as Proposed by Chen et al. (2002)

Tx	Rx	Idle	Sleeping
1400 mW	1000 mW	830 mW	130 mW

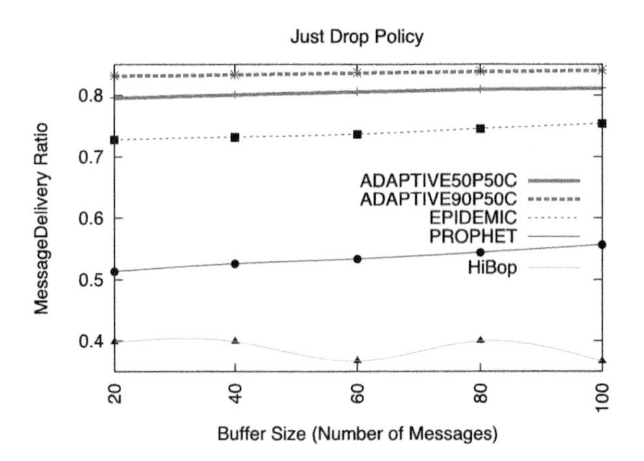

FIGURE 6.2
MDR for traffic type 1 with just drop buffer management policy.

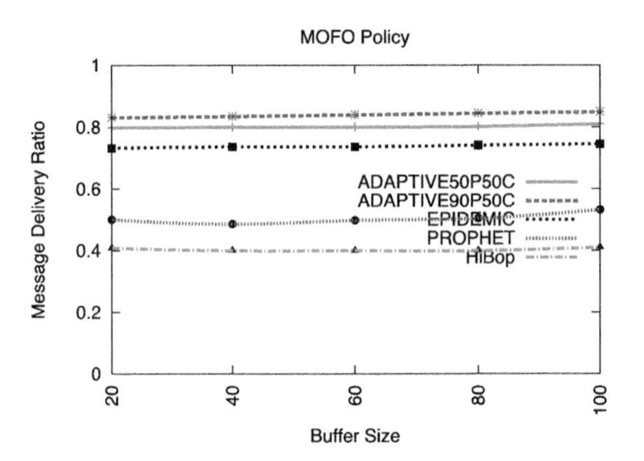

FIGURE 6.3
MDR for traffic type 1 with MOFO buffer management policy.

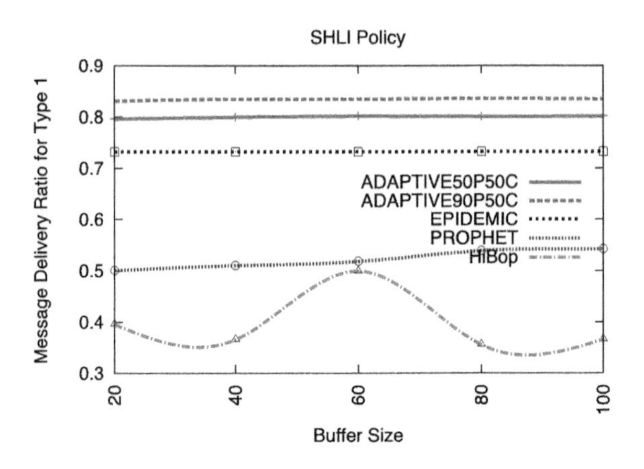

FIGURE 6.4
MDR for traffic type 1 with SHLI buffer management policy.

conditions, the epidemic routing protocol achieves a higher MDR than the PRoPHET routing protocol. We could also observe that the adaptive and epidemic routing protocols both achieve a higher MDR than the PRoPHET protocol.

From Figures 2.2 through 2.4, we can also observe that the MOFO policy outperforms the NOPO and SHLI policies in terms of the MDR. When there is congestion, the MOFO policy ensures that the historically least forwarded messages in the buffer are kept while the most forwarded messages are evicted from the buffer. This enables the least forwarded messages to more likely get to their destinations. The SHLI policy, which evicts messages based on their time to live values, performs similarly to the NOPO policy.

6.3 Adaptive Routing Protocol with Congestion Avoidance

The protocol described in Kathiravelu et al. (2010) estimates the two values of predictability and connectedness heuristically for each node based on its past history of contacts and then makes forwarding decisions for messages (Kathiravelu et al., 2010; Lakkakorpi et al., 2010). The protocol assumes that when a chosen forwarder receives messages as described above, it will have adequate storage to accommodate the arriving messages and will also try its best to forward them (Kathiravelu et al., 2010).

This can result in a situation where a few nodes will have to devote most of their resources to others in order to increase the overall delivery rate of messages. Even if properties such as the number of remaining hops toward the destination and the probability of reaching a given destination are considered by the adaptive routing protocol, some nodes can become more popular and will have to devote their precious resources, such as battery power, for mutual forwarding. This is because the heuristics considered so far do not pay attention to a node's position of being so popular among the other nodes. This ultimately results in the dropping of many messages by popular nodes and the degradation of final message delivery. Typical MANET approaches for avoiding congestion just discard arriving messages when a node is not ready to accommodate such messages, resulting in messages being lost in transit (Sharma and Bhadauria, 2011; Jain et al., 2012).

When nodes become overloaded, they employ buffer management policies to avoid congestion and to accommodate new messages. As a result, a significant number of messages can get lost in transit. In opportunistic networks, as there is no effective feedback path at the link layer between the source and the destination nodes, source nodes are often unaware of the congestion in intermediate nodes and the message losses. Congested nodes often refuse to receive messages and, as a result, the network becomes partitioned. Therefore, the typical approaches that are developed based on the end-to-end connectivity are not applicable for opportunistic networks. In our empirical studies, we have observed that the conventional approaches for congestion avoidance have impacted the overall performance of the network significantly. We have also observed that a better, more controlled approach is needed to avoid congestion.

6.3.1 Congestion-Aware Adaptive Routing Protocol

Our proposal for a CAARP considers the amount of buffer space needed for an incoming message at the node while making a forwarding decision. In a node with limited buffer space, a decision has to be made on which of the messages to accept for forwarding.

A message will be accommodated on a priority basis if the arriving message has yet to travel a larger number of hops before reaching its destination. Messages that have yet to travel a larger number of hops to their destinations should be handled with priority in order to counter the fact that the predictability and connectedness values for those messages are found to be lower than those of other messages, and therefore nodes will not prefer to receive them by default.

A message that has already been forwarded many times implies that the message is already at its destination or is closer to its destination. A proposal to the contrary is that keeping such messages at the end of the list with a lower priority will not do much harm to the message delivery (Lindgren and Phanse, 2006). We argue otherwise. A message that has already been forwarded by many nodes could be accommodated in the buffer in a priority manner. The MOFO (evict the most forwarded packet first) approach adopts this policy, and the effect of this approach is discussed in Lindgren and Phanse (2006).

On the other hand, considering messages that have been forwarded already by many nodes to be given a higher priority than messages that have been just forwarded by a few nodes can result in relatively new messages getting dropped because of the lower number of hops they have traversed. This could seriously affect the final delivery of such messages. Therefore, we have adopted a policy that gives a higher priority to a message that still needs to traverse many nodes to reach its intended destination.

When nodes come into contact with each other, summary vectors are exchanged, containing a list of messages that a node is willing to accept based on its congestion avoidance policy. The sending node will accordingly forward a number of messages from its buffer to the receiving node. The receiving node will keep those messages that need to be forwarded toward their final destinations through other nodes until a suitable forwarder is met while forwarding those that can be immediately sent.

When a message is received by a node and if the buffer occupation is less than 50% of the total space, the message is accepted without any issues (Lakkakorpi, 2011). If the space available in the buffer is more than 50%, then the congestion avoidance algorithm will be enabled to decide which messages to store in the buffer. The congestion avoidance algorithm first considers the rate at which the node itself is generating messages. It gives priority to those messages in case the space available in the buffer is less than a safety threshold and the algorithm will allow only the node's or nodes' own messages that cannot be immediately forwarded to occupy the buffer if the threshold remains exceeded (Figure 6.5).

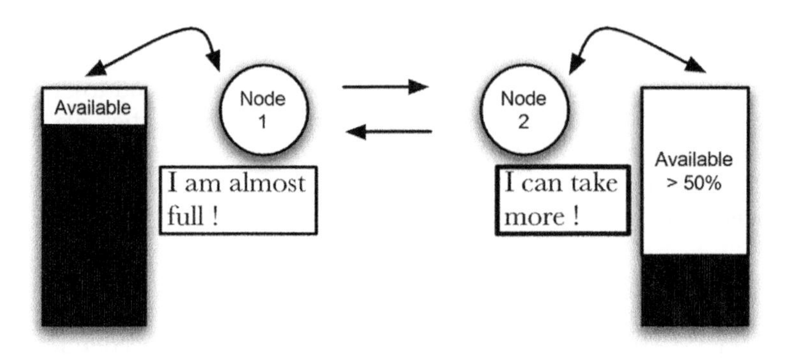

FIGURE 6.5
Message delivery mechanism of the adaptive routing protocol.

When the occupied buffer space is less than 50% but above the safety threshold, the congestion avoidance algorithm decides to accommodate messages that need be forwarded through a number of hops to reach their respective destinations. If the occupied space is below the threshold, only own messages will be accepted. The algorithm is able to determine which messages are to be accepted since each node maintains a routing table, as described earlier.

Algorithms 6.4 and 6.5 illustrate how congestion avoidance is in operation when the remaining buffer space in a node is less than 50% of the total space. Each node lists the nodes that it has a higher predictability of meeting and advertises them to its neighbors. Receivers of such advertisements shall not forward any messages to the sender that does not have a higher probability of reaching their destinations. Algorithm 6.4 illustrates how a node informs its neighbors about congestion at itself and refrains from receiving further messages from its neighbors. Algorithm 6.5 illustrates how a sender of messages prepares a list of messages to be sent to a neighbor that has indicated its willingness to receive messages. Our proposed approach for congestion avoidance is incorporated into the adaptive routing protocol. The enhanced protocol is named the congestion-aware adaptive routing protocol, which fully incorporates the functionalities described in Algorithms 6.4 and 6.5.

Algorithm 6.4: Message Acceptance Logic - *Receiver*. The node which is in contact with a new node will execute this algorithm and shall exchange its buffer space availability to accept messages from the node it has just met.

Input: Buffer size-$B_{capacity}$
Input: Current buffer occupancy-B_{occ}
Input: Traffic generation rate-T_{rate} given as the number of packets generated by a node per unit time
Input: The ordered list of predictability information of node's neighbors
Input: δt the worst case expected time duration till the next contact
Define: Available free space in the buffer B_{free};
Define: Expected occupancy requirement B_{req};
Define: List of nodes reachable in the descending order of the number hops to reach $List_{best\_nodes}$;
begin
  set $B_{req} \Leftarrow min(B_{occ} + (T_{rate} * \delta t),\ B_{capacity}\ )$;
  if $B_{occ} \leq 1/2 * B_{capacity}$ **then**
    send the list of messages that can be
    accepted within the buffer based on $B_{free} \Leftarrow (B_{capacity} - B_{req})$;
    set $B_{occ} \Leftarrow B_{req}$;
    Receive messages from the neighbor;
    set $B_{occ} \Leftarrow B_{occ} +$ *number of messages received*;
    set $B_{free} \Leftarrow B_{free}$ $-$ *number of messages received*;
    set $List_{best\_nodes} \Leftarrow$ *nodes in the descending order of hops*;
  else if $(B_{occ} > 1/2 * B_{capacity})$ *and* $(B_{occ} < B_{req})$ **then**
    while $B_{occ} < B_{capacity}$ **do**
      receive a message from the neighbor;
      accommodate this message in to the buffer;
      set $B_{occ} \Leftarrow B_{occ} + 1$;
      set $B_{free} \Leftarrow B_{free} - 1$;
      set $List_{best\_nodes} \Leftarrow$ *nodes in the descending order of hops*;

Simulations are carried out to compare the performance of CAARP and the epidemic, PRoPHET, and HiBOP routing protocols (Vahdad and Becker, 2000; Lindgren et al., 2003; Boldirini et al., 2007) under low to moderate congestion.

Simulations are carried out to compare the performance with the proposed connectivity model implemented to model the inherent opportunistic network properties (Kathiravelu et al., 2006, 2009).

Algorithm 6.5: Message Sending Logic-*Sender*. This algorithm will be executed by the intermediate node which receives the willingness information from a neighbor it had just met and shall prepare messages to be sent.

Input: the received B_{free} information form the neighbor just met. If there are two neighboring nodes that wants to send at the same time then choose one of them randomly.
Input: the $List_{best\_nodes}$ just received from a neighbor
begin
    Fetch buffered messages to be forwarded according to B_{free} and $List_{best\_nodes}$;
    When two or more messages have the same hop value, choose the node with the minimum predictability value;
    while $List_{best\_nodes}$ *is not Empty* **do**
        Forward the message to the neighbor;
        $List_{best\_nodes} \Leftarrow List_{best\_nodes} - 1$;

6.3.2 Results and Discussion

Figure 6.6 presents the MDR for the adaptive protocol and CAARP for the cases of 50P50C and 90P50C. It can be observed that CAARP shows better performance than the adaptive protocol when the buffer size is 100, but the adaptive protocol performs better for small buffer sizes.

One more interesting observation that can be made is that when the buffer sizes are smaller, the MDR is very low for CAARP in the cases of both 50P50C and 90P50C. The

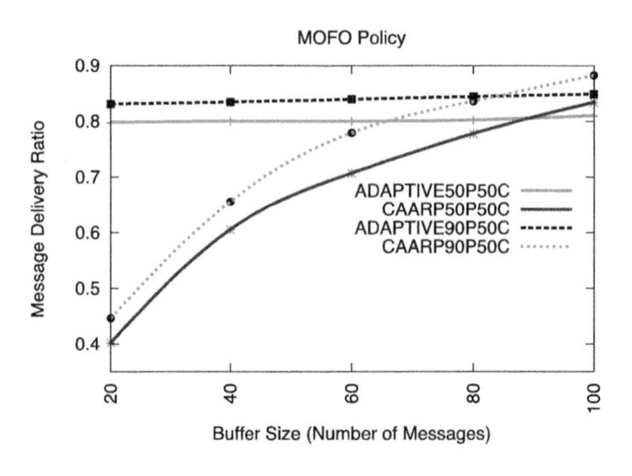

FIGURE 6.6
A comparison of the MDR for the adaptive routing protocol and CAARP.

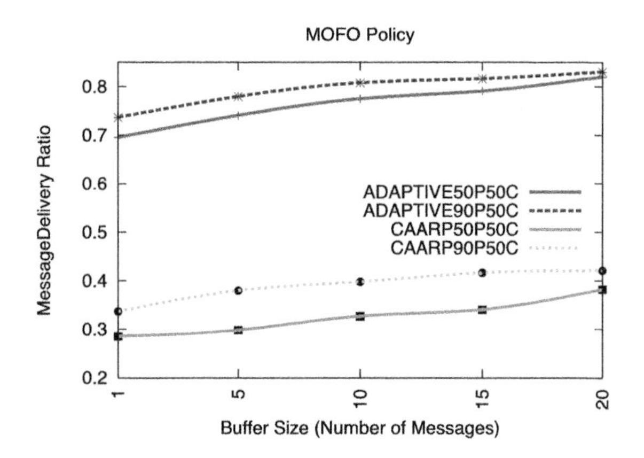

FIGURE 6.7
A comparison of MDR for the adaptive routing protocol and CAARP when the buffer sizes are smaller.

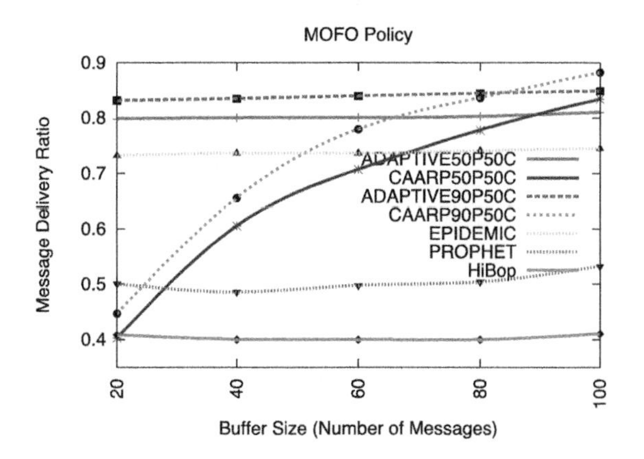

FIGURE 6.8
A comparison of the MDR for CAARP and the adaptive, PRoPHET, HiBOP, and epidemic routing protocols.

reason for this is that CAARP allocates at least half of its buffer space to store its own messages and only then will accommodate the messages it receives from its neighbors.

When the buffer sizes are comparably smaller, the MDR becomes low. To investigate this effect, the buffer sizes were varied from 1 to 20 and the adaptive protocol and CAARP behaviors were investigated. From Figure 6.7, we could clearly observe that the adaptive routing protocol performs over and above CAARP well when the buffer sizes are smaller.

Figure 6.8 presents the MDR for adaptive (Kathiravelu et al., 2009), our proposed CAARP, epidemic (Vahdad and Becker, 2000), PRoPHET (Lindgren et al., 2003), and HiBOP (Boldirini et al., 2007) routing protocols. It can be clearly observed that CAARP exhibits a sharp increase in the MDR when the buffer size increases. For all other protocols, MDR mostly remains invariant of the buffer size.

In conclusion, in this chapter we have proposed, designed, and developed a CAARP, which can perform better in opportunistic networking environments. This proposed protocol exhibits better performance than well-known protocols in the field, especially when the buffer sizes are larger. As future exploration in this field, we envision that more advanced routing protocols that incorporate the inherent properties of opportunistic networks, along with the recent trending approaches to solve problems, such as machine learning, could be proposed and developed.

References

Barr, R., Haas, Z. J., and van Renesse, R. (2005). JiST: An efficient approach to simulation using virtual machines. *Software Practice & Experience*, 35(6): 539–576.

Boldrini, C., Conti, M., Jacopini, I., and Passarella, A. (2007). HiBOP: A history based routing protocol for opportunistic networks. In *Proceedings of the (WoWMoM 2007)*, Helsinki, Finland, pp. 1–12.

Bruno, R., Conti, M., and Gregori, E. (2005). Mesh networks: Commodity multihop ad hoc networks. *IEEE Communications Magazine*, 43(3): 123–131.

Calegari, R., Musolesi, M., Franco, R., and Mascolo, C. (2007). CTG: A connectivity trace generator for testing the performance of opportunistic mobile systems. In *Proceedings of the ESEC and ACM SIGSOFT FSE07*, Dubrovnik, Croatia, pp. 415–424.

Camp, T., Boleng, J., and Davies, V. (2002). A survey of mobility models for ad hoc network research. *Wireless Communications & Mobile Computing (WCMC)*, Special Issue on Mobile Ad Hoc Networking Research Trends and Applications, 2(5): 483–502.

Cerf, V., Burgleigh, S., Hooke, A., Togerson, L., Dust, R., Scott, K., Fall, K., and Weiss, H. (2004). Delay tolerant network architecture. draft-irtf-dtnrg-arch-02.txt.2. https://tools.ietf.org/html/rfc4838.

Chaintreau, A., Hui, P., Crowcroft, J., Diot, C., Gass, R., and Scott, J. (2005). Pocket switched networks: Real-world mobility and its consequences for opportunistic forwarding. Technical report UCAM-CL-TR-617. University of Cambridge, Computer Lab.

Chaintreau, A., Mtibaa, A., Massoulie, L., and Diot, C. (2007). The diameter of opportunistic mobile networks. In *Proceedings of the 3rd International Conference on Emerging Networking Experiments and Technologies (CONEXT '07)*. New York, NY, pp. 1–12.

Chen, B., Jamieson, K., Balakrishnan, H., and Morris, R. (2002). SPAN: An energy-efficient coordination algorithm for topology maintenance in ad hoc wireless networks. *Wireless Networks*, 8(5): 481–494.

Conti, M. and Kumar, M. (2010). Opportunities in opportunistic computing. *Computer*, 43(1): 42–50.

Fall, K. (2003). A delay-tolerant network architecture for challenged internets. In *Proceedings of the SIGCOMM'03*. Karlsruhe, Germany, pp. 27–34.

Feeney, L. M. and Nilsson, M. (2001). Investigating the energy consumption of a wireless network interface in an ad hoc networking environment. In *IEEE Infocom*, Anchorage, AK, pp. 1548–1557.

Girvan, M. and Newman, M. (2002). Community structure in social and biological networks. *Proceedings of the Nationals Academy of Sciences, USA*, 99: 7821–7826.

Hsu, W.-J., Merchant, K., Shu, H., Hsu, C., and Helmy, A. (2005). Weighted waypoint mobility model and its impact on ad hoc networks. *Mobile Computing and Communication Review*, 9(1): 59–63.

Hui, P., Chaintreau, A., Gass, R., Scott, J., Crowcroft, J., and Diot, C. (2005a). Pocket switched networking: Challenges, feasibility, and implementation issues. In *Proceedings of the Workshop on Autonomic Communications*, Athens, Greece, pp. 1–12.

Hui, P., Chaintreau, A., Scott, J., Gass, R., Crowcroft, J., and Diot, C. (2005b). Pocket switched networks and the consequences of human mobility in conference environments. In *Proceedings of the 2005 ACM SIGCOMM First Workshop on Delay-Tolerant Networking and Related Topics (WDTN '05)*, Philadelphia, pp. 244–251.

Hui, P., Crowcroft, J., and Yoneki, E. (2008). Bubblerap: Social-based forwarding in delay tolerant networks. In *MobiHoc '08: Proceedings of the 9th ACM International Symposium on Mobile Ad Hoc Networking and Computing*, New York, pp. 241–250.

Hui, P., Yoneki, E., Chan, S.-Y., and Crowcroft, J. (2007). Distributed community detection in delay tolerant networks. In *Proceedings of the MobiArch'07*, Kyoto, Japan, pp.1–8.

Jain, S., Kokate, S., Thakur, P., and Takalkar, S. (2012). A study of congestion aware adaptive routing protocols in MANETs. *Computer Engineering and Intelligent Systems*, 3(4): 63–74.

Jung, S., Lee, U., Chang, A., Cho, D.-K., and Gerla, M. (2007). Bluetorrent: Cooperative content sharing for Bluetooth users. *Pervasive Mobile Computing*, 3: 609–634.

Kathiravelu, T. (2007). Towards content distribution in opportunistic networks. Licentiate degree thesis, Comprehensive summary, Uppsala University, Uppsala, Sweden.

Kathiravelu, T. and Pears, A. N. (2006). What & when: Distributing content in opportunistic networks. In *Proceedings of the International Conference on Wireless and Mobile Computing (ICWMC 2006)*, Bucharest, Romania, pp. 64–69.

Kathiravelu, T., Pears, A., and Ranasinghe, N. (2009). Evaluation of the impact of opportunistic networking on command and control system performance. In *Proceedings of the Next Generation Wireless Networks (NGWS 2009)*, Melbourne, Australia.

Kathiravelu, T., Ranasinghe, N., and Pears, A. (2010). Towards designing a routing protocol for opportunistic networks. In *Proceedings of the International Conference on Advances in ICT for Emerging Regions (ICTer2010)*, Colombo, Sri Lanka, pp. 56–61.

Konishi, K., Maeda, K., Sato, K., Yamasaki, A., Yamaguchi, H., Yasumoto, K., and Higashino, T. (2005). Mobireal simulator-evaluating MANET applications in real environments. In *Proceedings of the 13th IEEE International Symposium on Modeling, Analysis and Simulation of Computer and Telecommunication Systems (MASCOTS)*, Washington, DC, pp. 499–502.

Lakkakorpi, J., Pitkänen, M., and Ott, J. (2010). Adaptive routing in mobile opportunistic networks. In *Proceedings of the MSWiM'10*, Bodrum, Turkey, pp. 101–109.

LeBrun, J. and Chuah, C. N. (2006). Bluetooth-based content distribution stations on public transit systems. In *Proceedings of the ACM Mobishare'06*, Los Angeles, pp. 63–65.

Leguay, J., Lindgren, A., Scott, J., Friedman, T., and Crowcroft, J. (2006). Opportunistic content distribution in an urban setting. In *Proceedings of the ACM SIGCOMM 2006—Workshop on Challenged Networks (CHANTS '06)*, Pisa, Italy, pp. 205–212.

Lindgren, A., Doria, A., and Schelen, O. (2003). Probabilistic routing in intermittently connected networks. In *Proceedings of the Fourth ACM International Symposium on Mobile Ad Hoc Networking and Computing (MOBIHOC 2003)*, Annapolis, MD, pp. 19–20.

Lindgren, A. and Phanse, K. S. (2006). Evaluation of queuing policies and forwarding strategies for routing in intermittently connected networks. In *Proceedings of the First International Conference on Communication System Software and Middleware (COMSWARE 2006)*, Delhi, India, pp. 1–10.

Pareschi, L., Riboni, D., and Bettini, C. (2008). Protecting users' anonymity in pervasive computing environments. In *Proceedings of the Sixth Annual IEEE International Conference on Pervasive Computing and Communications (Per-Com2008)*, Hong Kong.

Pelusi, L., Passarella, A., and Conti, M. (2006). Opportunistic networking: Data forwarding in disconnected mobile ad hoc networks. *IEEE Communications Magazine*, 44(11): 134–141.

Pietilinen, A.-K., Oliver, E., LeBrun, J., Varghese, G., and Diot, C. (2009). Mobiclique: Middleware for mobile social networking. In *WOSN'09: Proceedings of the 2nd ACM SIGCOMM Workshop on Online Social Networks*, Barcelona, Spain, pp. 49–54.

Sharma, V. K. and Bhadauria, S. S. (2011). Agent based congestion control performance in mobile adhoc network: A survey paper. *International Journal of Advanced Computer Science and Applications*, Special Issue on Wireless and Mobile Networks, 197: 324–333.

Song, L. and Kotz, D. F. (2007). Evaluating opportunistic routing protocols with large realistic contact traces. In *Proceedings of the CHANTS '07 Workshop*, Montreal, Quebec, Canada.

Vahdad, A. and Becker, D. (2000). Epidemic routing for partially connected ad hoc networks. Technical report, Duke University.

Wang, Y., Jain, S., Martonosi, M., and Fall, K. (2005). Erasure-coding based routing for opportunistic networks. In *Proceedings of the 2005 ACM SIGCOMM Workshop on Delay-Tolerant Networking (WDTN '05)*, Philadelphia, PA, pp. 229–236.

7

Vehicular Ad Hoc Networks

Sara Najafzadeh

CONTENTS

7.1 Introduction 134
7.1.1 Overview of Dedicated Short-Range Communication (DSRC) Standard 135
7.2 Applications of VANETs 136
7.2.1 Safety Applications 136
7.2.2 Non-Safety Applications 136
7.2.3 Commercial Applications 137
7.3 Packet Forwarding and Dissemination in VANETs 137
7.4 Homogeneous and Heterogeneous Vehicular Ad Hoc Networks 137
7.4.1 Unicast Communication 138
7.4.2 Multicast Communication 139
7.4.3 Broadcast Communication 139
7.5 Broadcasting in Vehicular Ad Hoc Networks 140
7.6 Dissemination Protocols for Connected Vehicular Ad Hoc Networks 141
7.6.1 Heuristic 141
7.6.1.1 Probability-Based 141
7.6.1.2 Counter-Based 142
7.6.1.3 Delay-Based 142
7.6.2 Topology-Based 143
7.6.2.1 Local-Decision 143
7.6.2.2 Imposed-Decision 144
7.7 Routing Approaches for Fragmented Vehicular Ad Hoc Networks 145
7.7.1 Epidemic 145
7.7.2 VADD 145
7.7.3 Spray and Wait 146
7.7.4 MobySpace 146
7.8 Broadcasting Protocols for Connected and Fragmented Vehicular Ad Hoc Networks 147
7.8.1 IVG 147
7.8.2 DRG 147
7.8.3 DV-CAST 148
7.8.4 Mobicast 148
7.8.5 DECA 148
7.8.6 POCA 149
7.8.7 EDB 149
7.8.8 SRD 149
7.8.9 EAEP 150

7.8.10 Ack-PBSM ... 150
7.8.11 UV-CAST ... 151
7.8.12 Streetcast ... 152
7.9 Comparison of Broadcasting Protocols in VANETs ... 152
7.10 Conclusion ... 153
References ... 155

7.1 Introduction

This chapter starts with an overview of vehicular ad hoc networks (VANETs). It includes VANET characteristics and similarities, and the differences between VANETs and mobile ad hoc networks (MANETs). This chapter is about the diverse applications of VANETs and the requirements of these applications. In addition, it reviews VANET standards, heterogeneous VANETs, a variety of studies focused on different broadcasting methods, and also associated issues with data dissemination in vehicular networks and vehicular mobility models. The discussion will be about the challenges and solutions with respect to the related issues, based on the literature.

Vehicular ad hoc networks have been established as a potential candidate to connect vehicles to other vehicles which are traveling on the road and provide ubiquitous communication. VANETs can facilitate generic applications such as internet access, multiplayer games, and content distribution as well as some novel, purposely designed, applications. Even though VANETs is considered as a subcategory of MANETs, there are some unique features that make a clear distinction between VANETs and MANETs (Moustafa and Zhang, 2009). These features, including high mobility, dynamic topology, mobility modeling and prediction, localization, large-scale networks, and hard delay constraints, are described briefly in this section.

There is a potential for vehicles to travel at high speeds, and, therefore, the period of communication between them can be very short. Since vehicles have the characteristic of high mobility, the topology of the network changes quickly and unexpectedly, which leads to the frequent and unpredictable break down of wireless links (Moustafa and Zhang, 2009). Another feature of VANETs is that the network topology is highly dynamic (Bako and Weber, 2011). For example, late at night or on rural roads, the traffic density is very low; on the other hand, on huge highways or at midday, a very dense network can be experienced. Accordingly, in the communication range, the number of neighbor vehicles may differ from zero up to hundreds.

The mobility of MANETs is random and without a certain control, while in VANETs, nodes move throughout the network under strict rules. This control strategy causes the position of vehicles to be predictable. One of the most important features of the vehicular scenarios is the fact that sensor nodes are not allowed to freely move around an area, and the vehicles have to be completely respectful of the movements of other vehicles and the road layout. That is, vehicles have a propensity to drive around the forming groups and the radio coverage of the wireless interface of the VANET is usually smaller than the distance between the groups. Furthermore, based on the kind of the road that vehicles pass on, the traffic patterns vary. The road topology also puts a severe restriction on the movement of the vehicle. In other words, while moving around, the nodes have to comply with those mobility patterns which the road network has imposed. Roads can be categorized into three groups: rural roads, urban roads, and highways.

Because of its high potential, this communication platform is going to be a significant part of intelligent transportation systems (ITS). Allocating a unique spectrum to the dedicated short-range communication (DSRC), the wide adaption of IEEE 802.11, the popularity of Global Positioning System (GPS) and local regularity bodies, such as European Telecommunication Standard (ETSI), Industry Canada (IC) and the Federal Communications Commission (FCC) of the U.S, are major issues for the rapid development of VANETs.

For the majority of VANET applications, it is important to have access to real-time updated positioning information (Barani and Fathy, 2008; Boukerche et al., 2008). Unlike the other networks, in VANETs, the position information is easily accessible as they can be applied to the GPS systems in the vehicles, which can be simply installed. With the application of GPS, locating the position of the vehicles is possible in VANETs.

In VANETs, there are thousands of nodes traveling at speeds of up to tens of kilometers per hour. This is not a feature of any other mobile networks, such as wireless sensor networks (WSNs) or mobile ad hoc networks (Olariu and Weige, 2009).

The main objective of VANETs is the delivery of safety messages. These messages must have the highest priority and need to be delivered on time. Since most of the safety applications rely on mechanisms of broadcast, they are delay-critical. These mechanisms allow information to be distributed with minimal delay (Hafeez et al., 2010). Therefore, critical safety information has to be forwarded very quickly by broadcast mechanisms that are designed for this purpose (Chen et al., 2010a).

7.1.1 Overview of Dedicated Short-Range Communication (DSRC) Standard

Network operators are the service providers of VANETs. They can also be implemented with the collaboration of governmental authority. Recently, 75 MHz of the DSRC spectrum at the frequency of 5.9 GHz has been allocated for V2I and V2V communications by the Federal Communications Commission (FCC) of the U.S. (Kenney, 2011). The DSRC spectrum is divided in seven wide channels of 10 MHz. This type of architecture should allow communication between roadside equipment and vehicles and among vehicles that are nearby (Intl, 2003). It means that there are no supports from any infrastructures. One alternative is technology which is infrastructure based (V2I). In fact, the architecture of V2V is included in the V2I architecture. The approaches of V2I depend on the infrastructure that, due to its high cost, may not become a reality in the early stages of VANET development. Within the car to car communication consortium (C2C-CC) a reference architecture is proposed for vehicular networks, this reference helps to distinguish between two domains: the infrastructure domain and in-vehicle ad hoc. The standards of IEEE P1609.1, P1609.2, P1609.3, and P1609.4 have been completed by the IEEE for trial testing with vehicular ad hoc networks (Yin, 2004).

The standard P1609.1 is for the resource manager of wireless access for vehicular environment (WAVE) (IEEE, 2010). The standard IEEE 802.11 is mainly designed for networks with low mobility. However, considerable issues such as frequent disconnection and handoff are addressed by the working group of the standard of IEEE 802.11p (Jiang and Delgrossi, 2008). In order to support the vehicular network applications, alternatives to the IEEE 802.11 standard are defined by the IEEE 802.11p standard. CSMA/CA is used by the IEEE802.11p for link sharing, as the basic scheme for media access. The IEEE 802.11s protocols utilize multi-stage request to send, clear to send (RTS/CTS) handshake followed by an acknowledgment to guarantee the delivery of a unicast packet. Broadcast messages, on the other hand, cannot use the RTS/CTS exchange because it causes a collision in the network (Schmidt et al., 2011).

7.2 Applications of VANETs

In the context of VANET applications, DSRC has been employed to develop new types of applications (Olariu and Weige, 2009), which benefit from a combination of various hardware components (input and output devices, CPUs, navigations systems, sensors, and wireless transceivers) that will be employed in future vehicles. These categories are driver safety enhancement, non-safety utilization, and commercial intentions.

7.2.1 Safety Applications

The safety applications are aimed at improving public safety and protecting individuals against events that may cause loss of life. In safety applications, the safety data is required to be delivered to the intended receivers (e.g., those vehicles that are moving toward the dangerous zone). The most significant group of the VANET applications, the active safety applications, are aimed at decreasing the number of fatalities or injuries in road accidents. Safety applications broadcast information about risky positions and conditions, like the Road Caution Hazard Notification (RCHN) or Post Crash Notification (PCN), to vehicles that are in a position to benefit from the information, for example, keeping away from an accident or any unwanted events. The active safety applications depend upon the information broadcasting into a certain geographical region (Yin et al., 2004). Active safety applications are highly dependent on the information distribution into a particular region of interest (RoI) (Viriyasitavat et al., 2011).

Although the messages sent by this application are relatively few, it is essential that they are delivered and distributed immediately (Elbatt et al., 2006). The minimum frequency for these kinds of applications is 1 Hz (Olariu and Weige, 2009).

7.2.2 Non-Safety Applications

The convenience and efficiency of the driver are improved by the non-safety applications. Unlike safety applications that do not deal with a lot of data, non-safety applications need to process a higher volume of information (Chen et al., 2010b). Travel time can be minimized using driving efficiency applications. They distribute information about the roads and traffic conditions so that the driver can avoid roads with high traffic density or traffic jams. If a driver has information about the roads which lead to their destination, they can save time by choosing the best route (by the application of the car navigation system), which has the least traffic. There are many situations such as merging into the flow of traffic or finding free parking, that comfort applications, like the applications for driver efficiency, can help the driver. Information is periodically exchanged by non-safety applications. This is aggregated with the information received by the vehicle's sensors from neighboring vehicles and, after all these processing stages, the information is distributed to other vehicles. Therefore, unlike safety applications, comfort applications do not impose tight time constraints. However, they also need to periodically exchange information and data in their direct neighborhood.

7.2.3 Commercial Applications

There are some services that are provided by commercial applications such as advertisement, web access, and entertainment. In addition, services such as map downloading (for the navigation systems), video streaming and remote vehicle diagnostics are also included in these applications. Unlike the safety and non-safety applications, commercial applications are heavily dependent on unicast communications (Schoch and Kargl, 2010).

7.3 Packet Forwarding and Dissemination in VANETs

Monitoring the different applications that rely on the VANET reveals that a unicast communication needs to be established by a number of applications. Since the number of the nodes is very high and the topology of the network is rapidly changing, the message sender can send messages without knowing the receiver and messages can be sent without changing the network topology. In many cases, receivers can be defined, such as those located in front of or behind the sender, those in a particular spot, and those receivers that are able to offer a particular service. In addition, some of the applications depend on the local broadcast which can be considered as a type of one-to-all application: emergency signal pre-emption, SoS services, and post-crash warnings (Olariu and Weige, 2009). Therefore, there are three categories for the communications in VANETs: unicast, multicast and broadcast.

7.4 Homogeneous and Heterogeneous Vehicular Ad Hoc Networks

There is an assumption for the simplification of routing in the wireless networks that most of the research in this area has been based on. This assumption is that all the vehicles have the same communication range (Homogeneous) and this range is represented by *R*, which is a fixed number. Therefore, all the existing links in the network, called Bidirectional Links (BL), have two directions and they can operate perfectly well in both directions. It means that if the node R is able to reach the node S with one hop, the node S can also reach the node R with one hop and this has been shown in Figure 7.1. However, in real life, vehicles with different features also have a heterogeneous communication

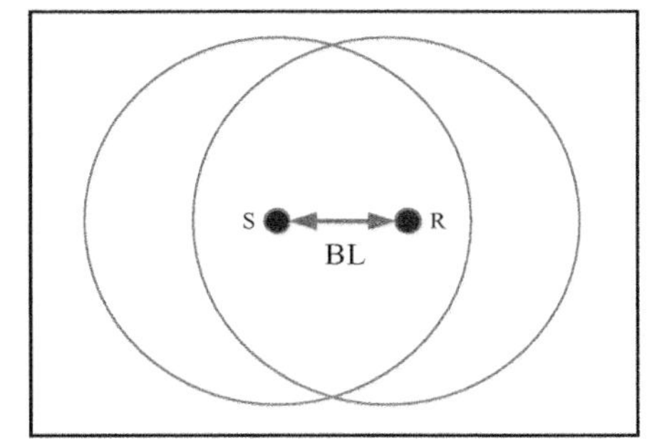

FIGURE 7.1
Bidirectional link.

range. This may be the result of the differences in the algorithms of topology control, heterogeneity of the transmitter and receiver hardware, or the vehicle height, such as buses, trucks, and cars (Li et al., 2004).

There is another reason for the links with one direction and that is the interference, which can be closer to one node than the other. This prevents nearby nodes from obtaining some of the packets of data that are sent to them. The cause of this interference may be the other node's transmissions or transmissions that originate from other devices such as jammers, which are operating at an overlapping frequency. Although it is more likely for this interference to have short-lived one direction links, a network that is made up of different wireless range devices can also have permanent unidirectional links for the given positions of nodes (Zhang et al., 2004).

If the node R is not able to reach the node S, but the node S is able to reach the node R, the link between these two nodes has one direction (unidirectional). A network that is considered as unidirectional consists of links with one direction. An instance of a Unidirectional Link (UL), which is the result of transmission power difference, is shown in Figure 7.2.

7.4.1 Unicast Communication

The protocols of unicast or one-to-one communications are based on the geographical position [such as distance routing effect algorithm for mobility (DREAM) (Basagni and Chlamtac, 1998) and greedy perimeter stateless routing (GPSR) (Karp and Kung, 2000)] or on the topology [such as ad hoc on-demand distance vector (AODV) (Perkins and Royer, 1999) and optimized link state routing (OLSR) (Clausen et al., 2004)]. In a routing protocol which is topology-based (such as AODV and OLSR), the network topology helps the messages to be routed. In a protocol that is based on geographical position, knowledge of the positions of neighbors affects the forwarding decision. The position information is accessible to neighbors due to some location services, which are normally dependent on the one-to-all communication mechanism. However, when the duration of the communication between two vehicles is long, one-to-one communication can be applied. Nevertheless, this type of long communication between two vehicles is not common in highly dynamic networks such as VANETs. Additionally, one-to-one communication is rarely needed in VANET applications (Tonguz et al., 2010).

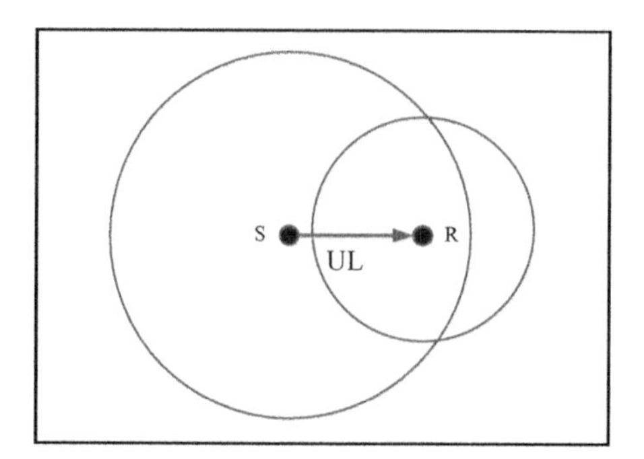

FIGURE 7.2
Unidirectional link.

7.4.2 Multicast Communication

In a multicast routing protocol, a mesh or tree has been constructed to connect senders and receivers. Some vehicles serve as "forwarders" for the multicast message. Although these vehicles are not interested in the multicast message, they act as a router. In tree-based multicasting [such as multicast ad hoc on-demand distance vector (MAODV) (Souza et al., 2013) and ad hoc multicast routing protocol (AMRoute) (Xie et al., 2002)], a sender floods a message to all vehicles in the VANET and only vehicles that are interested in receiving the message reply via a reverse path. To solve the robustness problem of tree-based multicasting, mesh-based multicast protocols have been proposed [such as on-demand multicast routing protocol (ODMRP) (Zhao et al., 2003), group header-based multicasting (GHM) (Hsieh and Wang, 2011)), topological multicast routing protocol (ToMuRo) and geographical multicast routing protocol (GeMuRo) (Santos et al., 2007)]. They use alternative paths to avoid the effect of frequent topology changing.

7.4.3 Broadcast Communication

This type of communication is the most convenient way of information distribution within a VANET. A message received by a vehicle is retransmitted to the adjacent vehicles. This assures that the message is received by as many vehicles as possible. This type of communication can be properly employed in location services (Peng and Cheng, 2007). In the mechanism of one-to-all communication, occasionally, each vehicle disseminates information about itself. As soon as a message is delivered to a vehicle, the vehicle stores the message and forwards it instantly. Since there are a lot of information and message flows in the network, it is clear that this mechanism cannot be scaled, especially when there is a scenario with a high traffic density. By means of IEEE 802.11 standard, Neves et al. (2011) conducted a research on the diffusions on a convoy of vehicles and found that flooding leads to considerable contentions and collisions because of the high quantity of redundant broadcasts.

Each vehicle estimates global topology to observe neighbors or local topology information, which has been collected in a particular time period. Thus each vehicle has the information of its one-hop neighbor which is called neighbor-awareness. On the other hand, in order to apply this method, it is required that the frequency of the hello beaconing update be at a suitably high level to achieve measurements with good accuracy.

Data such as the global topology, which is considered as the density or the volume of the traffic or even a more complete topology of the k-hop neighbors, where $k > 1$, may be applicable to design a protocol, which by its hierarchal structure will result in a much lower overhead. For instance, one of the methods that might be applicable is utilizing the one-hop neighbor information for identification and then adding it to the local data, which is achievable through the broadcast of the periodic hello messages. In a dense traffic system, the application of the periodic hello messages is eliminated or reduced by the coarse information cloud and, as a result, the bandwidth is saved.

For the purpose of traffic density estimation, it is possible for the vehicles to cooperate with each other and exchange information about the topology. However, this method may use a lot of bandwidth and bring a high level of overhead. For this reason, when there are no available smart infrastructures, it may not be appropriate to utilize global topology information.

7.5 Broadcasting in Vehicular Ad Hoc Networks

For the purpose of comfortable and safe driving for vehicles, data can be exchanged among them in vehicular ad hoc networks. There are numerous applications that have been developed and they depend on the distribution of data over long distances or in a geographical zone. Routing is about data packets delivery from the origin to the target over long distances via multi-hop steps (intermediate nodes). However, data dissemination refers to data distribution to all of the nodes in a particular zone. The main focus of data dissemination is on the delivery of safety-related data to the safety applications, especially real-time warning, and collision avoidance data. Although trying not to overload the network is one of the main goals of the distribution, it is essential issues to ensure the delivery of the information to all of the necessary recipients in the RoI.

Another way of looking at the broadcasting in VANETs is to see it as a controlled flooding in the network. Suppose that there is a network with high density and in this high-density network an event has been detected by the vehicles. Then vehicles try to inform the other vehicles about this event by broadcasting the data to them. Now, when there are numerous candidates to forward and broadcast this data, an overload of the shared wireless channel will occur. Therefore, there has to be a well-designed forwarding strategy so that the congestion of the wireless channel will not take place. In addition, safety messages are of a broadcast nature and the on-time availability of them needs to be ensured. Hence, in order to avoid overloading the channel, the number of unnecessary rebroadcasts needs to be minimized by the adopted techniques of data dissemination.

The classification of the protocols of broadcasting in VANETs based on the network density assumption is shown in Figure 7.3. On this basis, there can be three different categories: broadcasting protocols that consider networks that are well-connected, broadcasting protocols that consider fragmented vehicular ad hoc networks and the protocols that consider both the fragmented and connected networks. A classification of these categories is shown in Figure 7.3. This classification is founded on the assumed condition of the network density. By using the particular assumed condition, each category corresponds to the protocols.

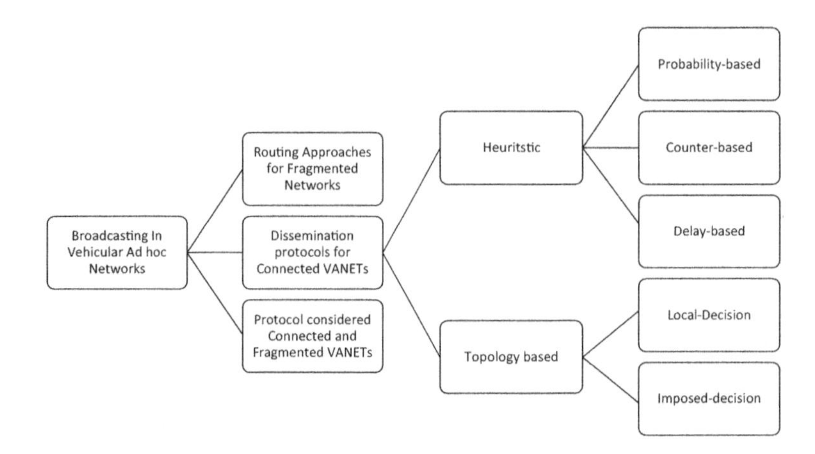

FIGURE 7.3
Taxonomy broadcasting protocol in vehicular ad hoc networks.

7.6 Dissemination Protocols for Connected Vehicular Ad Hoc Networks

For the broadcast packets, there are different protocols in VANETs. Flooding is the simplest technique. In the flooding technique, the packets are rebroadcasted by each node at the time it has been first received. Here, the broadcast total number is equal to *N*-1 and *N* refers to the total number of the vehicles. Although flooding is a simple technique, it may cause some issues. First, when a node tries to forward a data packet to the neighbors and they have received the packet beforehand, then redundant rebroadcast occurs and this would be a redundant transmission.

Second, there will be a contention at a medium level when a packet is received by a neighbor and that neighbor tries to rebroadcast the packet. This will lead to collision, redundancy, and contention in mobile networks. This phenomenon is referred to as the broadcast storm problem. In this category, the protocol's main objective is restricting the rebroadcasting numbers, which will lead to broadcast storm problem mitigation. Topology-based and heuristic-based protocols are the categories of the protocols in this section.

7.6.1 Heuristic

The methods of heuristic broadcasting need parameter selection and thresholds selection, which are mostly associated with environments of ad hoc networks. The performance of these methods depends on the thresholds in the heuristic and the parameters that have been selected (Williams and Camp, 2002).

7.6.1.1 Probability-Based

In the probability-based method, in order to reduce the redundancy of the packets and collision avoidance, the rebroadcast of the messages are decided by the vehicles with some probability. Static gossiping is one of the schemes that is probabilistic-based and is used to enhance the flooding. In order to forward the messages, it applies a probability that is globally defined (Haas et al., 2002). If the characteristics of the network are known in advance and static, all of these variants properly work. Otherwise, the result will be a low delivery ratio or a high number of messages that are redundant. Adaptive gossiping methods have been developed in order to solve these problems. A scheme with two thresholds was proposed by Haas et al. (2002). For static gossiping, this scheme is an expansion that is based on the number of neighbors. If there are n neighbors for a node, this node will forward the data packet with the probability of *P1*. The n is the threshold and if the node neighbors number becomes less than this threshold, then a higher probability of *P2* is used to forward the messages. A great advantage of this improvement is that it prevents messages from dying in networks with sparse connectivity, because in these networks the forwarding probability is greater than the forwarding probability in dense networks. Haas et al. (2002) have also proposed a second improvement, which may prevent the dying out of a message. If there are n neighbors for a node and the probability is p, then each message is received by each node $p.n$ times from its neighbors.

Optimized adaptive probabilistic broadcast (OAPB) (Alshaer and Horlait, 2005) is a probability-based protocol designed to mitigate the broadcast storm problem. In OAPB a rebroadcast probability is assigned to a node based on the density of vehicles in its zone. For this purpose, OAPB utilizes two-hop neighbor information. This information can only be accessed through one-hop neighbors.

AutoCast protocol (Wegener et al., 2007) functions similarly to OAPB. Rebroadcast probability in AutoCast is calculated from a number of nodes around the vehicle area. The only difference between Autocast and OAPB is the use of a different equation to determine broadcast probability.

7.6.1.2 Counter-Based

In this method, there is a defined counter referred to as C and each time that the same data packet is received by the node, the number of this counter increases. When the defined variable C becomes greater than a threshold, the node is dropped off the packet. During the time that the packet is dropped off and the first packet is received, the node begins the rebroadcast of the packet and it is followed by a little delay for each of the retransmissions (Wu et al., 2010b).

A counter-based scheme was proposed by Tseng et al. (2001). The mechanism of this scheme is that a random timeout is set when a message is received by a node for the first time. During the period of the timeout, a counter is increased for every duplicate message received. When the timeout expires, the message is forwarded only if the number of the counter has not passed a value for the threshold which is predetermined.

7.6.1.3 Delay-Based

In this method, in order to omit retransmissions of unnecessary information, smart flooding algorithms are used. In an effort to maximize the nodes that are reachable, in order to forward the message, a set of nodes or a relay node are chosen instead of selecting all of the nodes to distribute the information to all of the neighbors. Delay-based methods are able to deal with the problem of scalability of the nodes with high density.

Urban multi-hop broadcast (UMB) is a V2V delay-based broadcasting protocol. This approach comprises two phases: the intersection and directional broadcast (Korkmaz et al., 2004). The road section that is within the communication range of the origin node is divided into subdivisions with equal lengths. The road that is in the direction of the distribution is the only road that is divided. The forwarding task is assigned to the vehicle from the farthest subdivision. However, in scenarios with high density, there might be more than one vehicle in the most distant segment. In such cases, the farthest subdivision is divided into narrower sub-segments. Then, in order to choose the vehicle that is in the farthest sub-segment, a new iteration starts. When a request to forward the received information is received by the vehicles that are in the distribution direction, the distance of the vehicles to the source node is calculated by the vehicles themselves. According to the calculated distance, each of the vehicles transmits a jamming signal (black burst signal) in the period of the short interframe space (SIFS).

Transmission range adaptive broadcast (TRAB) is another broadcast algorithm for VANETs that is delay-based (Wu et al., 2010b). This algorithm considers the communication range of the vehicles together with the inter-vehicle distances. The waiting time is calculated by the TRAB algorithm in order to decide on the relay vehicles compliant with the additional coverage area of neighboring nodes. This is to guarantee that there will be a reduced number of relay nodes for forwarding emergency packets. In addition, it adopts two types of mechanisms for answering to ensure the reliability of the distribution. These mechanisms are adaptively called explicit acknowledgement and implicit acknowledgement. The packets are forwarded by these mechanisms, founded on vehicles in the two-way lane.

7.6.2 Topology-Based

Topology-based broadcasting can be classified into two sub-categories. These sub-categories are imposed decision-based and local decision-based methods. In local decision-based methods (which are referred to as receiver based or reactive methods as well), the decision making of each node is on its own. The node decides whether to broadcast or forward a particular message or not. In contrast, in the approaches that are imposed decision-based (which are referred to as sender-based methods or proactive methods as well), it is the other nodes that determine whether to forward a message or not. These other nodes can be the previous relay nodes or cluster head.

7.6.2.1 Local-Decision

This approach is fundamentally based on the idea that the node exploits neighborhood connectivity and the history of the nodes that have been already visited by the message; this way, it can decide whether it is a forward node or not. A generic scheme has been proposed by Wu et al. (2010a). Most of the local decision-based methods that exist today are covered by this scheme. The history of visited nodes and neighborhood connectivity are the base of this scheme. The k-hop neighbors of each node have some information and this information will be built up by exchanging information with each node. This information will be exchanged via the periodic hello messages between one-hop neighbors. The node's property information such as node degree, list of the previously visited nodes, and the node ID are added to the broadcast message. According to this type of information, the decision of whether a message should be forwarded or not will be made. Flooding with self-pruning or a neighbor coverage scheme is the most straightforward local decision-based method (Chiang et al., 2005). The list of one-hop neighbors of a sender gets piggybacked by the sender itself. This piggyback occurs on each one of the broadcast messages that gets transmitted. The message immediately gets forwarded, if some additional nodes can be covered by a receiver. The additional nodes are the nodes that are addition to those of the sender. A strategy for forwarding is used by the scalable broadcast algorithm (SBA), which is similar to the scheme of the neighbor coverage (Khan et al., 2008a). However, there are two differences as follows: first, the list of the nodes' one-hop neighbors are not inserted into data messages, but in the hello packets. Secondly, the messages are not immediately forwarded by the nodes and a random assessment delay (RAD) is initiated by the nodes. During the period of waiting, the additional coverage is recalculated by the node for each neighbor forward. At the time that the random assessment delay expires, if the recalculated additional coverage has not reach the zero value, then that node is considered as a forwarder node. In the SBA, the adaptation of RAD is according to the neighbor degree of the node. One of the variants of the protocol of SBA is the Scoped Flooding. The condition of the forwarding is changed upon the expiration of the RAD. There are fixed ratios for each RAD and if the uncovered neighbors are more than this fixed ratio, then the message is forwarded by the node.

An algorithm introduced by Stojmenovic (2004) requires only two-hop neighbors. If there are two neighbors that are not connected, then the dominating set includes the node. The only nodes that forward the message are the ones that belong to the CDS. In order to detect if there are any connections between neighbors, the information of one-hop neighbor is enough. But this information will only be enough if the positions of the nodes are known to the nodes themselves (Stojmenovic, 2004).

7.6.2.2 Imposed-Decision

In these protocols, a broadcast message from a sender specifies which neighbors have to execute a rebroadcast. These types of protocols are called deterministic broadcast approaches. Deterministic approaches clearly select a subclass of neighbors as the forwarding nodes. These selected neighbors can get to the expected destinations, which were supposed to be reached by all the nodes together. Therefore, there is a need for a relaying node to know at least its one-hop neighbors. Since finding a minimal sized optimal subset is considered as NP-hard, heuristic approaches are used. Therefore, the problem of the broadcast storm can be dealt with. That is the reason that there are different types of deterministic broadcasting protocols in the literature background.

There are some examples of deterministic methods (approaches) such as cluster-based methods (Lou and Wu, 2004), total dominant pruning (Lou and Wu, 2003), multi-point relaying (MPR) (Qayyum et al., 2002), and dominant pruning (Lim and Kim, 2001). Although deterministic broadcast is highly efficient, it also has a considerable disadvantage. The disadvantage is that a single point of failure is represented by the relaying nodes. If the job of forwarding a message is failed by a relay, for any reason such as node failure, wireless losses, or not being in the communication range, then it is possible that the message reception rate will significantly drop. Therefore, there is a lack of robustness in these types of protocols and their performance is poor in dynamic environments, such as VANETs. As a result, they are not suitable to be applied for robust and safety-critical applications in VANETs.

The main idea behind the MPR (multi-point relay) is a policy for a message to be forwarded (Plesse et al., 2005). In this policy, a subset of one-hop neighbors of a node is selected by that node in order for the broadcast message to be forwarded. This process has to be performed in such a way that the two-hop neighbors of the node can be reachable with this subset. In the multi-point relay, the list of the one-hop neighbors is inserted by the nodes into the hello packets of the nodes. As a result, the awareness of the nodes from their two-hop neighbors is assured. The forwarder nodes are chosen from the one-hop neighbors of the sender node. Therefore, the set covers all of the two-hop neighbors (Lou and Wu, 2003). The forwarding list is piggybacked by the nodes in their hello beacons. The broadcast message is forwarded only by the nodes that exist in this list. Similar to MPR, by using hello beacons, the nodes that exist in the dominant pruning obtain the knowledge about the two-hop neighbors. In addition, by utilizing the same rule of MPR, the designated forwarders are selected by the senders. Different from MPR, the forwarding set is selected by the receivers, based on the selection rule of MPR. Besides, one other base for this selection is the knowledge of the neighbors previously covered by the broadcast of the sender. The selection of the forwarding set is out of the one-hop neighbors that are not included in the previous relay node's neighbors. The forwarding list is piggybacked on the broadcast message. Therefore, for a particular node, the forwarding message may be different from a message to another one.

Double-covered broadcast (DCB), which is a broadcast scheme that was proposed by Lou and Wu (2004), includes a specific policy for the forwarder node selection. In this policy, first the two-hop neighbors of the sender are covered and then the one-hop neighbors of a sender are either a non-forwarding node or a forwarding node, but in any case, at least two forwarding neighbors cover them. Simulation results indicate that DCB provides fine performance for the operation of broadcasting in an environment with a high rate of transmission error (Lou and Wu, 2004).

7.7 Routing Approaches for Fragmented Vehicular Ad Hoc Networks

Since there are frequent partitions in the VANETs, vehicle connectivity may not be present between most of the node pairs. In cases like this, failure of many of the traditional broadcasting protocols is unavoidable. This part of the study will be assigned to the review fragmented network's routing schemes.

In rural areas or at late at night when there is light traffic, it is possible for the VANETs to become partitioned. These scenarios occur when there is a large distance between the closest vehicles and this distance is greater than the communication range. Therefore, for some of the nodes the termination of broadcasting will occur (Chen et al., 2010a). In such scenarios of networking, as a capable scheme, it has been proposed to apply the store carry forward scheme. With the scheme of store carry forward, a packet that has been received by a node is stored and carried while in motion and the packet is then forwarded to other nodes when they are encountered.

7.7.1 Epidemic

The protocol of epidemic routing is for the delivery of messages in a network that is disconnected most of the time and has mobile nodes (Hayel and Tembine, 2007). Each message has an ID and the summary vector of this ID is maintained by each node. But only the previously received messages are maintained in each node. When contact is initiated by two nodes, the first exchanges are these summary vectors that exist in the session of anti-entropy. In this contact, the nodes compare the message IDs and then identify the messages that have not been received yet and determines whether or not the message have to be drawn from the other node or not. There is also a second phase of contact, in this phase the messages are exchanged by the nodes. There is a limit for every message and this limit is referred to as the time-to-live (TTL) field. The number of the contacts that a message can go through is limited by this field. When the value of the TTL for a message is "1", this message is only forwarded to the destination. The main problem of epidemic routing is the flooding of the whole network with the messages that need to get to the destination. This will result in contentions for the transmission time and the buffer space (Chigira and Higaki, 2011).

7.7.2 VADD

Vehicle-assisted data delivery (VADD) is a protocol for sharing, storing, and forwarding the information of data packets (Zhao and Cao, 2008). If the neighbor is not promising enough, they wait for another one, which is more reliable and is in their communication range. However, their intention is to forward the messages at the earliest time possible. In addition, the decision of which road the packet needs to follow is made using the road and the vehicle information, including maximum allowed speed, next junction distance, and current speed.

The main objective of the vehicle-assisted data delivery is to choose the path that has the minimum delay in the packet delivery. The node that holds the message has a position and this position affects the protocol's behavior. There have been two cases under the consideration: when the nodes, which route the messages, are in the middle of a road and when a junction is the location for those nodes. There are fewer alternatives for the first case, which is also referred to as straightway routing. The alternatives either forward the data packet to the previous junction or to the next one. However, the second case, which is also referred to as intersection routing, is much more complicated than straight roads. The reason for this complication is that there are different roads to be considered at the junctions and this leads to a higher number of options. In both cases, the applied approach is the same. This approach is to determine which road is the next one that the message needs to follow and after that, among the current neighbors, which relay has to be selected. The authors of VADD have proposed a common way to determine the next road, by selecting the outgoing road the lowest delay.

7.7.3 Spray and Wait

Spray and wait is a routing protocol with zero knowledge, it means this protocol does not need neighbor information. To decrease the useless messages that are flooded in the delay-tolerant network (DTN), this protocol was introduced. Like epidemic routing, copies of messages are forwarded to nodes by this protocol. The main difference between these two protocols is that the total number of the distributed message copies are restricted to a number *N* by spray and wait protocol and this number is constant. In the phase of spray, the nodes receiving the message (total number of *N* relays) and the source forwards *N* copies for every message that originates from the same source. In the phase of wait, direct transmission is performed by all the nodes that have a stored copy of the message.

At the start, the *N* copies of a particular message are spread by the spray and wait protocol and this takes place in an epidemic fashion. This to increase the possibility of direct contact with at least one relay node with the node of the destination. All the *N* copies of the message are forwarded by the source node to the first *N* encountered nodes. This can take place in a simple heuristic of the source spray and wait. The optimal policy for forwarding is the binary spray and wait. In this policy, the movements of the nodes are random and they have their own independent and identical distribution of probability. The storage of a message is physical and it will be transmitted only once, even at the times that multiple copies may be involved in a transmission. There is a header field for every message and the number of copies is indicated by this header. A binary tree, the root of which is in the source node, can represent the paths that the copies follow.

7.7.4 MobySpace

In the MobySpace, it is more likely for two nodes that are closer to have contact than two nodes that are farther apart. The forwarding algorithm decides if a message is to be forwarded, during contact with a node that is closer to the destination of the message (Leguay et al., 2006). Messages take paths through the MobySpace in order to bring them closer to their destination. There have been several proposed functions of distance for the similarity measurement in the patterns of mobility. If stable patterns of mobility are shown by the nodes then the approach of MobySpace can be effective. When there is a similar pattern of mobility for a current node with the destination, it is possible for the Moby Space to be ineffective. However, as a result of trajectory synchronization, it is rare to find a direct contact with the destination (Leguay et al., 2007).

Although there might be similarity in the patterns of mobility in two nodes, it does not mean that there are frequent contacts. In addition, an effective path is not provided for the transmission of the messages by this similarity of mobility. The application of the frequency or the probability of direct contact with the other nodes as the distributions in the MobySpace might be a solution for this problem. Complementing the MobySpace by converting the spatial visit patterns to the domain of frequency is another method to cope with the temporal variability of mobility patterns. In this case, the frequency domain represents the phase and the main frequency of visitation. The other matters concerning the MobySpace include the effective distribution of location probabilities.

7.8 Broadcasting Protocols for Connected and Fragmented Vehicular Ad Hoc Networks

Just a few broadcasting protocols have been developed to function in both connected and fragmented conditions. This section will review these protocols.

7.8.1 IVG

Inter-vehicle geocast (IVG) is a timer-based distribution protocol for safety messages in VANETs (Bachir and Benslimane, 2003). If there are any incidents or any accidents on a highway, all the vehicles get informed by the IVG. The position and the direction of a vehicle are the factors that determine the areas of risk. For any incident or accident, the relevant areas are determined by this protocol. The broadcast group that is restricted is referred to as the multicast group. The direction of driving, velocity, and the location are the parameters that dynamically define a multicast group. The message received by the vehicles is not rebroadcasted by them immediately. Before rebroadcasting, there has to be a defer time. If the vehicle has not received a message that has the same ID upon the expiration of the defer time, that vehicle appoints itself as a relay and commences to rebroadcast the message in order to inform the other vehicles. Equation 7.1 is used to calculate the defer time.

$$\text{Defertime}(x) = \text{Maxdefertime}.\left(\frac{R^{\varepsilon} - Dsx^{\varepsilon}}{R^{\varepsilon}}\right) \quad (7.1)$$

where Dsx is the gap between nodes s and x, and R is the communication range. The assumption of the IVG is that there is an equal communication range for all vehicles. The number of unnecessary safety messages is decreased by IVG. IVG performs this task by dynamically maintaining a relay in every driving direction. In addition, safety messages are periodically rebroadcasted by IVG in order to address the fragmentation of the network.

7.8.2 DRG

Distributed robust geocast(DRG) is a fully-distributed broadcasting approach, which considers the fragmentation of the network (Joshi et al., 2007). The zone of forwarding is defined by DRG and the region of interests is surrounded by this zone. The zone of forwarding is defined as a series of geographic criteria that need to be satisfied by the vehicles in order to forward a geocast message. The base of DRG is a back-off scheme and it is for the relay node selection. In addition, DRG is a protocol which is thoroughly distributed.

After a backoff time that is distance based, for each node, a transmission time is scheduled by a vehicle when it receives a safety message.

$$Bo_d(R_{tx}, d) = MaxBo_d.S_d\left(\frac{R_{tx} - d}{R_{tx}}\right) \tag{7.2}$$

In Equation 7.2, S_d is the distance sensitivity factor, BO_d is the backoff time, $maxBO_d$ is the maximum backoff time and R_{tx} is the communication range. In order to deal with temporary fragmentations of the network, DRG applies packet periodic retransmission. This task continues until it is transmitted by a new relay. This is in fact the acknowledgement of the previous relay. When a node transmits a message at the time t, a retransmission is scheduled by that node at the time $t + maxBO_d$.

7.8.3 DV-CAST

DV-Cast applies three types of light-weight suppression techniques, slotted p-persistence, slotted 1-persistence, and weighted p-persistence (Tonguz et al., 2010). In these techniques in order to calculate the probability of forwarding and/or the waiting time before the rebroadcast, instead of threshold values, a light-weight distributed algorithm has been used. The technique of weighted p-persistence uses the relative distance between two vehicles to calculate the rebroadcasting probability. The probability of forwarding P_{ij} is determined by Equation 7.3, where D_{ij} is distance from vehicle i and vehicle j, and R is the communication range.

$$P_{ij} = \frac{D_{ij}}{R} \tag{7.3}$$

In contrast with the gossip-based scheme or the p-persistence scheme, weighted p-persistence assigns a higher probability to nodes that are more distant from the sender. This method does not account for density and, as a result, when there is a high density in the network, the messages are rebroadcasted by the more distant nodes.

7.8.4 Mobicast

In the case of highways, there is a broadcasting protocol called mobicast. This protocol supports the applications of convenience and safety (Chen et al., 2010b). In the network, zero infrastructure is assumed by mobicast. At the time t, a message is distributed by the mobicast to all of the vehicles in a particular zone from a particular vehicle. It is possible to divide the mobicast into two different mechanisms, store carry forward and multiple forwarding. There are two different zones defined by mobicast, the zone of forwarding (ZoF) and the zone of relevance (ZoR). The zone of forwarding indicates which vehicles have to carry the packet forward and the zone of relevance indication vehicles that are message receiver candidates.

7.8.5 DECA

Density-aware reliable broadcasting for vehicular ad hoc networks (DECA) is designed for urban and highway scenarios (Na Nakorn and Rojviboonchi, 2010b). By utilizing periodic beaconing, the local density is gathered by this protocol. There are two lists for the DECA, the broadcast list and the neighbor list. For all of the one-hop neighbors, the identifier is the neighbor list. It also identifies their local density. The waiting time of the broadcast messages and the broadcast messages themselves are maintained by the broadcast list. DECA chooses a vehicle with the highest local density to send a message. If the selected vehicle is the source vehicle, another vehicle is chosen. The waiting time for each node is a random number.

7.8.6 POCA

Position-aware reliable broadcasting protocol (POCA) is a broadcasting protocol to eliminate the broadcast storm problem in VANETs (Na Nakorn and Rojviboonchi, 2010a). Also, it is designed to function in intermittent connected networks. It utilizes adaptive beaconing technique to obtain one-hop neighbor velocity and position information. POCA assumes all the vehicles in the networks have an homogeneous communication range. Relay selection in POCA is based on the distance between vehicles and the selected node. The selected node instantly rebroadcasts the packet. If the selected node does not rebroadcast the packet, other nodes will be chosen as an alternative. In POCA, waiting time is calculated based on the distance between the precursor node and the vehicle.

7.8.7 EDB

Efficient Directional Broadcasting protocol (EDB) is a directional- and distance-based broadcasting protocol for the urban vehicular ad hoc networks and it applies directional antennas (Li et al., 2007). In EDB, the furthest receiver has the responsibility of distributing the message when it arrives in the opposite direction of the highway. Fixed directional antennas are the equipment of each vehicle in EDB and the beam width of these antennas is about 30 degrees. There are two points for these antennas to be mounted, one at the back and one at the front. Since the vehicular ad hoc networks have a highly dynamic mobility, receiver-based decisions are made by the EDB in order to forward packets in the opposite direction of the highway. EDB calculates waiting time from Equation 7.4.

$$\text{Waiting Time} = \left(1 - \frac{D}{TR}\right) * maxWT \tag{7.4}$$

where TR is the communication range, D is distance from the source, and maxWT is the maximum waiting time. The last vehicle sends an acknowledgement to notify the sender.

7.8.8 SRD

Simple and robust dissemination (SRD) is a protocol for broadcasting in highway cases (Schwartz et al., 2011). Vehicle-to-vehicle communication is assumed by the simple and robust dissemination protocol. In SRD, a time slot assignment is the optimized slotted 1-persistence. This technique operates as follows, when the vehicle j is moving in the direction of the message, a message is received by this vehicle from the vehicle i. Then the $PDij$ distance, which is the distance between the two vehicles, is calculated based on Equation 7.5. SRD assumes the communication range of the vehicles is the same.

$$PD_{ij} = \left\lceil \frac{min(D_{ij}, R)}{R} \right\rceil \tag{7.5}$$

where R is the communication range and D_{ij} is the distance between vehicles i and j. The number of the time slot S_{ij} assigned to vehicle j is calculated by Equation 7.6.

$$S_{ij} = NS * (1 - PD_{ij}) \tag{7.6}$$

where NS is the whole number of time slots. The vehicles are divided into two different categories by the SRD. Tail states: the vehicles that do not have any connection with the other vehicles, and those which are located at greater distances in the direction of the message are referred to as the cluster tail. Non-tail states: there is at least one neighbor for the vehicles that are categorized in this state. There are two responsibilities for the vehicles, which are classified in the second category. When a message is received, if the receiving vehicle is in the direction of the message, the vehicle rebroadcasts it and, on the contrary, if the vehicle is not in the direction of the message, the message is dropped.

7.8.9 EAEP

Edge aware epidemic protocol (EAEP) (Nekovee, 2009) is designed for VANETs to solve the broadcast storm problem in highway scenarios. EAEP is an epidemic protocol which assumes end to end connection between vehicles. It decreases the overhead by omitting beacon exchange. EAEP utilizes GPS to determine the location information of each vehicle. Each node piggybacks its geographical location upon receiving a new packet in broadcast message to withdraw hello packets. Each node is assigned a random waiting time. This waiting time is selected exponentially based on the distance from the source. This random waiting time is chosen from the interval $[0, T_{max}]$ with Equation 7.7.

$$T_{max} = \min \begin{cases} \dfrac{T_0}{u} \exp\left(\dfrac{|x_{rec} - x_{sou}|}{L}\right) \\ T_{min} = \dfrac{T_0}{2u} \end{cases} \tag{7.7}$$

In Equation 7.7, U is used to indicate the "Urgency" of the packet and T_0 and L are parameters related to the protocol. While the assigned waiting time expires, the vehicle counts the number of received packet from nodes in the front and the back. Then this vehicle makes a decision to rebroadcast the packet or not based on the difference between count numbers.

7.8.10 Ack-PBSM

Acknowledgement parameterless broadcast in static to highly dynamic mobile (Ack-PBSM) (Ros et al., 2009) is an extension of the protocol into highly mobile and static scenarios (PBSM) (Khan et al., 2008b). In order to broadcast in networks with good connections, the dominating set that is connected is used by Ack-PBSM. It can be applied in both urban and highway scenarios. The broadcast packet's acknowledgment is handled by this protocol. In periodic beacons, these acknowledgments are piggybacked. The function t_{oev} assigns a waiting to each vehicle before there is any retransmission possibility. The value of t_{oev} can be calculated from Equation 7.8.

$$t_{oev} = \frac{1}{|N|} \tag{7.8}$$

$|N|$ is the number of elements in N and indicates whether the node is in the CDs or not. In order to deal with the temporary network fragmentation, the store carry forward approach is applied by this protocol.

7.8.11 UV-CAST

Urban vehicular broadcast (UV-CAST) is broadcast protocol designed for urban scenarios (Viriyasitavat et al., 2011). The assumption of this protocol is the vehicle-to-vehicle communication and there are no supports of any infrastructures involved. There are two ranges of transmission for communications, non-line of sight and line of sight. In this protocol, communication can only occur between two vehicles that are in the corresponding range of communication. The node that has the shortest healing time is assigned with the task of store carry forward. When a message is received for the first time, the angle θ is computed by the node for all of its neighbors. Maximum (θ^{+}) and Minimum (θ^{-}) angles are then computed from Equations 7.9 and 7.10.

$$\theta^{-} = \min\left(\min i(\theta_i), 0\right) \tag{7.9}$$

$$\theta^{+} = \max\left(\max i(\theta_i), 0\right) \tag{7.10}$$

$$\text{If } |\theta^{+}| + |\theta^{-}| < \pi \text{ then A=SCF Task} \tag{7.11}$$

The boundary vehicles are selected by Equation 7.11 for UV-CAST from θ^{+} and θ^{-} angles in order to be assigned to the task of store carry forward. An overhead and a high complexity is given to this protocol by this process. The task of store carry forward is assigned to many vehicles in high and medium densities of traffic. UV-CAST uses a timer-based approach when a new packet is received, the vehicle computes the waiting time based on Equation 7.12. UV-CAST uses two independent equations for highways and intersections.

$$T_i = \begin{cases} \frac{1}{2}\left(1 - \frac{d_{ij}}{R}\right) T_{\max} & \text{If } i \text{ is at intersection} \\ \frac{1}{2}\left(2 - \frac{d_{ij}}{R}\right) T_{\max} & \text{otherwise} \end{cases} \tag{7.12}$$

where $T_{\max}$ is the maximum waiting time which is set to 400 ms, R is the communication range, and d_{ij} is the distance that sits between the vehicle j and vehicle i. If the timer expires and a duplicate packet has been not received by the vehicle i, the rebroadcast is performed, and otherwise the packet is dropped.

7.8.12 Streetcast

Streetcast is an urban broadcast protocol for VANETs to solve the broadcast storm problem, which assumes a homogeneous communication range for vehicles (Yi et al., 2010). The three components of streetcast are adaptive beacon control, multicast request-to-send (MRTS) handshaking, and relay node selection. For the relay node selection, the information of one-hop neighbor and the information of the digital street map are applied. The mechanism of MRTS is used for the protection of the transmissions of the messages. For the information exchange between the neighbors, hello beacons are utilized. Meanwhile, there is a proposed adaptive beacon control heuristic for the dynamic adjustment of the number of the transmitted beacons. For the redundancy reduction, multi-point relay (MPR) is applied as the strategy of broadcast for the reduction of the number of relay nodes. Since the distribution of the vehicles is along the streets, the MPR selection can be simplified by applying the digital street map. A neighbor table is maintained by each road side unit (RSU) and on-board unit (OBU). For the direction of each road, a neighbor is maintained by the RSU and only two lists of neighbors is maintained by an OBU for the directions of forward and backward. This study has assumed that a GPS is provided for each vehicle to gain the information of the position. A "Hello" message is periodically broadcasted by each node in the VANET. This beacon comprises of the ID of the node, time stamp, and the location. At the time that a "Hello" beacon is received by a node, the digital street map is checked by the node and then the information of the neighbors is updated on the neighbor list.

7.9 Comparison of Broadcasting Protocols in VANETs

In the previous section, a variety of broadcasting protocols for dissemination of information in vehicular ad hoc networks has been reviewed. A classification of these protocols is illustrated in Table 7.1.

All of the protocols discussed assume homogeneous communication range and bidirectional link. The reduction of redundant broadcast is done through the distance between sender and relay node. Although there are already different broadcasting protocols for VANETs, most of them can function only in specific network scenarios such as on highways or only in urban areas. There is a great need to have an ultimate protocol with no assumptions about network scenarios, which can function in different road topology, such as highway and urban.

Table 7.2 shows strength and weaknesses of these broadcasting protocols in VANETs.

Most of these broadcasting protocols choose a relay node based on the distance from the source because of greatest additional coverage. This is not always true in heterogeneous vehicular networks as a vehicle which is closer to the sender but has a larger communication range may have more additional coverage than a vehicle which is far from the sender and has small communication range. While most of the protocols take into account distance from sender to select a relay node, in heterogeneous communication range VANETs, the farthest vehicle does not necessarily have maximum coverage area. Therefore, aside from the distance, the algorithm for choosing the next hop will also

TABLE 7.1

Comparison of Broadcasting Protocol in VANETs

Existing Protocol	Network Scenario	V2V	Node Selection Parameter	Link Assumption	Mechanism for Network Fragmentation
IVG	Highway	*	Distance	Bidirectional	Periodic broadcast
DRG	Highway	*	Distance	Bidirectional	Periodic broadcast
DV-CAST	Highway	*	Distance	Bidirectional	Store carry forward
MobiCast	Highway	*	Distance	Bidirectional	Store carry forward
SRD	highway	*	Distance Direction	Bidirectional	Store carry forward
EAEP	highway	*	Random	Bidirectional	Epidemic
UV-CAST	Urban	*	Distance Angle	Bidirectional with two different power level	Store carry forward
EDB	Urban		Distance	Bidirectional	Repeater
Streetcast	Urban		Distance	Bidirectional	RSU
POCA	Both	*	Distance	Bidirectional	Store carry forward
DECA	Both	*	Density information around node	Bidirectional	Store carry forward
Ack-PBSM	Both	*	Connected Dominating Set	Bidirectional	Store carry forward

require the radius of communication, which will be in the beacon messages. For instance, once vehicle A broadcasts a packet, vehicle B and vehicle C receive it. Among these two vehicles, B is the farthest as illustrated in Figure 7.4. The decision to select vehicle B as a relay to alert may not be wise, as vehicle B could have a very short communication range. The communication range of vehicle C, which is between the source and B, span farther than that of vehicle B.

Additionally, these broadcasting protocols adjust the waiting delay and the rebroadcast probability based on the vehicle location and the physical characteristics, such as vehicle density of the network.

7.10 Conclusion

In this chapter, an extensive literature review is discussed. The main parts of this chapter include an overview of VANETs and DSRC standard, VANET characteristics and applications, and their requirements. Other sections present the heterogeneity of the communication range in VANETs. The current research challenges of VANETs broadcasting protocols are focused on issues such as the broadcast storm problem and network fragmentation. The disseminating protocols of VANETs and their approaches, strengths, and weaknesses for handling these problems are discussed.

TABLE 7.2

Strength and Weakness of Broadcasting Protocol in VANETs

Existing Protocol	Strength	Weakness
IVG	1. Mitigates broadcast storm problem 2. Efficient in fragmented networks 3. Distributed algorithm	1. Only functions in highway scenario 2. Assumes homogeneous communication range 3. Requires accurate GPS information 4. Periodically rebroadcasts safety message
DRG	1. Distributed algorithm 2. Mitigates broadcast storm problem	1. Data dissemination may be slow because ZoF 2. Mitigates network fragmentation periodically causing high reception overhead 3. Assumes homogeneous communication range
DV-CAST	1. Distributed framework 2. Mitigates broadcast storm problem and network fragmentation in a single framework 3. Efficient for safety emergency applications	1. Only functions in straight highways 2. Highly dependent on the position and direction information of vehicles gathered from GPS 3. Assumes homogeneous communication range
MobiCast	1. Mitigates broadcast storm problem 2. Efficient in fragmented network	1. Only functions in highway scenario 2. Complicated mechanism to select vehicles in ZoF and ZoR 3. Assumes homogeneous communication range
SRD	1. Simplicity 2. Mitigates broadcast storm problem and fragmented network problem simultaneously	1. Only functions in highway scenario 2. Not reliable for safety messages 3. Assumes homogeneous communication range
EAEP	1. No beacon exchange 2. Mitigates broadcast storm problem	1. Only functions in highway scenario 2. Assumes end to end connection between vehicles 3. Assumes homogeneous communication range
UV-CAST	1. Mitigates broadcast storm problem and network fragmentation 2. Considers two different levels for communication range which is a more realistic assumption	1. High complexity because of gift-wrapping algorithm 2. Only functions in urban scenario 3. Assigns task of store carry forward to different vehicles
EDBww	1. Receiver-based decision	1. Only functions in highway scenario 2. Fixed antenna direction with beam width of about 30 degree
Streetcast	1. By utilizing digital map, streetcast is a fast and accurate broadcast protocol	1. Homogeneous communication range assumed 2. No specific method for fragmented network condition
POCA	1. Eliminates broadcast storm problem and network fragmentation 2. Functions in different network scenarios such as highway and urban	1. Utilizes 2-hop neighbor information 2. Assumes homogeneous communication range 3. Very high reception overhead
DECA	1. Functions in different network scenarios such as highway and urban 2. Mitigates broadcast storm problem and network fragmentation problem	1. Selects relay vehicle based on random waiting time 2. Assumes homogeneous communication range 3. Requires knowledge of 2-hop neighbors
Ack-PBSM	1. Functions in different network scenarios such as highway and urban	1. Data dissemination speed may be slow because of using CDs 2. Not efficient for safety emergency messages 3. Assumes homogeneous communication range

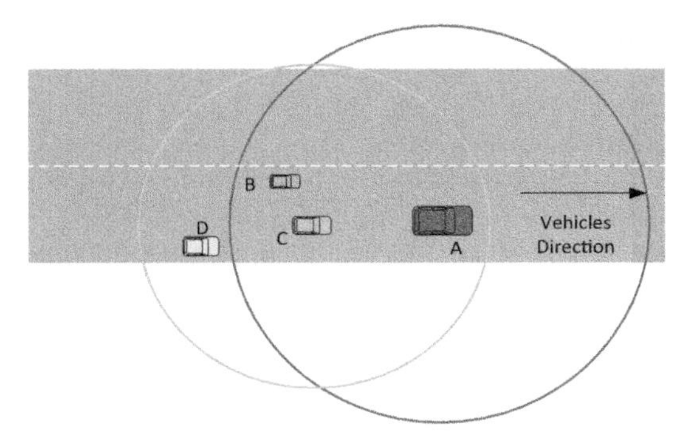

FIGURE 7.4
Broadcasting in heterogeneous communication range.

References

Alshaer, H. and Horlait, E. (2005). An optimized adaptive broadcast scheme for inter-vehicle communication. *Proceedings of the IEEE 61st Vehicular Technology Conference -VTC 2005 - Spring Stockholm: Paving the Path for a Wireless Future*, May 30–June 1, Stockholm, Sweden, 2840–2844.

Bachir, A. and Benslimane, A. (2003). A multicast protocol in ad hoc networks inter-vehicle geocast. *Proceedings of the 57th IEEE Semiannual Vehicular Technology Conference (VTC2003)*, April 22–25, Jeju, Korea, Republic of, 2456–2460.

Bako, B. and Weber, M. (2011). Efficient information dissemination in VANETs. In: Almeida, M. (Ed.). *Advances in Vehicular Networking Technologies*, InTech, London, pp. 45–65.

Barani, H. and Fathy, M. (2008). Overview on ad hoc networks localization. *Proceedings of the International Conference on Wireless Networks, ICWN 2008*, July 14–17, Las Vegas, NV, United States, 407–413.

Basagni, S. and Chlamtac, I. (1998). A distance routing effect algorithm for mobility (DREAM). *Proceedings of the 4th Annual ACM/IEEE International Conference on Mobile Computing and Networking*, Dallas, TX, United States, ACM, 76–84.

Boukerche, A., Oliveira, H. A., Nakamura, E, F., and Loureiro, A. A. F. (2008). Vehicular ad hoc networks: A new challenge for localization-based systems. *Computer Communications* 31(12): 2838–2849.

Chen, R., Jin, W.-L., and Regan, A. (2010a). Broadcasting safety information in vehicular networks: Issues and approaches. *IEEE Network* 24(1): 20–25.

Chen, Y.-S., Lin, Y.-W., and Lee, S.-L. (2010b). A mobicast routing protocol with carry-and-forward in vehicular ad-hoc networks. *5th International ICST Conference on Communications and Networking in China*, ChinaCom 2010, August 25–27, Beijing, China, IEEE Computer Society, 1–5.

Chiang, T.-C., Jiunn-Yin, J. L., and Huang, Y.-M. (2005). Adapted DP based algorithm with distance-based approach in ad hoc networks. *Proceedings of the 16th International Workshop on Database and Expert Systems Applications, DEXA 2005*, August 22–26, Copenhagen, Denmark, 123–127.

Chigira, Y. and Higaki, H. (2011). DTN routing of data messages with epidemic distribution of mobility plans of mobile wireless nodes. *Proceedings of the 10th IASTED International Conference on Parallel and Distributed Computing and Networks, PDCN 2011*, February 15–17, Innsbruck, Austria, 163–169.

Clausen, T., Baccelli, E., and Jacquet, P. (2004). OSPF-style database exchange and reliable synchronization in the optimized link-state routing protocol. *Proceedings of the First Annual IEEE Communications Society Conference on Sensor and Ad Hoc Communications and Networks, IEEE SECON 2004*, October 4–7, Santa Clara, CA, United States, 227–234.

ElBatt, T., Goel, S. K., Holland, G., Krishnan, H. and Parikh, J. (2006). Cooperative collision warning using dedicated short range wireless communications. *Proceedings of the 2006 VANET - Third ACM International Workshop on Vehicular Ad Hoc Networks, September 29*, 2006, September 29, Los Angeles, CA, United states, 1–9.

Haas, Z. J., Halpern, J. Y., and Li, L. (2002). Gossip-based ad hoc routing. *Proceedings of the IEEE Infocom 2002*, June 23–27, New York, NY, United States, 1707–1716.

Hafeez, K. A., Zhao, L., Liao, Z., and Ma, B. N.-W. (2010). Performance analysis of broadcast messages in VANETs safety applications. *Proceedings of the 53rd IEEE Global Communications Conference, GLOBECOM 2010*, December 6–10, Miami, FL, United States, 1–5.

Hayel, Y. and Tembine, H. (2007). Information dissemination using epidemic routing with delayed feedback. *Proceedings of the IEEE International Conference on Mobile Adhoc and Sensor Systems, MASS*, October 8, 2007–October 11, 2007, Pisa, Italy, 1–4.

Hsieh, Y.-L. and Wang, K. (2011). Road layout adaptive Overlay Multicast for urban vehicular ad hoc networks. *Proceedings of the IEEE 73rd Vehicular Technology Conference, VTC2011-Spring*, May 15–18, Budapest, Hungary, 1–5.

IEEE. (2010). *802.11p-2010 - IEEE Standard for Local and Metropolitan Area Networks - Specific requirements Part 11: Wireless LAN Medium Access Control (MAC) and Physical Layer (PHY) Specifications Amendment 6: Wireless Access in Vehicular Environments*, http://standards.ieee.org/findstds/standard/802.11p-2010.html.

Intl, A. (2003). Standard specification for telecommunications and information exchange between roadside and vehicle systems-5 GHz band Dedicated Short Range Communications (DSRC). *Medium Access Control and Physical Layer Specifications*, 213–303.

Jiang, D. and Delgrossi, L. (2008). IEEE 802.11p: Towards an international standard for wireless access in vehicular environments. *Proceedings of the IEEE 67th Vehicular Technology Conference-Spring, VTC*, May 11–14, Marina Bay, Singapore, 2036–2040.

Joshi, H. P., Sichitiu, M., and Kihl, M. (2007). Distributed robust geocast multicast routing for inter-vehicle communication. *In WEIRD Workshop on WiMax, Wireless and Mobility*, May, 9–21.

Karp, B. and Kung, H. T. (2000). GPSR: Greedy perimeter stateless routing for wireless networks. *Proceedings of the 6th Annual International Conference on Mobile Computing and Networking (MOBICOM 2000)*, August 6–11, Boston, MA, USA, 243–254.

Kenney, J. B. (2011). Dedicated short-range communications (DSRC) standards in the United States. *Proceedings of the IEEE* 99(7): 1162–1182.

Khan, A. A., Stojmenovic, I., and Zaguia, N. (2008a). Parameterless broadcasting in static to highly mobile wireless ad hoc, sensor and actuator networks. *Proceedings of the 2008 22nd International Conference on Advanced Information Networking and Applications, AINA 2008*, March 25–28, Ginowan, Okinawa, Japan, 620–627.

Khan, I. A., Peng, E. Y., Javaid, N., and Qian, H. L. (2008b). Angle-aware broadcasting techniques for Wireless Mobile Ad Hoc Networks. *Information Technology Journal* 7(7): 972–982.

Korkmaz, G., Ozguner, F., Ekici, E., and Ozguner, U. (2004). Urban multi-hop broadcast protocol for inter-vehicle communication systems. *Proceedings of the VANET - Proceedings of the First ACM International Workshop on Vehicular Ad Hoc Networks, Held in Conjunction with MOBICOM 2004*, October 1, Philadelphia, PA, United States, 76–85.

Leguay, J., Friedman, T., and Conan, V. (2006). Evaluating mobility pattern space routing for DTNs. *Proceedings of the 25th IEEE International Conference on Computer Communications*, April 23, Barcelona, Spain, 1–10.

Leguay, J., Friedman, T., and Conan, V. (2007). Evaluating MobySpace-based routing strategies in delay-tolerant networks. *Wireless Communications and Mobile Computing* 7(10): 1171–1182.

Li, D., Huang, H., Li, X., Li, M., and Tang, F. (2007). A distance-based directional broadcast protocol for urban vehicular ad hoc network. *Proceedings of the International Conference on Wireless Communications, Networking and Mobile Computing, WiCOM 2007*, September 21–25, Shanghai, China, 1520–1523.

Li, X.-Y., Song, W.-Z., and Wang, Y. (2004). Localized topology control for heterogeneous wireless ad-hoc networks. *Proceedings of the IEEE International Conference on Mobile Ad-Hoc and Sensor Systems*, October 25–27, Fort Lauderdale, FL, United States, 284–293.

Lim, H. and Kim, C. (2001). Flooding in wireless ad hoc networks. *Computer Communications* 24(3–4): 353–363.

Lou, W. and Wu, J. (2003). A reliable broadcast algorithm with selected acknowledgements in mobile ad hoc networks. *Proceedings of the IEEE Global Telecommunications Conference GLOBECOM'03*, December 1–5, San Francisco, CA, United States, 3536–3541.

Lou, W. and Wu, J. (2004). A K-hop zone-based broadcast protocol in mobile ad hoc networks. *Proceedings of the* 2004 *GLOBECOM'04 - IEEE Global Telecommunications Conference*, November 29–December 3, Dallas, TX, United States, 1665–1669.

Moustafa, H. and Zhang, Y. (2009). *Vehicular Networks Techniques, Standards and Applications*. CRC Press, London.

Na Nakorn, N. and Rojviboonchai, K. (2010a). POCA: Position-aware reliable broadcasting in VANET. *Proceedings of the Asia-Pacific Conference of Information Processing APCIP2010*, Nanchang, China, 420–428.

Na Nakorn, N. and Rojviboonchai, K. (2010b). DECA: Density-aware reliable broadcasting in vehicular ad hoc networks. *Proceedings of the 7th Annual International Conference on Electrical Engineering/Electronics, Computer, Telecommunications and Information Technology, ECTI-CON 2010*, May 19–21, Chiang Mai, Thailand, 598–602.

Nekovee, M. (2009). Epidemic algorithms for reliable and efficient information dissemination in vehicular ad hoc networks. *IET Intelligent Transport Systems* 3(2): 104–110.

Neves, F., Cardote, A. and Sargento, S. (2011). Real-world evaluation of IEEE 802.11p for vehicular networks. *Proceedings of the 2011 17th Annual International Conference on Mobile Computing and Networking, MobiCom'11 and Co-Located Workshops – 8th ACM International Workshop on Vehicular Inter-Networking, VANET'11, September 19, 2011*, September 23, Las Vegas, NV, United states, 89–90.

Olariu, S. and Weige, M. C. (2009). *Vehicular Networks from Theory to Practice*. CRC Press, London.

Peng, J. and Cheng, L. (2007). A distributed MAC scheme for emergency message dissemination in vehicular ad hoc networks. *IEEE Transactions on Vehicular Technology* 56(61): 3300–3308.

Perkins, C. E. and Royer, E. M. (1999). Ad-hoc on-demand distance vector routing. *Proceedings of the 2nd IEEE Workshop on Mobile Computing Systems and Applications, WMCSA'99*, February 25–26, New Orleans, LA, United States, 90–100.

Plesse, T., Adjih, C., Minet, P., Laouiti, A., Plakoo, A., Badel, M., et al. (2005). OLSR performance measurement in a military mobile ad hoc network. *Ad Hoc Networks* 3(5): 575–588.

Qayyum, A., Laouiti, A., and Viennot, L. (2002). Multipoint relaying technique for flooding broadcast messages in mobile wireless networks. *Proceedings of the 35th Annual Hawaii International Conference on System Sciences*, Big Island, HI, United States, 3866–3875.

Ros, F. J., Ruiz, P. M., and Stojmenovic, I. (2009). Reliable and efficient broadcasting in vehicular ad Hoc networks. *Proceedings of the VTC Spring 2009 - IEEE 69th Vehicular Technology Conference*, April 26–29, Barcelona, Spain, 1–5.

Santos, R. A., Villasenor, L., Sanchez, J., Gallardo, J. R., and Edwards, A. (2007). Performance of topological multicast routing algorithms on wireless ad-hoc networks. *Proceedings of the Electronics, Robotics and Automotive Mechanics Conference, CERMA 2007*, September 25–September 28, Cuernavaca, Morelos, Mexico, 3–8.

Schmidt, R. K., Kloiber, B., Schuttler, F., and Strang, T. (2011). Degradation of communication range in VANETs caused by interference 2.0 - Real-world experiment. *Proceedings of the 3rd International Workshop on Communication Technologies for Vehicles, Nets4Cars and Nets4Trains 2011*, March 23–24, Oberpfaffenhofen, Germany, 176–188.

Schoch, E. and Kargl, F. (2010). On the efficiency of secure beaconing in VANETs. *Proceedings of the 3rd ACM Conference on Wireless Network Security, WiSec'10*, March 22–March 24, Hoboken, NJ, United States, 111–116.

Schwartz, R. S., Barbosa, R., Meratnia, R. R., Heijenk, G., and Scholten, H. (2011). A directional data dissemination protocol for vehicular environments. *Computer Communications* 34(17): 2057–2071.

Souza, A. B., Celestino, J., Xavier, F. A., Oliveira, F. D., Patel, A., and Latifi, M. (2013). Stable multicast trees based on Ant Colony optimization for vehicular Ad Hoc networks. *Proceedings of the 27th International Conference on Information Networking, ICOIN 2013*, January 27–30, Bangkok, Thailand, 101–106.

Stojmenovic, I. (2004). Comments and corrections to dominating sets and neighbor elimination-based broadcasting algorithms in wireless networks. *IEEE Transactions on Parallel and Distributed Systems* 15(11): 1054–1055.

Tonguz, O. K., Wisitpongphan, N., and Bai, F. (2010). DV-CAST: A distributed vehicular broadcast protocol for vehicular ad hoc networks. *IEEE Wireless Communications* 17(2): 47–57.

Tseng, Y. C., Ni, S. Y., and Shih, E. Y. (2001). Adaptive approaches to relieving broadcast storms in a wireless multihop mobile ad hoc network. *Proceedings of the 21st IEEE International Conference on Distributed Computing Systems, April 16, 2001–April 19, 2001*, Mesa, AZ, United States, 481–488.

Viriyasitavat, W., Tonguz, O. K., and Bai, F. (2011). UV-CAST: An urban vehicular broadcast protocol. *IEEE Communications Magazine* 49(11): 116–124.

Wegener, A., Hellbruck, H., Fischer, S., Schmidt, C., and Fekete, S. (2007). AutoCast: An adaptive data dissemination protocol for traffic information systems. *Proceedings of the IEEE 66th Vehicular Technology Conference, VTC 2007-Fall*, September 30–3, Baltimore, MD, United States, 1947–1951.

Williams, B. and Camp, T. (2002). Comparison of broadcasting techniques for mobile ad hoc networks. *Proceedings of the 3rd ACM International Symposium on Mobile Ad Hoc Networking and Computing, MOBIHOC 2002*, June 9–11, Lausanne, Switzerland, 194–205.

Wu, X., Yang, Y., Liu, J., Wu, Y., and Yi, F. (2010b). Position-aware counter-based broadcast for mobile Ad Hoc networks. *Proceedings of the 5th International Conference on Frontier of Computer Science and Technology, FCST 2010*, August 18–22, Changchun, China, 366–369.

Wu, X.-W., Yan, W., Song, S.-M., and Wang, H.-B. (2010a). A transmission range adaptive broadcast algorithm for vehicular ad hoc networks. *Proceedings of the 2nd International Conference on Networks Security, Wireless Communications and Trusted Computing, NSWCTC 2010*, April 24–25, Wuhan, Hubei, China, 28–32.

Xie, J., Talpade, R. R., McAuley, A., and Liu, M. (2002). AMRoute: Ad Hoc multicast routing protocol. *Mobile Networks and Applications* 7(6): 429–439.

Yi, C. W., Chuang, Y.-T., Yeh, H.-H., Tseng, Y.-C. and Liu, P.-C. (2010). Streetcast: An urban broadcast protocol for vehicular ad-hoc. *Proceedings of the IEEE 71st Vehicular Technology Conference*, Taipei, Taiwan, 1–5.

Yin, J., Elbatt, T., Yeung, G., Ryu, B., Habermas, S., Krishnan, H., et al. (2004). Performance evaluation of safety applications over DSRC vehicular ad hoc networks. *Proceedings of the First ACM International Workshop on Vehicular Ad Hoc Networks, Held in Conjunction with MOBICOM 2004*, October 1, Philadelphia, PA, United States, 1–9.

Zhang, L., Wang, X., and Dou, W. (2004). A distributed topology control algorithm for heterogeneous ad hoc networks. *Proceedings of the 5th International Conference, PDCAT 2004*, December 8–10, Singapore, 681–684.

Zhao, J. and G. Cao (2008). VADD: Vehicle-assisted data delivery in vehicular ad hoc networks. *IEEE Transactions on Vehicular Technology* 57(3): 1910–1922.

Zhao, Y., Xu, L., and Shi, M. (2003). On-Demand Multicast Routing Protocol with Multipoint Relay (ODMRP-MPR) in mobile ad-hoc network. *Proceedings of the International Conference on Communication Technology, ICCT 2003*, April 9–11, Beijing, China, 1295–1300.

8

Energy Management in OppNets

Itu Snigdh and K. Sridhar Patnaik

CONTENTS

8.1 Introduction 159
8.1.1 OMN Characteristics 160
8.1.2 Categories of OMNs 160
8.2 Comparison of OMNs with MANETs and WSNs 161
8.2.1 Mobility of Nodes 161
8.2.2 Energy Conservation 161
8.2.3 Density of Nodes 161
8.3 Energy Constraints 162
8.3.1 Factors Affecting Energy Depletion 162
8.4 Energy Conservation Methodologies 163
8.4.1 Existing Methodologies 163
8.4.1.1 Encounter Frequency 163
8.4.1.2 Strength of Social Relationships between the Users 164
8.4.1.3 Regularity of Encounter 164
8.4.1.4 Interest in the Type of Content 165
8.4.1.5 Intelligent Infrastructure-Based Energy Conservation 165
8.5 Customized Routing Protocols for Optimizing Energy Consumption 165
8.5.1 Forwarding-Based Approach 166
8.5.2 Flooding-Based Approach 167
8.6 Performance Analysis of Existing Routing Protocols in Context to Energy Consumption 167
8.7 Existing Research and Milestones 168
8.8 Ongoing Research Challenges 168
References 169

8.1 Introduction

With the proliferation of sensor networks and mobile ad hoc networks (MANETs), we have a plethora of intelligence-equipped smart devices that are heterogeneous in nature and need to communicate frequently over the network. This will increase the network capability requirements in the near future. With the spreading of innumerable heterogeneous devices, the quality of service will degrade due to the insufficient bandwidth coverage provisions of wired and wireless networks. Also, we cannot just rely on the wireless infrastructure unconditionally to service our demands. Moreover, with such devices, the basic requirement is supporting mobility and ubiquitous computing. Opportunistic mobile (self-organizing) networking (Pelusi et al., 2006) may be considered a remarkable concept and the first step toward realizing these requirements.

8.1.1 OMN Characteristics

Opportunistic mobile networks (OMNs) generally invalidate the assumptions of general network architecture. Though there is no common definition to suffice its characteristics, it is essentially a typical wireless network that is based on the direct communication of nodes with each other. Sometimes it may require additional support from infrastructure, although primarily it is based purely on ad hoc connections. An OMN essentially consists of a source node that generates a message and forwards the message to its intermediate nodes. The forwarding decision may be random or intelligent, so as to choose a candidate intermediate node that ensures bringing the message closer to the destination node.

The protocol stack of OMNs considers an additional layer in the conventional TCP/IP architecture, commonly referred to as the bundle (Huang et al., 2008) layer. These networks use nodes as an entity to implement aggregation of data (Socolofsky and Kale, 1991). Since these networks are characterized by frequent disconnections, the nodes also need a buffer storage to store the message until the connection can be reestablished. The general OMN characteristics can be summarized as

- Mobility
- Resource constraints
- High heterogeneity
- Resource failures
- Dynamic contacts

8.1.2 Categories of OMNs

An OMN can be visualized as emerging out of the following special cases of network connections:

1. *Human-centered mobile phones*: People carry smartphones that can be used to set up this type of network. Hence, such networks are called pocket switched networks (PSNs), that is, networks formed by mobile devices carried by people. A subset of opportunistic networks (OppNets) is therefore called PSNs (Lilien et al., 2006).
2. *VANETs*: Vehicles are equipped with wireless connection–enabled devices and are thus ideal to connect to the other similar devices whenever possible. Vehicular ad hoc networks (VANETs) have wider application in OppNets as they can be coupled by infrastructure nodes at the roadside, mostly known as roadside units (RSUs), for 3G/2G cellular connectivity or serve as Wi-Fi access points to facilitate communication. Hence, an OppNet finds vivid applications in the VANET domain.
3. *Hybrid networks*: Since networks today comprise heterogeneous devices like smartphones, Wi-Fi interface–enabled devices in vehicles, or Wi-Fi adapters, any combination of infrastructure-based or infrastructure-less wireless devices also form a type of OppNet connecting whenever possible for the necessary communication.

8.2 Comparison of OMNs with MANETs and WSNs

OMNs are considered a subset of delay-tolerant networks (DTNs) and an interesting derivative of MANETs. Some literature also assumes that OppNets have emerged from the concept of opportunistic sensor networks (Pelusi et al., 2006). Nodes in such networks are usually the handheld devices carried by people, which discover each other automatically, without user intervention, for the necessary communication.

The main concepts that differentiate an OppNet from a MANET are discussed below.

8.2.1 Mobility of Nodes

In the case of a MANET, the intermediary devices, if mobile, would incur battery wastage, an essential overhead, to enable communications between nodes. So, both MANETs and wireless sensor networks (WSNs) employ an almost fixed routing strategy that may or may not be updated when the nodes move. Thus, stable connectivity is compromised with the unpredicted mobility of nodes in these networks. Mobility is therefore treated as a constraint, and it needs to be controlled to a certain extent.

An OMN eliminates the prime concern of connectivity establishment of MANETs and WSNs, which is mobility. In fact, it uses mobility to establish a path for data transmission. It leverages the mobility of devices and takes it as an opportunity to carry the message while moving and forwarding data to the destination via whichever path and whenever it deems possible.

8.2.2 Energy Conservation

The main concept that differentiates an OppNet from a MANET is that in the case of a MANET, the intermediary devices incur battery wastage to enable communications between nodes. On the contrary, in the case of an OppNet, human mobility is exploited to move information rather than depending on connections to dedicated or ad hoc gateway nodes.

Energy efficiency in the implementation of routing protocols for MANETs is much needed to prolong the operational time of the network (Eu et al., 2010; Rodrigues et al., 2011). Similarly, energy also constrains the operating of an OMN but affects it only in the sense that faster energy depletion requires frequent charging intervals. Since most of the mobile devices in OMN environments are usually equipped with an energy-limited battery, the energy-efficient protocols are an obligation in these networks too.

8.2.3 Density of Nodes

Internode or intercluster interference tends to increase with the increase in the number of nodes deployed or visiting an area. On the contrary, with the increase in the number of available mobile devices in an environment, contact opportunities greatly increase, which is very useful in establishing an OppNet. Thus, the adverse conditions are turned into advantages with OMNs.

An OMN is a challenged network, as it is characterized by long and variable delays, high latency, frequent disconnections, long queuing times, limited longevity, and limited resources (Dhurandher et al., 2013). In DTNs, too, applications report experiencing long or variable delays, high error rates, low reliability, security issues, and greatly heterogeneous scenarios.

8.3 Energy Constraints

Energy is expedited in any ad hoc network when establishing and maintaining the proactive or reactive network routing paths, transmitting and receiving messages, scheduling active and sleep cycles to reduce collisions, and in idle listening and overhearing. In addition, there is considerable loss in energy due to interference among the nodes. In the case of OMNs, the energy expenditure accounts for frequent SCAN and OFF cycles, in addition to the aforementioned factors. Energy is also wasted due to unnecessary or out-of-context forwarding and processing of information, which is a requisite of OMNs.

8.3.1 Factors Affecting Energy Depletion

- *Neighbor discovery*

  Since most of the devices are handheld or incessantly mobile, continuous or frequent scanning is required for establishing opportunistic encounters. Thus, neighbor discovery requires sending and receiving requests for association. For example, in Bluetooth-enabled OMNs (which are present in most smartphones sold today), the neighbor discovery procedure uses node discovery requests (NDREQs) and node discovery replies (NDREPs). This drains the battery on existing personal mobile wireless devices. Furthermore, there is a great deal of uncertainty about when encounters between devices carried by humans will take place (Orlinski and Filer, 2015).

- *Data forwarding and connectivity (routing)*

  Connectivity in OppNets can be either human centric or data centric and is facilitated by either Bluetooth or Wi-Fi, the most popular. Communication happens only when the nodes are in reciprocal range of each other (Conti et al., 2010). Mobility is essential to identify the features that affect the data delivery schemes used in OMNs; for example, the intercontact time between the node or a certain pair of nodes can be studied or used to generate a pattern for exploiting in robust communication. Mobility information or knowledge is essential to estimate the delay that will be incurred, and hence the extent of buffer or cache that will be required by the devices. These in turn require evaluating the performance of the proposed protocols in the realistic domain.

- *Speed of mobile nodes*

  Node speed affects the probability of detecting encounters between mobile devices in OMNs. Due to the limitations of short communication ranges to save energy and avoid interference, rapid and unpredictable changes occur in the connectivity due to unprecedented mobility patterns. Energy is also consumed due to burden of irregular cellular congestion. Highly mobile devices continuously scan for fluctuations in the wireless signals to initiate the neighbor discovery interval. This continuously probing and listening for wireless signals is a waste of energy in case no new devices are added.

- *Data processing*

  Irrespective of the other features, every node in an OMN is expected to accept a message and forward it to the most suitable candidate. Therefore, energy is expedited in initial computations related to whether to accept the process and forward or to reject the packet received.

8.4 Energy Conservation Methodologies

Energy conservation can be achieved by the following ways:

- Data reduction
- Protocol overhead reduction
- Energy-efficient routing
- Duty cycling
- Topology control

8.4.1 Existing Methodologies

OppNets can be formed using user mobility when there is a potentially exploitable regularity in patterns. The basic ideology behind reducing energy is to limit the number of message copies that need to be generated in case of the usual flooding-based data delivery approaches. Thus, the existing methodologies focus on deciding when and how much to replicate messages. The decisions may solely be dependent on the speed at which the nodes move rather than the probability of successful packet delivery. The following section outlines some of the methodologies based on the following factors that are used to leverage the characteristics of nodes and their connections. They can be listed as

- Frequency at which the nodes are encountered
- Strength of social relationships that exist between the users
- Places regularly visited
- Type of content they are interested in
- Intelligent infrastructure

8.4.1.1 Encounter Frequency

This factor is used for optimizing the degree and extent to which data forwarding needs to be implemented. It can broadly be categorized as

- Human centric (Chaintreau et al., 2007; Perino and Varvello, 2011)

  For example, a human-centric approach would be data forwarding that happens between nodes A and B if they explicitly know each other's identity and the condition is favorable. For example, if A wants to communicate with B, the end points are generated, and the route is established to carry on the communication.
- Group communication (Phuong et al., 2016)

  This is a data-centric approach where the knowledge of the nodes' identity is not important, but rather the emission of the context reaching the destination is more important. In such cases, the content is generated by any user and is absorbed by other users that find interest in the data. It resembles the publish–subscribe architecture where information is pushed to users that belong to the same interest domain.

8.4.1.2 Strength of Social Relationships between the Users

Many routing algorithms, rather than blindly forwarding packets to nodes encountered, incorporate intelligent decision-making strategies to choose the best candidate to ensure message delivery. These are usually coupled with a prior knowledge of contacts, or the history of last successful data delivery, or simply the mobility of the nodes. Thus, many routing algorithms use Markov-based approaches or Bayesian belief networks to estimate the success probability of each node, and then initiate communicating with them or forwarding messages with the hope that the message would successfully be delivered. These approaches work similar to social networks where nodes compute and assign a confidence value to each node and decide accordingly. This method also reduces the total number of message replications as well as the energy expedited in the random scan operations. The data-forwarding schemes under this category may be classified as (Pelusi et al., 2006; Waltari and Kangasharju, 2016; Bulut and Szymanski, 2010; Conti and Kumar, 2010).

1. *Social context oblivious*: Also called randomized. In this case, a certain number of copies are generated and replicated with the hope that one of the replicas would reach the destination.
2. *Social content aware*: Also called utility based. These try to estimate the probability of nodes entering the destination. This probability is computed on the basis of the frequency of contact, intercontact time, and the probability of successful delivery, which is the basis of choosing the most eligible candidate to forward the data. This usually is an inference-based scheme on the past encounters and experiences of the nodes. These schemes are more accurate in reaching the destination but have a larger requirement of storage of contextual information and are more delay prone.

8.4.1.3 Regularity of Encounter

For nodes that are frequently moving along the same region of interest as well as there is a probable fixed interval or duration associated to the place of visit, the contact time can be synchronized. The literature mentions adoption of symmetric and asymmetric neighbor discover intervals for the above. In addition to this, the symmetric discovery interval class can further be categorized as synchronous and asynchronous.

This can be understood by a simple example. If user A frequently boards a bus to reach office from the same location and meets approximately the same group of people destined to the same or different places, then their devices may use this regularity of meeting to transfer information to different locations where the device is being carried. In such cases, the message does not need to be redundantly broadcasted for delivery, and it would likely improve the probability of message reception with conservation of battery life. This concept has been widely researched and adopted for guaranteed message delivery.

The adoption of synchronized symmetric neighbor discovery intervals would enable the personal mobile wireless devices to save their probable neighbors as well as discover other devices autonomously by initiating the symmetric neighbor discovery simultaneously on every device. For example, IMPALA (Liu and Martonosi, 2003) and CENWITS (Huang et al., 2005) use GPS-aided time calibration and a regular operation schedule to synchronize symmetric neighbor discovery intervals. So, they simultaneously turn on their receivers to initiate communication without the burden of scanning the network. The disadvantage, however, is that if two devices are in each other's range, but their symmetric discovery interval does not overlap, their encounters will be missed completely.

The other method used is the asynchronous–symmetric approach. These two approaches allow devices to choose the length of time between symmetric neighbor discovery intervals. For example, STAR (Wang et al., 2009) and DWARF (Izumikawa et al., 2010) allow different participants to have different encounter patterns. Their discovery intervals are called interprobe time.

Likewise, autonomous neighbor discovery is altogether a different strategy where autonomous neighbor discovery devices keep their radio off for most of the time but can still guarantee that new encounters between devices will be discovered.

8.4.1.4 Interest in the Type of Content

Energy can also be saved by intelligent routing (Pelusi et al., 2006) based on the amount of knowledge about the context of users they connect to. Context may contain attributes like buffer size, density, remaining energy, and network bandwidth. Information traditionally pertaining to lower levels of the stack (such as link availability, contact opportunities, and communication costs) also becomes fundamental for the middleware operations, and lower layers can also benefit from context information managed by the middleware. Social-aware middleware is a content-sharing service that exploits user social relationships to optimize the distribution of data so as to maximize the probability of delivering content to interested users (Conti et al., 2010). This strategy looks for nodes that show an increasing match with known context attributes of the destination. A high match means high similarity between the node's and destination's contexts, and therefore high probability for the node to bring the message in the destination's community. They also consider that people are not likely to move around randomly. Rather, they move in a predictable fashion based on repeating behavioral patterns at different timescales (day, week, and month). If a node has visited a place several times before, it is likely to visit this location again in the future (Boldrini et al., 2007, 2010). Middleware like CAMEO (Arnaboldi et al., 2014) extend the paradigm of online social networks with additional interaction opportunities generated by user mobility and opportunistic wireless communications among users who share interests, habits, and needs. The data-forwarding schemes under this category are context oblivious, partially context aware, and fully context aware.

8.4.1.5 Intelligent Infrastructure-Based Energy Conservation

Similar to exploiting node mobility and contact patterns, user mobility can also be used for packet transport, packet routing, and radio operation decisions, with the help of wireless hotspots, rather than depending solely on the cellular networks. Ideally, nodes extend their radio signal coverage to connect nodes and form links. The MANETs formed by these devices could be used for sending and routing latency-insensitive packets. This would provide mobile devices rich communication capabilities for longer periods of time, using low-power radios and thereby saving energy (Su et al., 2004).

8.5 Customized Routing Protocols for Optimizing Energy Consumption

As the nodes do all the computations for next hop selection in infrastructure-less OMNs, a lot of battery power gets consumed. which reduces the network lifetime (Lilien et al., 2006). Thus, a proper energy-efficient routing protocol should be selected for message passing in these types of networks. The ad hoc nature of the links and participation of the nodes in communication impact the battery reserve of the nodes themselves. Existing algorithms force energy conservation through optimal routing schemes. Thereby routing protocols may be categorized on the basis of how they plan to achieve energy optimization. A number of routing protocols have been proposed to date to deal with the highly dynamic environment of opportunistic networking. The two main categories in which we can divide routing protocols are

- Forwarding-based approach
- Flooding-based approach

8.5.1 Forwarding-Based Approach

This scheme uses a single-copy approach in which only one copy of message is created and forwarded to destination node using intermediate nodes. In this approach, based on certain knowledge, nodes select the best route to send the packet to the destination node. The forwarding-based approach can be classified into three main categories: direct transmission, location based, and knowledge based.

Direct transmission: Direct transmission (Spyropoulos et al., 2005) is one of the simplest routing protocols. In this, a single copy of message is created and the source node forwards the packet only if it encounters the destination node. Hence, the delay latency is very high in this approach, but the buffer requirement gradually becomes much less.

Location-based approach: In this approach, the node chooses the neighbor that is closest to it and forwards the packet to the nearest neighborhood node. The motion vector of mobile nodes (MoVe) (LeBrun et al., 2005) uses the relative velocity of nodes to find the closest distance between the nodes to predict their future location. Based on the predicted future location, the message is forwarded to the node that is moving closest to the destination node. In this approach, the bandwidth requirement is less than that for epidemic routing, but the delivery latency is more.

Knowledge based: In the knowledge-based strategy, the source and intermediate nodes decide which node must forward the message and whether it should transmit the message immediately or hold the message until it meets a better node; this decision is based on certain knowledge about the network. Jain et al. (2004) proposed the knowledge-based routing strategy, where the basic idea is to apply the traditional shortest-path routing techniques to OMN by using the network knowledge.

Other routing strategies include the context-aware routing (CAR) strategy (Musolesi et al., 2005), where the delivery probability of a node is calculated and the message is forwarded to the node with a higher delivery probability. The delivery probabilities are exchanged periodically among nodes in order to compute the best carrier for each destination node. MaxProp (Boldrini et al., 2010; Burgess et al. 2006) is a routing protocol based on prioritizing the schedule of packets transmitted to other nodes, as well as scheduling the packets to be dropped when the buffer is full. Here, the scheduling of packets is based on the path likelihoods to peers according to historical data. The shortest-expected-path routing (SEPR) strategy (Tan et al., 2003) estimates the link-forwarding probability based on history data. It computes the effective path length (EPL) and updates the local EPL value if it receives a smaller EPL value whenever two nodes meet.

8.5.2 Flooding-Based Approach

The flooding-based routing protocol is a multi-copy-based strategy. In this approach, the node holding the packet broadcasts its packet to all of its neighborhoods. This process continues until the packet is delivered to the destination node.

Epidemic routing (Vahdat and Becker, 2000) is used for forwarding data in an OMN where the connections are highly intermittent. Epidemic routing is similar to the flooding approach, as in this approach the message is broadcasted by the node to all its neighborhoods, but with some limitations. In the dissemination process, each message generated is assigned a hop count and the buffer space of each node is bounded in order to achieve the limitations. Epidemic routing uses the epidemic algorithm (Rodrigues et al., 2011) proposed for synchronizing duplicate databases.

Another similar technique, called spray and wait (Spyropoulos et al., 2004), is the variance of flooding approach, and it limits the level of flooding. Basically, it contains two phases: the spray phase and the wait phase. A number (N) of copies are generated by the node in the spray phase and are spread randomly to its neighbors present at one hop count distance. In the wait phase, if the destination is not found nodes perform direct transmission. The major limitation of this approach is that the intermediate nodes are selected randomly, and they fail to implement the social behavior of nodes. Hence, to enhance the performance of the routing strategy the future prediction of contacts plays a vital role.

The probabilistic routing protocol (PROPHET) uses the history of encounters and transitivity (Lindgren et al., 2003) for prediction. In this, nodes estimate the probability of each available route or existence of linked nodes to the destination and use this information to decide which node must transfer its message and whether it should store the packet and wait for a better chance to forward the message. The history-based opportunistic protocol (HiBOp) not only uses a predefined set of context information, but also uses any information users want to provide to describe their current context (Boldrini et al., 2007). It entails that each node should automatically learn its current context and remember it. The context data would then feed algorithms to decide next hops toward eventual destinations.

8.6 Performance Analysis of Existing Routing Protocols in Context to Energy Consumption

All protocols have similar behavior for sparse networks: the delivery ratio decreases and the delay increases following the connectivity. The message delivery relies on a predicted sequence of communication opportunities, which is defined as a contact. Most protocols aim to maximize the data delivery and minimize the delay.

The major drawback of the forwarding-based approach is the latency delay and the low delivery ratio, but the buffer required is decreased as well as the traffic, and hence the bandwidth required is less than that of the flooding-based approach. Likewise, in the flooding-based approach, if the network traffic increases, the bandwidth requirement increases, and so does the requirement of the buffer space when compared with the forwarding-based approach. In contrast, the delivery latency decreases in the flooding-based approach as multiple copies of message are created and broadcasted to neighbors.

8.7 Existing Research and Milestones

Current research applications (Xi and Chuah, 2009) that are successfully employing OppNets can be summarized as

- The ZebraNet system that has been implemented at the Mpala Research Centre and is currently under test.
- The Shared Wireless Infostation Model (SWIM), which is based on monitoring whales.
- The DakNet project in India. The OppNet is used to provide intermittent Internet connectivity to rural and developing areas and emerges as the only affordable way to help bridge the digital divide.
- The Saami Network Connectivity (SNC) project, which aims to provide network connectivity to the nomadic Saami population of the reindeer herders who live across the Sápmi region (also known as Lapland) in the northwest part of Sweden, Norway, and Finland. Providing network connectivity to the Saami population enables them to continue to live according to their traditions and, at the same time, have economic sustenance through distance work and net-based businesses.

8.8 Ongoing Research Challenges

An OMN therefore considers almost all of the wireless network's peculiarities, like mobility of the user, ad hoc node connections and links, frequent disconnections, network partitions, and link instability, which have to date been dealt with in the wireless network scenario as "exceptions." These peculiarities require specialized routing protocols, and energy management techniques. Though OppNets seem to be a viable solution to the incessant mobility-based communication, deploying an OppNet in the real environment poses several challenges. Some of the areas that are still under rigorous development are

- Security and privacy
- Data caching
- Content diffusion
- Context- and location-aware routing

Several models representing infrastructure requirements as well as optimal routing have been proposed, capturing different characteristics of node behavior in different kinds of OMNs. Yet, much remains to be done, to incorporate intelligent behavior in nodes as well as in leveraging the network and human-centric statistics to predict and optimize the network and routing requirements.

References

Arnaboldi, V., Conti, M., and Delmastro, F. (2014). CAMEO: A novel context-aware middleware for opportunistic mobile social networks. *Pervasive and Mobile Computing*, 11, 148–167.

Boldrini, C., Conti, M., Delmastro, F., and Passarella, A. (2010). Context-and social-aware middleware for opportunistic networks. *Journal of Network and Computer Applications*, 33(5), 525–541.

Boldrini, C., Conti, M., Jacopini, J., and Passarella, A. (2007, June). HiBOp: A history based routing protocol for opportunistic networks. In *IEEE International Symposium on a World of Wireless, Mobile and Multimedia Networks, 2007 (WoWMoM 2007)* (pp. 1–12).

Bulut, E. and Szymanski, B. K. (2010, December). Friendship based routing in delay tolerant mobile social networks. In *Global Telecommunications Conference (GLOBECOM 2010), 2010 IEEE* (pp. 1–5).

Burgess, J., Gallagher, B., Jensen, D. D., and Levine, B. N. (2006, April). MaxProp: Routing for vehicle-based disruption-tolerant networks. In *Infocom*. 2006. 25th IEEE International Conference on Computer Communications. Proceedings (pp. 1–11). IEEE.

Chaintreau, A., Hui, P., Crowcroft, J., Diot, C., Gass, R., and Scott, J. (2007). Impact of human mobility on opportunistic forwarding algorithms. *IEEE Transactions on Mobile Computing*, 6(6), 606–620.

Conti, M., Giordano, S., May, M., and Passarella, A. (2010). From opportunistic networks to opportunistic computing. *IEEE Communications Magazine*, 48(9), 126–139.

Conti, M. and Kumar, M. (2010). Opportunities in opportunistic computing. *Computer*, 43(1), 42–50.

Dhurandher, S. K., Sharma, D. K., Woungang, I., and Bhati, S. (2013). Routing protocols in infrastructure-less opportunistic networks. In *Routing in Opportunistic Networks* (pp. 353–382). Springer, New York.

Eu, Z. A., Tan, H. P., and Seah, W. K. (2010). Opportunistic routing in wireless sensor networks powered by ambient energy harvesting. *Computer Networks*, 54(17), 2943–2966.

Huang, C. M., Lan, K. C., and Tsai, C. Z. (2008, March). A survey of opportunistic networks. In *22nd International Conference on Advanced Information Networking and Applications—Workshops, 2008 (AINAW 2008)* (pp. 1672–1677).

Huang, J. H., Amjad, S., and Mishra, S. (2005, November). Cenwits: A sensor-based loosely coupled search and rescue system using witnesses. In *Proceedings of the 3rd International Conference on Embedded Networked Sensor Systems* (pp. 180–191).

Izumikawa, H., Pitkänen, M., Timm-Giel, A., and Bormann, C. (2010, February). Energy-efficient adaptive interface activation for delay/disruption tolerant networks. In *12th International Conference on Advanced Communication Technology (ICACT), 2010* (Vol. 1, pp. 645–650).

Jain, S., Fall, K., and Patra, R. (2004). Routing in a delay tolerant network. *ACM SIGCOMM Computer Communication Review*, 34(4), 145–158.

LeBrun, J., Chuah, C. N., Ghosal, D., and Zhang, M. (2005, May). Knowledge-based opportunistic forwarding in vehicular wireless ad hoc networks. In *Vehicular Technology Conference, VTC 2005—Spring, IEEE 61st* (Vol. 4, pp. 2289–2293).

Lilien, L., Kamal, Z. H., Bhuse, V., and Gupta, A. (2006). Opportunistic networks: The concept and research challenges in privacy and security. In *Proceedings of the WSPWN* (pp. 134–147).

Lindgren, A., Doria, A., and Schelén, O. (2003). Probabilistic routing in intermittently connected networks. *ACM SIGMOBILE Mobile Computing and Communications Review*, 7(3), 19–20.

Liu, T. and Martonosi, M. (2003, June). Impala: A middleware system for managing autonomic, parallel sensor systems. *ACM Sigplan Notices*, 38(10), 107–118.

Musolesi, M., Hailes, S., and Mascolo, C. (2005, June). Adaptive routing for intermittently connected mobile ad hoc networks. In *Sixth IEEE International Symposium on a World of Wireless Mobile and Multimedia Networks, 2005 (WoWMoM 2005)* (pp. 183–189).

Orlinski, M. and Filer, N. (2015). Neighbour discovery in opportunistic networks. *Ad Hoc Networks*, 25, 383–392.

Pelusi, L., Passarella, A., and Conti, M. (2006). Opportunistic networking: Data forwarding in disconnected mobile ad hoc networks. *IEEE Communications Magazine*, 44(11), 134–141.

Perino, D. and Varvello, M. (2011, August). A reality check for content centric networking. In *Proceedings of the ACM SIGCOMM Workshop on Information-Centric Networking* (pp. 44–49).

Phuong, D. T., Le, T. A., Nguyen, T. A. T., and Vo, P. L. (2016, November). LCD-based on probability in content centric networking. In *International Conference on Context-Aware Systems and Applications* (pp. 82–90). Springer, Cham.

Rodrigues, J. J. P. C., Soares, V. N. G. J., Farahmand, F., and Denko, M. K. (2011). Stationary relay nodes deployment on vehicular opportunistic networks. *Mobile Opportunistic Networks: Architectures, Protocols and Applications*, 56(6), 227–243.

Socolofsky, T. J. and Kale, C. J. (1991). TCP/IP tutorial (no. RFC 1180).

Spyropoulos, T., Psounis, K., and Raghavendra, C. S. (2004, October). Single-copy routing in intermittently connected mobile networks. In *First Annual IEEE Communications Society Conference on Sensor and Ad Hoc Communications and Networks, 2004 (IEEE SECON 2004)* (pp. 235–244).

Spyropoulos, T., Psounis, K., and Raghavendra, C. S. (2005, August). Spray and wait: An efficient routing scheme for intermittently connected mobile networks. In *Proceedings of the 2005 ACM SIGCOMM Workshop on Delay-Tolerant Networking* (pp. 252–259).

Su, J., Chin, A., Popivanova, A., Goel, A., and De Lara, E. (2004, December). User mobility for opportunistic ad-hoc networking. In *Sixth IEEE Workshop on Mobile Computing Systems and Applications, 2004 (WMCSA 2004)* (pp. 41–50).

Tan, K., Zhang, Q., and Zhu, W. (2003, December). Shortest path routing in partially connected ad hoc networks. In *Global Telecommunications Conference, 2003 (GLOBECOM'03)* (Vol. 2, pp. 1038–1042).

Vahdat, A. and Becker, D. (2000). Epidemic routing for partially connected ad hoc networks. Tech Report CS-2000-06, Duke University, Durham, NC.

Waltari, O. and Kangasharju, J. (2016, January). Content-centric networking in the Internet of things. In *Consumer Communications & Networking Conference (CCNC), 2016 13th IEEE Annual* (pp. 73–78).

Wang, W., Motani, M., and Srinivasan, V. (2009). Opportunistic energy-efficient contact probing in delay-tolerant applications. *IEEE/ACM Transactions on Networking*, 17(5), 1592–1605.

Xi, Y. and Chuah, M. C. (2009). An encounter-based multicast scheme for disruption tolerant networks. *Computer Communications*, 32(16), 1742–1756.

9

Network Coding Schemes

Amit Singh

CONTENTS

9.1 Introduction 172
9.1.1 Traditional Routing vs. Network Coding 172
9.1.2 Classification of Network Coding 174
9.2 Inter-Session Network Coding 175
9.2.1 COPE 175
9.2.1.1 Overview 175
9.2.1.2 Packet Coding 177
9.2.1.3 Packet Decoding 178
9.2.1.4 Packet Structure 178
9.2.1.5 Performance 179
9.2.2 BEND 181
9.2.2.1 Overview 181
9.2.2.2 Packet Mixing and Queuing Strategy 182
9.2.2.3 Packet Decoding 183
9.2.2.4 Packet Header 183
9.2.2.5 Performance 184
9.2.3 AONC 184
9.2.3.1 Overview 184
9.2.3.2 Protocol Mechanism 185
9.2.3.3 Packet Header 186
9.2.3.4 Performance 186
9.3 Intra-Session Network Coding 187
9.3.1 CodeOR 187
9.3.1.1 Overview 187
9.3.1.2 Protocol Mechanism 187
9.3.1.3 CodeOR Motivation 187
9.3.1.4 Sending Window and Acknowledgment 188
9.3.1.5 Performance 188
9.3.2 SlideOR 189
9.3.2.1 Overview 189
9.3.2.2 Encoding and Decoding Mechanism 189
9.3.2.3 Sliding Window Process 190
9.3.2.4 Performance 190
9.4 Research Aspects 190
References 190
References for Advanced Reading 191

9.1 Introduction

Network coding (NC) is a technique where network elements like routers, switches, etc. mingle the packets before putting it over the communication channel, which ultimately reduces the number of packet transmissions. The packets to be mixed at the node may belong to the same or different sources depending upon the immediate authoritative node. NC breaks the traditional network traffic assumption that information is separate but may share network resources. Rather, information mixing provides higher network throughput, greater robustness to link failures as well as resilience to attacks and eavesdropping. NC is used to attain maximum possible information flow to approach the Shannon capacity limit by dispersing and uniting information at intermediate nodes. It was first proposed for wired networks, to deal with the bottleneck problem, but link diversity and broadcast nature make it more attractive in wireless networks.

On the other hand, Opportunistic Networks (OppNets) (Pelusi et al., 2006; Huang et al., 2008; Rodriguez Aranguren, 2013) are based on spontaneous interaction and collaboration among devices equipped with short-range wireless transmission technologies like Bluetooth and Wi-Fi. It takes advantage of mobile nodes whenever they are in communication range to discover and asynchronously exchange information with each other. Each node exploits any opportunity that brings data closer to the actual destination. OppNets are only suitable for delay tolerant networks due to the store-carry-and-forward paradigm. An individual node's intention is to disseminate data at best possible next-hop among the neighborhoods, with the intention of successful delivery at the actual destination, where end-to-end paths between communicating nodes are rare or unpredictable if they exist. The link reliability among nodes in OppNets is unpredictable; therefore, each node keeps the data in a buffer during the nodes' separation and resumes transmission when the connection is reestablished.

In such a highly dynamic network, mixing multiple packets and transmitting them as a single may better exploit nodes' contact and coding opportunity, which in turn provides high throughput gain. Although coding schemes never wait for an additional codable packet to opportunistically overload each transmission if it is permitted (as the node may delay packet transmissions in OppNets), undoubtedly network performance will be lifted high. The application of NC changes the OppNets' traditional store-carry-and-forward paradigm with the store-carry-code-and-forward model.

9.1.1 Traditional Routing vs. Network Coding

In the traditional packet transmission, the source node splits the message into packets and transmits it over the medium. The packets route along different paths, but they all must arrive at the destined node where the packets reassemble into the original message. The most common problem with the approach in a high traffic network was a bottleneck, causing a packet drop, delay, etc., which drastically drops network throughput. Moreover, very few routes and nodes between the communicating ends might be congested, and others remain under-utilized.

NC was first introduced by Yeung and Zheng in late 1999 as an alternative of routing among various mutually independent sources (Ahlswede, 2000). It is a transmission technique used to accumulate various packets at the source node and de-accumulate the received packet at the destination node. In NC, traditional routers and switches are equipped with a coder/decoder that makes it opportunistic to snoop on the communication medium and encode multiple packets instead of just forwarding as it arrives. The merging of multiple packets depends on the coding opportunity at the node, which itself depends on its output buffer and the neighborhoods' information at that moment. If the opportunity exists, the node encodes multiple data packets and transmits it to the downstream in a single transmission for the actual recipients or the next-hops, i.e. closer to the actual recipients.

Let V denote the set of nodes that can function as an encoder and decoder and E represent the set of connections among the nodes, then graph $G=(V, E)$ in Figure 9.1 illustrates a point-to-point communication network of noiseless information exchange. Each node in the graph can collect information from all the intensive and mutually independent sources, encode it and forward it to the output link (for wireless networks) or set of output links (for wired networks).

Node s_i represents the ith source station and d_i denotes the jth destination station of the sub-network. Nodes 1, 2, 3 and so on are the intermediate relay routers or switches capable of the encoding and decoding of outgoing and incoming packets respectively. The arrows are representing the data flow in the network, with the different bit rate capacity of each link. The information exchange between s_i's and d_j's is accomplished with the relay nodes, which are in the communication range of both, and thus exploit the coding opportunity before putting it in the outgoing link.

The advantages of network coding can also be seen in Figure 9.2 with one source and two destination nodes in a single communication range; this is popularly known as butterfly network.

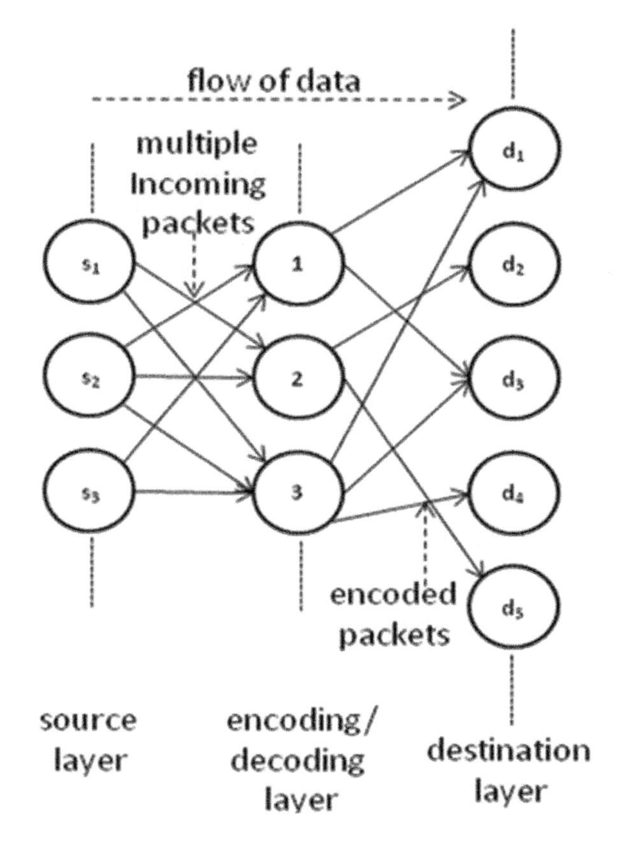

FIGURE 9.1
Graph representing flow network.

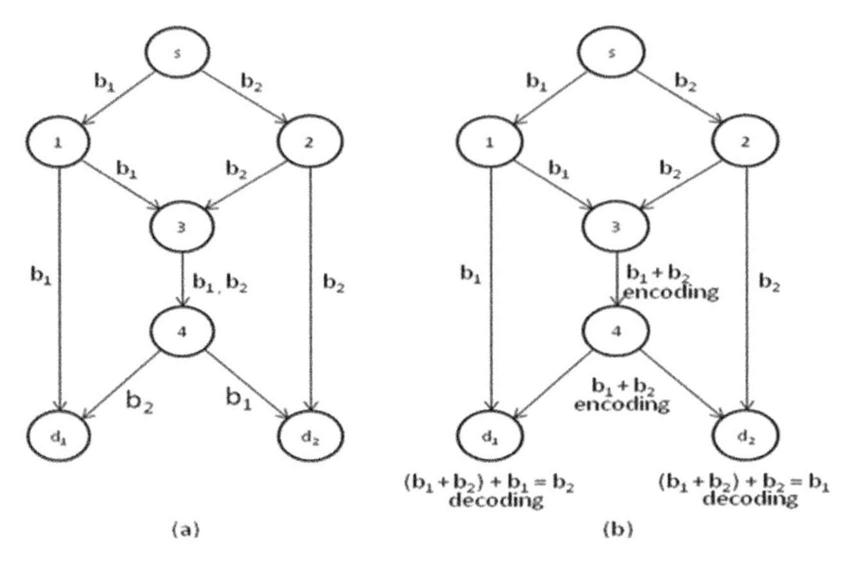

FIGURE 9.2
Butterfly network (a) without and (b) with network coding.

In Figure 9.2 (a) packets p_1 and p_2 are intended for destinations d_2 and d_1 respectively. Nodes 3 and 4 are functioning as a relay for the packets p_1 and p_2. Without the coding scheme, nodes 3 and 4 will transmit both the packets in separate transmissions and, hence, take *four* transmission cycles collectively to accomplish, whereas in Figure 9.2 (b) with NC, nodes 3 and 4 will encode the packets before sending it out for the destinations. NC will take only *two* transmission cycles, i.e. one transmission at node 3 of the encoded packet $b_1 + b_2$ and another of the same encoded packet from node 4 for the same communication. Therefore, NC shows the opportunistic and intelligent behavior of mixing packets to defend 50% bandwidth in a small topology, compared to without coding. The '+' denotes the NC operator which will be covered in the next section.

In recent years, as communication services become more relevant in real life, NC plays a significant role in efficient information exchange. OppNets provide short-range connectivity among highly mobile devices, and the NC alliance guarantees maximum data transfer, as nodes are in close contact with each other. The prominent networks like cellular, Wi-Fi, Wireless Sensor and Mesh, Mobile Ad-hoc (MANETs), Vehicular Ad-hoc (VANETs), Wi-max, etc., are more influenced by coding-aware OppNets.

9.1.2 Classification of Network Coding

Figure 9.3 shows the classification of network coding schemes which depends on several parameters. The first category is based on the previous hop(s) of the incoming packets at the midway nodes. At any instant, multiple information flows exist among different communicating pairs in the network. The first category of NC is inter-session, which allows the mixing of packets from different sessions (sources). It solves the hindrance of bottleneck and cuts down the number of transmissions, which in turn raises network throughput. However, intra-session is another category, which allows the fusion of packets from a single session (source). It contributes the reliable conversation between the communicating ends which, in contrast, consumes bandwidth in the form of a feedback mechanism in the conventional way of transmission.

Another viewpoint from which to classify NC is the fusion of multiple packets (coding) and the separation of a packet into multiple packets (decoding). The mathematical operation-based random linear (RL) coding creates a complex coded packet of the form $\sum_{i}^{k} \boldsymbol{\alpha}_i \boldsymbol{p}_i$ where the random coefficient α_i is chosen over a finite field and $\boldsymbol{p}_i$ denotes the native or coded packets. In contrast, the binary operation-based exclusive-OR (XOR) coding executes a logical exclusive exercise ($\oplus$), which results in a simple coded packet.

From another aspect, we can classify the coding technique based on the processing of coded packet at midway hops of the source-to-destination route. While transmitting between the communicating nodes, the coded packet may be decoded at each intermediate hop or at the destination hop only, and depending on this the coding schemes belong either to *local* or *global* respectively. Indeed, binary XOR prefers hop-by-hop *(local)* encoding and decoding while RL adopts the host-by-destination *(global)* coding scheme.

9.2 Inter-Session Network Coding

Inter-session NC recognizes the broadcast nature of the wireless medium as an opportunity if midway nodes adopt encoding and decoding capability. The nodes of long delay tolerant OppNets may afford more time to exploit the coding opportunity in a more profitable manner before relaying packets at intermediate hops. Some relevant techniques compatible with opportunistic networks will be discussed in this section along with its working concept, implementation detail and performance analysis.

9.2.1 COPE

In the literature, the idea of an opportunistic wireless NC named COPE (Katabi et al., 2006; Katti et al., 2006) was introduced, which bridges the theoretical and practical concepts of bursty and unicast flow in multi-hop wireless mesh networks. Binary XOR-based COPE raises the information content of each transmission which largely increases network throughput from a few percent to several folds depending on the traffic pattern, degree of congestion and transport layer protocol. COPE inserts a coding shim between the Internet Protocol (IP) and Medium Access Control (MAC) layers, which identifies the coding opportunities at midway hops of the route and exploits them to forward multiple packets in a single transmission, which in turn leads to a large bandwidth saving (Figure 9.3).

9.2.1.1 Overview

To exploit coding opportunities in OppNets, COPE integrates a new layer in the network stack. It consolidates three main techniques:

1. *Opportunistic listening:* Each node overhears the wireless channel and buffers the packets for a limited period of time (0.5 sec.), irrespective of whether or not they are destined for them. Owing to the broadcast characteristic of the wireless medium, immediate next-hops in the communication range can sense the data transmitted by the relay node and keep it in custody for further coding opportunity. Therefore, an OppNet's node needs to snoop on the medium regularly.

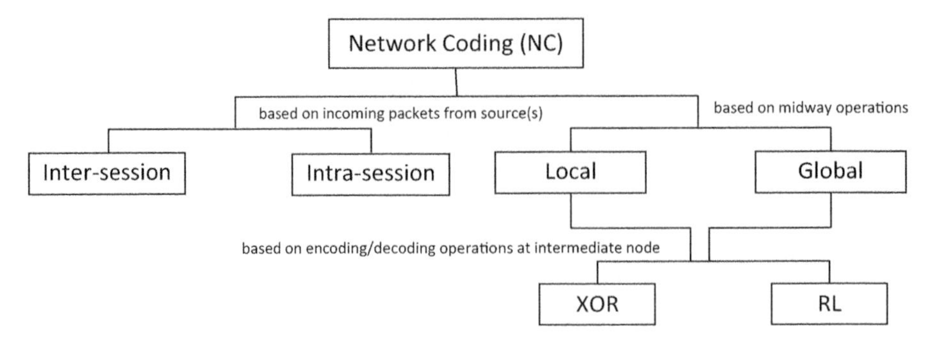

FIGURE 9.3
Classification of network coding.

2. *Opportunistic coding*: After sensing a new data packet from the medium, the node finds the opportunity to combine with the buffer packets before transmitting it into the air. If the opportunity does not exist, it transmits the packet alone to avoid packet delay.
3. *Learning neighbor's state*: The packet encoding depends on the degree of its decoding at the intended destinations. Thus, each node explores its neighbor's buffer pool. Indeed, the network elements convey the reception reports, i.e. knowledge of recently received packets, to its neighbors at regular intervals.

An essential prerequisite of network coding in opportunistic networks is to have at least three nodes in one communication range, as shown in Figure 9.4. It represents a small part of an OppNet and explains the functioning of opportunistic COPE. The nodes s_1 and s_2 want to exchange data packets p_1 and p_2 to destination nodes d_2 and d_1 respectively. The wireless communication range can be clearly observed in the figure. Destination node d_2 is not in the range of s_1 and, similarly, d_1 is not in the transmission range of node s_1 while all four nodes, i.e. communicating pairs s_1, d_2 and s_2, d_1 are in one radiocast range of node r. Hence, the communication between s_1, d_2 and s_2, d_1 can be relayed through node r, which sends forth the packet from respective source-destination pairs.

Further, the packet transmissions required to accomplish the communication without a network coding scheme are as follows:

1. Source node s_1 hands over packet p_1 to relay node r *(1st transmission).*
2. Source node s_2 hands over packet p_2 to relay node r *(2nd transmission).*
3. Relay node r sends forth packet p_1 to destination node d_2 *(3rd transmission).*
4. Relay node r sends forth packet p_2 to destination node d_1 *(4th transmission).*

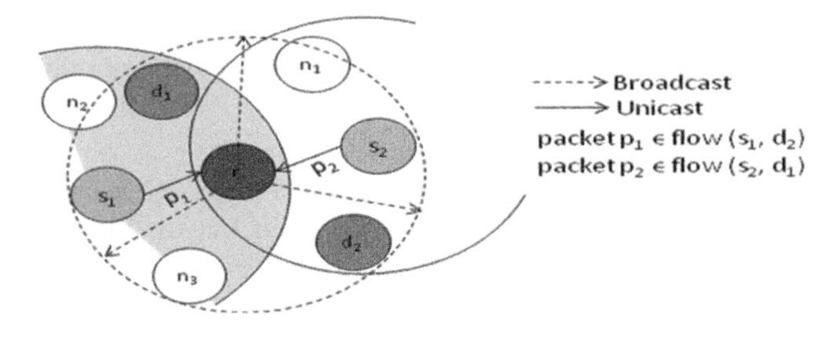

FIGURE 9.4
Understanding of COPE.

The cumulative transmissions required in the exchange of packets p_1 and p_2 between source-destination pairs s_1, d_1 and s_2, d_2 are *four.* Rather, consider the same communication scenario with the opportunistic COPE scheme:

1. Source node s_1 unicasts packet p_1 to relay node r, which can be opportunistically overheard by neighboring node, i.e. n_2, n_3 and d_1 *(1st transmission).*
2. Source node s_2 unicasts packet p_2 to relay node r, which can be opportunistically overheard by its surrounding neighborhoods, i.e. n_1 and d_2 *(2nd transmission).*
3. Further, relay node r combines the received packets p_1 and p_2 using an XOR operation and broadcasts it over the communication medium, which can be heard by its immediate next-hops e.g. d_1 and d_2 *(3rd transmission).*
4. Destination node d_1 and d_2 XOR the coded packet received from relay node r and the overheard packet from sources s_1 and s_2 to decode their respective packets p_2 and p_1.

The above demonstration of COPE requires only three transmissions and, self-evidently, with a small overhead of XOR operation. In fact, COPE defends more bandwidth saving, as appears in the example that will be discussed later in the subsection. Because of this, the overhearing and buffering of neighbors' transmissions for a short time span, as well as the updating of neighborhoods' buffer's knowledge, i.e. the capturing of reception reports regularly, makes opportunistic coding beneficial for OppNets.

9.2.1.2 Packet Coding

Some design issues have been taken care of in the COPE scheme to avoid its side effects while integrating with the existing network stack. First is the principle of transmitting packets without any delay. COPE is intelligent enough to encode multiple packets before putting them in the air but simultaneously ensures packet delay, due to the encoding and decoding process before each transmission. In the absence of coding opportunity, the node transmits the packet alone rather than waiting for a codable matching packet.

Second, COPE priorly encodes packets of identical length. An XOR-ing of unequal-length packets increases the bandwidth consumption. In turn, an individual node maintains one first in, first out (FIFO) output queue for storing its own data packets and two virtual queues per neighbor for referencing neighborhoods' packets: separate for small and long length. Whenever a coding opportunity exists, the node enquires the head of the virtual queue of matching packet length first. It is noted that unequal-sized packets can have zero padding to exploit COPE coding opportunity if equal-sized packets are not available, but that may cause underutilization of bandwidth.

Third, COPE never encodes packets together headed to the same next-hop. The encoding rule of native packets which can be decodable at the next-hop is:

To transmit n packets $p_1, p_2, \ldots p_n$, to next-hops $r_1, r_2, \ldots r_n$, a node can XOR n packets together only if each next-hop r_i has all n-1 packets p_j while $j \neq i$.

Packet reordering is another concern for COPE while it decodes at the recipient. In practice, reordering is quite rare due to the fact that most data packets of Transmission Control Protocol (TCP) flow are large enough to be queued in a proper packet queue. But packet retransmission leading causes reordering and hence to manage such scenarios, COPE tunes with an ordering agent to ensure harmonious delivery of TCP packets at recipients.

Thus, the strong reasons for scanning head packets from the hop's output and virtual queues during code fixing.

Finally, COPE ensures the packet decodability at the intended next-hops. The relay node estimates the probability of the buffer's packets being previously heard by its immediate next-hops. This probability assessment depends on the neighbor's reception report or the delivery probability between the packet's previous hop and the neighbors captured by the routing protocol intermittently.

Precisely, COPE maintains the following data structures at each network node:

1. Each node enriched with FIFO queue to buffer packets being forwarded, called output queue.
2. Each node is equipped with two virtual queues per neighbor for small and large packets, which take in pointers of neighbors' buffer packets. In COPE, a packet that is less than 100 bytes in size is considered to be small.
3. The node also maintains packet info, a hash table that is keyed on the packet ID and represents the buffer packets' probability of having it at neighborhoods.

9.2.1.3 Packet Decoding

Packet decoding performs a similar XOR operation as encoding. Each node maintains a log, i.e. a packet pool that keeps a copy of received and sent native packets. Whenever a node hears a coded packet, the scheduler scans the packet pool sequentially and retrieves the intended packet.

9.2.1.4 Packet Structure

The coding shim lies between the routing and the MAC header if the routing protocol comprises its own header; otherwise, it lies between the IP and the MAC header as shown in Figure 9.5.

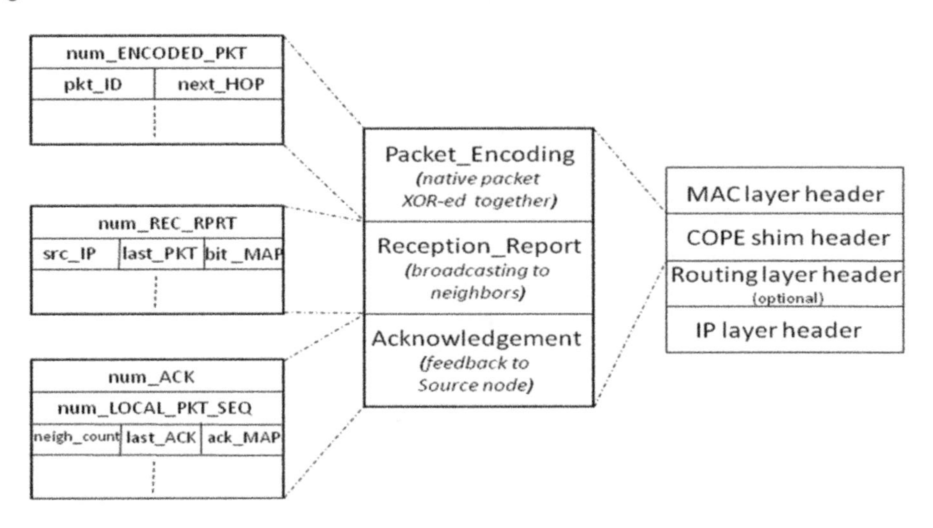

FIGURE 9.5
COPE header format.

The COPE header consolidates the following fields:

1. *Native packet information*: The first segment of the COPE header stocks the information about the native packets, the packets being encoded together. It assists in decoding at the recipient. The num_ENCODED_PKT holds the number of native packets being encoded at the node. The native packet's identity is stored at pkt_ID, which is a 32-bit hash of the IP address of the packet's origin and IP sequence number. next_HOP keeps the MAC address of packet's next-hop.
2. *Reception report*: The individual node shares ownership of its updated buffer information with its neighbors in order to exploit opportunistic coding with minimal computational overhead. The center block of the XOR header starts with the num_REC_RPRT; it contains a number of reports currently being shared. src_IP and last_PKT specify the source IP address and the most recently snooped packet from that source. The compact and robust bit_MAP field accommodates the binary sequence of bits that corresponds to the recently heard packets from the same source. For example, the report of the structure {128.0.16.28, 49, 10010001} is interpreted as the last packet heard from the source. 128.0.16.28 is sequenced as 49 and previously snooped packets are 41, 44 and 48. The duplicate report sending of a recently received packet guards against its dropping because of congestion and channel noise.
3. *Asynchronous acknowledgement*: The last section of the COPE header is dedicated to the acknowledgment part of packet delivery at the destination node. It starts with the number of the acknowledgement being accommodated in the report. A 16-bit-long counter is managed by each node for each neighbor, called neigh_count. The respective neighbor's counter is incremented whenever a packet is transmitted to that neighbor, and its value is assigned to the packet as the local sequence number, i.e. num_LOCAL_PKT_SEQ, which is used to identify the packet at communicating pairs. Similar to the reception report, this section practices the cumulative acknowledgement. Each acknowledgement contains the neighbor's MAC address followed by the acknowledgement pointer of the last packet received successfully. The ack_MAP field indicates the previously heard and missed packets. For example, an acknowledgement entry of {X, 25, 01111111} certifies packet 25, as well as the packet sequenced with 18–24. It also cites the loss of packet number 17. COPE retransmits the missing packet a few times (default is 2) and then gives up. If a packet is not acknowledged within a predefined time (Ta), which is slightly greater than the round-trip time of a single link, the packet is again inserted at the head of the output queue and retransmitted after exploiting the coding opportunity.

9.2.1.5 Performance

The benefits of COPE is measured by the coding gain, i.e. the number of transmissions saved against non-coded transmissions, which in turn leads to throughput gain.

As inferred from the following equation, the coding gain is greater than or equal to 1.

$$\text{Coding gain} = \frac{\text{Number of transmisions required by non-coded scheme}}{\text{Number of transmissions required with COPE scheme}}$$

Throughput gain is the ratio of network throughputs with and without COPE. The throughput improvement relies on the opportunistic coding, which itself depends on the traffic pattern.

The single source-destination scenario in Figure 9.5 shows *four* transmissions using a non-coding scheme, whereas only *three* transmissions are needed with the COPE scheme. Thus, the coding gain is 4/3 = 1.33, which means 33%, i.e. COPE requires 33% fewer transmissions to accomplish communication. In practice, the coding gain exceeded greatly due to the beneficial side effect of the MAC layer and coding shim interaction known as *coding + MAC gain*. The reason for being so is the fairness MAC channel allocation. In the previous single source-destination example, MAC partitions the available bandwidth in three equal parts. COPE, rather, allows the relay node to transmit packets twice as fast as s_1 and s_2 due to mixing before transmitting, i.e. exploiting the opportunity of coding, which in turn raises gain 2.

Similarly, other topologies shown in Figure 9.6 have greater coding + MAC gain. In all scenarios, the favorable effect of MAC demonstrates higher coding + MAC gain than theoretical coding gain given in Table 9.1.

It is well proved that the theoretical coding gain of the COPE scheme is upper bounded by 2. In a realistic scenario, the coding + MAC gain is lower bounded by 2 in the absence of opportunistic listening, and the maximum coding + MAC gain is unbounded. The energy consumption is upper bounded by 3 in unicast random networks (Liu et al., 2007).

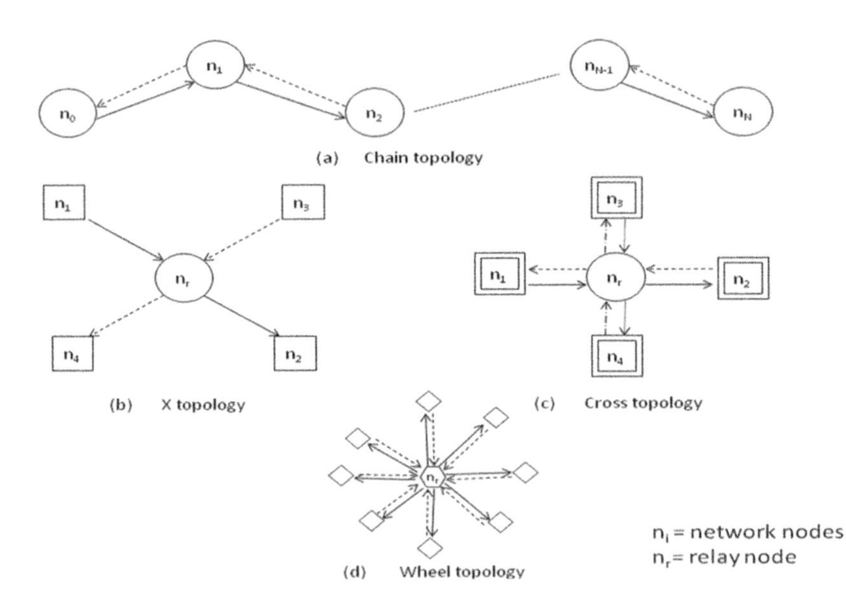

FIGURE 9.6
Basic topologies to understand the concept of coding and coding + MAC gain: (a) chain topology; (b) X topology; (c) cross topology; and (d) wheel topology.

TABLE 9.1
'Coding' and 'Coding + MAC' Gains for basic topologies

Gain ↓/Topologies →	Single Source - Dest	X	Cross	Infinite Chain	Infinite Wheel
Coding gain	1.33	1.33	1.6	2	2
Coding + MAC gain	2	2	4	2	∞

It does not show significant improvement in TCP traffic due to collision-related losses and claims an average gain of approx. 2%–3%, whereas high coding gain was achieved with the User Datagram Protocol (UDP) packets and the result of the average throughput gain was 3–4 fold.

COPE was designed for OppNets like wireless mesh networks with some constraints, like stationary network devices, sufficient memory for buffering overheard packets, an omni-directional antenna for better opportunistic listening and enough energy resources. COPE is an important step forward toward throughput gain beyond the traditional way. To satisfy the Quality of Service (QoS) requirement, a more efficient coding can be made possible with a few modifications in COPE architecture, such as a priority queue instead of FIFO, which assigns different weight to differentiate data packets, and introducing a fast scheduling algorithm to support the modified queuing system (Chi et al., 2009). Some research articles propose a virtual queue at the coding node to give higher priority to coded packets (Zhao & Médard, 2010).

9.2.2 BEND

Packet encoding in COPE can only be performed at the intersecting nodes of the path determined by the routing module, i.e. *traffic concentration*. Zhang et al. (2008, 2010) figured this problem out and in 2010 proposed BEND with the key foundation of scattering flows more evenly in the network, i.e. *traffic separation*, and thus minimizing the interference to achieving network capacity limit. BEND is a practical network coding approach in multi-hop wireless networks; it makes use of packet redundancy with a low overhead. It proactively captures coding opportunities without relying on fixed forwarders. In the opportunistic networks, an individual node in the neighborhood coordinates their packet transmissions and functions as potential coder and forwarder.

The mixing of packets in COPE relies on the focal nodes of the path determined by the routing module, which restricts the coding opportunity, which in turn prevents actual possible throughput gain of the network. Thus, some nodes in the network are favored by the routing mechanism whereas the rest are discouraged. Further, more traffic diverted through the focal nodes increases coding opportunity but may cause of packet drop, end-to-end delay, battery-power depletion etc.

9.2.2.1 Overview

BEND comes with the concept of splitting traffic among the forwarder nodes to achieve high throughput gain in opportunistic networks. It exploits over listening at the whole proximity of a node rather than only concentrating on the few joint nodes for a packet mixing opportunity as in COPE. Figure 9.7 depicts an example of two flows which are not intersecting each other, but are close enough to use the coding opportunity seized by the MAC layer.

In the figure, packets p_1 and p_2 correspond to network flows f_1 and f_2 for the communicating node pairs (A_1, B_1) and (A_2, B_2), which are two hops away and may accomplish transmission via intermediate nodes M and N respectively. Packet p_1 is broadcasted by node A_1 and intended for the node M while overheard by node A_2, X, Y and Z, as all are in the proximity of node A_1. Similarly, packet p_2 is broadcasted by node B_2 and intended for N but snooped by B_1, X, Y and Z, as they are a neighborhood of B_2. These two flows are not intersecting each other, but crossing node M and N closely. In this scenario, COPE does not find any coding opportunity while BEND coordinates the coding and forwarding of

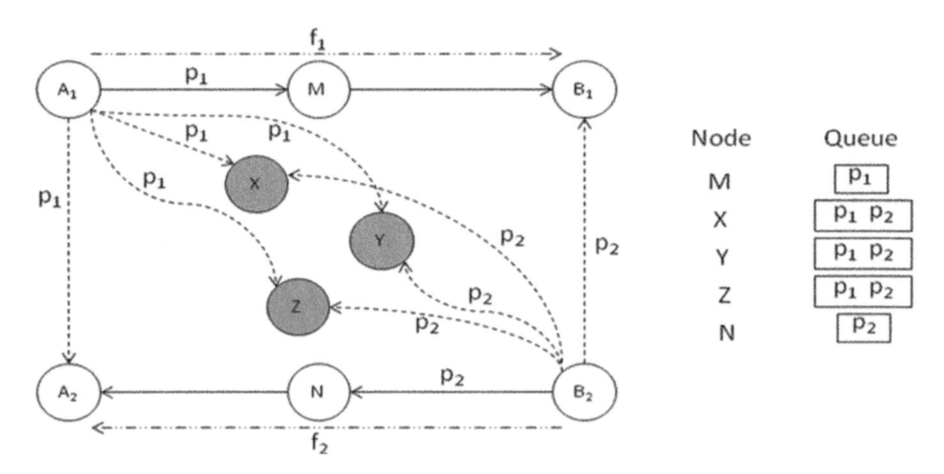

FIGURE 9.7
Conceptualization of BEND.

the packet queued at the neighborhoods, i.e. X, Y, Z. Instead of discarding these packets as in COPE, BEND stores them at the MAC layer and uses them to raise coding opportunities in the future. In the communication depicted in Figure 9.7, any one of nodes X, Y and Z that win the MAC layer medium contention encode packet p_1 and p_2 and transmit it in the medium. Further, destination nodes B_1 and A_2 can decode their intended native packets after receiving the encoded packet and its own buffer information.

BEND's objective is to increase coding gain and, hence, throughput gain using the per-packet and per-hop decision-making promoted by the MAC layer, i.e. the packets can be routed through distinct paths in order to find opportunistic coding with other packets of different flows in the network at various instances with different mixing nodes.

9.2.2.2 Packet Mixing and Queuing Strategy

BEND promotes the XOR operation to mingle multiple native packets in a single transmission. The coding condition ensures that the receiver of packet p_i has p_j (where $i \neq j$) in its buffer so that it can XOR p_j with coded packet to get the intended native packet.

Each node maintains two different FIFO queues i.e. Q_1 to keep packets that are intended for it during transmission and Q_2 to carry overheard transmission. The coding capable packets are shifted from Q_1 and Q_2 to a different queue, named *mixing-Q*. These three queues are responsible for packet matching for the mixing process, as follows.

When network layer passes a new packet p_1 to the MAC layer, BEND traverses all three queues for the coding pair. It always starts the exploration with *mixing-Q* to satisfy the pair-wise matching condition. If no suitable packet is found for coding, it searches Q_1 and Q_2 in turn. It starts with the head packet of the queues and removes the first matching packet to enqueue at the tail of the *mixing-Q* along with the packet p_1. If no match is found and the node is the intended forwarder of p_1, it will be enqueued at the tail of either Q_1 or Q_2. Q_1 and Q_2 hold the packet until they are transmitted. Similar to COPE, the node transmits the packet alone if coding is still unmatched till its scheduled turn.

BEND's packet matching mechanism is slow and complex compared to COPE. Instead of concentrating only on the foremost packets in the virtual queue, the matching process of BEND cares for the packet's next-hop receiver as well as the packet's previous forwarders. Hence, BEND is hungry for memory space and requires a complicated queuing structure.

BEND allocates higher priority to coded packet transmission in order to maximize the coding opportunity but simultaneously disallows starving non-coded packet transmissions. The *mixing-Q* is assigned higher priority than Q_1 and Q_2. The scheduler generates a random uniform number between *0* and *1*. If the number is greater than W_x, a predefined threshold value called tunable weight, and *mixing-Q* is not empty, the node generates a coded transmission with a matching mechanism or else it schedules a non-coded transmission.

9.2.2.3 Packet Decoding

A node in the network has its own buffer to store packets that had either been forwarded or originated earlier or overheard from the medium. These packets service the node to extract the native packet from the received coded packet. When a coded packet arrives at any node, it checks whether the node's MAC address is in the recipient list or not. If the node is the actual recipient, it determines the packet IDs from the header and looks up the recipient's buffer to retrieve the intended packet by XOR-ing the required packets. Further, the packets can be mixed with other packets in the queue in accordance of coding opportunity. This process repeats until the packet is delivered to the destination node.

9.2.2.4 Packet Header

BEND is managed by the MAC layer, and therefore packets adopt 802.11 specifications (IEEE Computer Society LAN MAN Standards Committee, 1997) with a few modifications in the data and control headers, as shown in Figure 9.8 The data frame header has distinct field arrangement for the native data packet or the encoded data packet. The native data packet accommodates the source and destination IP address (source and destination) along with the IP address of the second next-hop of the packet. This field acquires an IP address of the packet's second next-hop from routing protocols which is exploited by neighborhoods. Whenever a forwarder or neighborhood receives a native data packet, it checks the second next-hop field to identify the next intended node for that packet. The encoded packet contains the length of the encoded packet (code_len), i.e. the number of native packets in the encoded packet, all recipient lists (destinations) and the corresponding list packet_ID. The recipient address is 6 bytes long in Adaptive Opportunistic Network Coding (AONC). Similar to COPE, the hash of the host's IP address and the IP packet sequence number generates a 4-byte-long packet ID. The type and subtype bit of the frame control field corresponds to either the native packet, coded packet, acknowledgement (ACK), negative acknowledgement (NACK) or other 802.11 frame. If a coded packet arrives at the receiver, it checks the packet decodability by traversing its own buffer's packets. If the encoded packet is found to be decodable, the receiver sends either an ACK or a

NATIVE_PKT_MAC header	Frame control	Duration	destination	TA	2nd next-hop	
ENCODED_PKT_MAC header	Frame control	Duration	code_len	destination [code_len]	TA	packet_ID [code_len]
ACK/NACK_PKT_MAC header	Frame control	Duration	source	FCS	packet_ID	

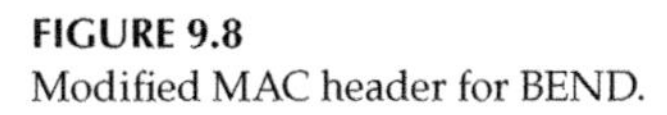

FIGURE 9.8
Modified MAC header for BEND.

NACK. This feedback frame accommodates the packet's source IP address and the packet ID of the acknowledged and negative-acknowledged packet. As BEND generates packet duplication at multiple forwarders, the ACK frame is used to clear all replicas of delivered packets at the previous hops (Figure 9.8).

9.2.2.5 Performance

BEND figures out the dilemma of *coding* and *diffusion gain,* which cannot be achieved simultaneously by coding-aware routing and COPE. The link layer dynamically scatters the flows through numerous forwarders in multi-hop OppNets, which leads to benefits of coding opportunity. Such dispersing of flows is termed "diffusion gain".

In a 3-tier network architecture, BEND has consistently higher-gain (approx. 55%–97%) over COPE by an increasing number of nodes at tier-2 level. The simulation results of BEND proved that the coding opportunity with more than two native packets is much higher than the existing COPE. BEND performance is highly appreciable with heavy multi-hop traffic flow, i.e. huge number of flows which pass multiple intermediate nodes in the network. Since BEND exploits packet redundancy in OppNets but packet duplication at UDP recipients is as identical as COPE.

9.2.3 AONC

Although COPE and BEND have demonstrated their capability of improving the network throughput gain, they cannot efficiently support video data communication. A bandwidth and energy efficient AONC (Shen et al., 2012) scheme was proposed for upgrading the transmission quality of the video stream in multimedia-based OppNets. The fundamental variation between traditional data transmission and real-time video data transmission inspired AONC. The traffic-aware data scheduling algorithm of AONC functions together with the special features of the existing network coding techniques.

9.2.3.1 Overview

AONC is motivated by COPE and performs a similar XOR operation for packet mixing, but in an asymmetric fashion. To realize the conceptual difference between and COPE and AONC, consider the simple example shown in Figure 9.9. The unequal-length packets buffered in the output queue are divided into three data sets, i.e. data 1, data 2 and data 3. Let us assume that any two data sets match the coding criteria. Figure 9.9 (b) shows the randomly encoded packets using the XOR operation in symmetric mode. Due to varying length, which is very common with video data traffic, packets are padded with zeros to match the largest-length packet, which, ultimately, is a wastage of the bandwidth medium. In this scenario, the node is required to transmit three times. Instead of symmetric mode coding, Asymmetric mode-based AONC apprehends the benefits of variable-length data packets, which leads to the effective utilization of bandwidth. In Figure 9.9 (c) video data packets of the same flow are spliced first and then combined with the packets of other flows according to the code matching condition. The Figure 9.9 (c) shows that only *one* transmission is required for the set of data shown in Figure 9.9 (a) while the traditional symmetrical coding method, COPE, has taken *three.*

AONC considers the dynamic priority assignment for the coded packets due to variable-length video data packet while COPE priority tags coded transmissions as higher priority

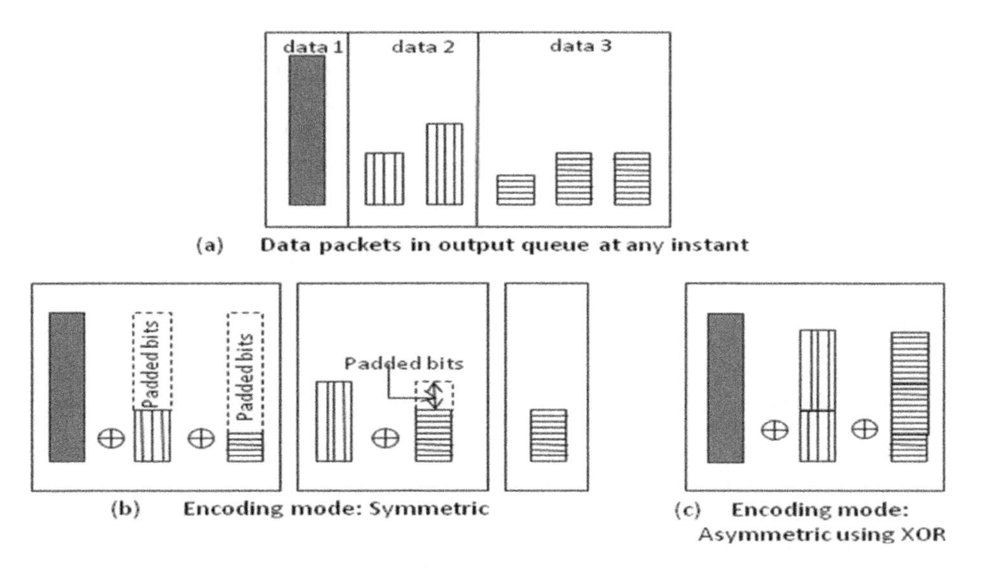

FIGURE 9.9
Comparisons of traditional coding scheme i.e. COPE vs. AONC.

in the scheduling. Another major difference between AONC and symmetric mode coding is the compression technique over the video data before transmission. The compressed video stream has very complex traffic characteristic and, hence, FIFO-based scheduling mechanism loses coding opportunities.

9.2.3.2 Protocol Mechanism

The traditional symmetric coding and data scheduling methods degrade the performance of OppNets due to the different data behavior of the video traffic over the network. Along with this, the priority assignment method of coded packets is also unreasonable. The fundamental mechanism of the AONC is comprised of the following:

1. An asymmetric coding approach is used to enhance the data exchange gain. The data exchange gain is the number of transmissions reduced by successful encoded packet interchange.
2. An opportunistic transmission strategy with dynamic priority is designed to improve network throughput.
3. A traffic-aware data scheduling algorithm is used to reduce coding opportunity loss.

AONC is a flow-oriented coding mechanism, i.e. it groups the data packets of the single previous forwarder and the next-hop receiver into one flow and stores it in the dedicated output queue of that flow. A flow is termed as "coding flow" when it transmits through intermediate nodes and the packets of flow have coding opportunities during transmission. AONC selects data packets from each flow of the pre-decided group of coding flows and slices them. These slices of data packets mingle in a single transmission, and therefore increase the information content per transmission.

The opportunistic coding-based forwarding strategy of AONC is based on dynamic priority assignment. The dynamic priority-setting of coded packets depends on data exchange gain and space utilization. Space utilization is a metric used in AONC to estimate the amount of actual data handled by an encoded packet. Unlike BEND, where higher priority is assigned to the encoded packets, AONC uses dynamic priority to indicate the degree of importance of the encoded packets. In this way, the coded packet with higher space utilization is transmitted with higher priority. Thus, we conclude that the node with coding opportunity and better space utilization can have a high probability to capture communication media.

Along with the above improvement over BEND, a traffic prediction mechanism is also used to predict the volume of incoming traffic at the intermediate nodes. AONC considerably enhances the transmission quality of video packets and attains a notable gain in bandwidth utilization and energy consumption.

9.2.3.3 Packet Header

AONC performs packet encoding/decoding and tagging at the MAC layer. It implements IEEE 802.11 with some modification in the DATA and ACK frame headers to provide real-time video transmission. Figure 9.10 shows the modified header fields of AONC.

The destination[CFN] contains the receiver addresses of the native packets, where CFN is the coding flows of coded packet. The NUM_packet[CFN] represents the number of original packets belonging to that corresponding coded flow. The INFO_packet[CFN] [NUM_packet[CFN]] records the list of the native packet IDs and lengths in the encoded packet. The packet IDs are computed similarly to COPE. The ACK frame is modified with a new field, i.e. INFO_packet[NUM_packet] for the efficient feedback of multiple slicing units at a time.

9.2.3.4 Performance

Compared to BEND and COPE, the dynamic nature of AONC provides enhancement in network throughput as well as stability in video transmission. The average peak signal-to-noise ratio (PSNR) value of AONC is higher than any of the other strategies like BEND, COPE or 802.11 that concludes its higher reliability and quality of video transmission. The video quality of AONC is close to the original sequence of non-encoded transmission, which is an almost error-free transmission. The coding mechanism and data scheduling scheme of AONC provides better network bandwidth utilization than the BEND and COPE that avoids packet loss problem in the communication medium. It delivers interrupt-free video signals at the recipient station in cross, 3-tier and mess network topologies. The enhancement in data exchange gain reduces energy consumption in data transmission. Since a wireless node consumes more energy in packet transmission and reception, AONC is more energy efficient than BEND and COPE.

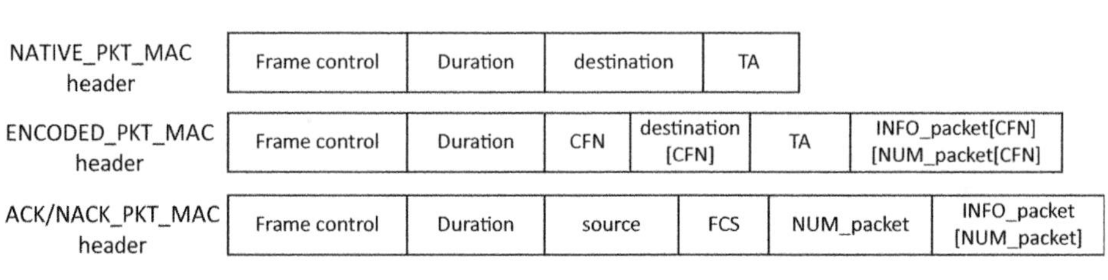

NATIVE_PKT_MAC header	Frame control	Duration	destination	TA		
ENCODED_PKT_MAC header	Frame control	Duration	CFN	destination [CFN]	TA	INFO_packet[CFN] [NUM_packet[CFN]
ACK/NACK_PKT_MAC header	Frame control	Duration	source	FCS	NUM_packet	INFO_packet [NUM_packet]

FIGURE 9.10
Modified MAC header for AONC.

9.3 Intra-Session Network Coding

The discussed coding techniques in the previous section were based on inter-session while ExOR (Biswas & Morris, 2005), MORE (Chachulski et al., 2007), MIXIT (Katti et al., 2008), Coding in Opportunistic Routing (CodeOR) (Lin et al., 2008) and SlideOR (Lin et al., 2010) are intra-session-based network coding techniques. Opportunistic routing improves the unicast network throughput of OppNets by exploiting the shared wireless broadcast medium. But, on the other hand it has the challenging issue of the unnecessary forwarding of duplicate packets by multiple intermediate nodes and overloading of one forwarding path between sources to destination. The packet duplicity can be avoided by the large number of control packets and the complex packet scheduling algorithm used in ExOR. However, a more feasible and practical approach referred to as MORE uses random network coding and stop-and-wait protocol in its sending window. Its successor MIXIT has attained better throughput in lossy wireless medium but uses the same stop-and-wait mechanism, and, therefore, shares the same drawbacks in large-scale networks.

The following sub-sections will present CodeOR and SlideOR encoding techniques, which are especially appropriate for unicast real-time multimedia data traffic in wireless mesh networks.

9.3.1 CodeOR

The computational complexity of random network coding increases as the number of data packets are increased. The constraints of computation complexity and bad utilization of bandwidth enlightens the CodeOR; this is a segmented network coding scheme in which the data stream is divided into segments and packet encoding is performed in the same segment. CodeOR allows the transmitting node to broadcast a sliding window of multiple segments using opportunistic routing protocol.

9.3.1.1 Overview

CodeOR improves network throughput for real-time multimedia traffic by using flow control in opportunistic routing and segmented network coding that partitions the data stream into multiple segments. The multiple small segments are transmitted concurrently, but in a sequential manner. CodeOR decreases the decoding delay and increases throughput by varying the segment size and the window size. It combines the benefits of opportunistic routing and segmented network coding.

9.3.1.2 Protocol Mechanism

CodeOR is inspired by MORE protocol. The protocol mechanism is described in Figure 9.11. The part of the network shown in the figure has *(k-1)th, kth,* and *(k+1)th* segments are in flight simultaneously. The detailed working of CodeOR is presented in the subsequent sections.

9.3.1.3 CodeOR Motivation

In MORE, nodes keep on transmitting irrespective of destination and obtain a sufficient number of coded packets to decode the segment and hence it starts transmitting next

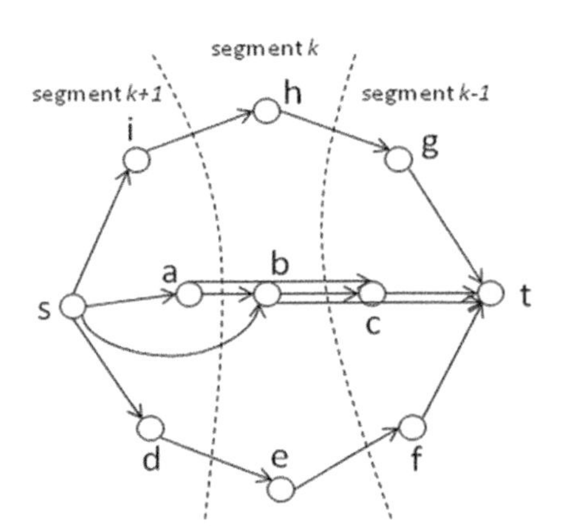

FIGURE 9.11
Working of CodeOR: concurrent but sequential transmission of segment *k-1, k,* and *k+1.*

segment much later. A motivation to CodeOR is to stop transmitting a segment as soon as the destination receives the required number of packets to decode that segment and to start transmitting the next segment much sooner than the CodeOR. MORE protocol transmits one segment and waits for its acknowledgment whereas CodeOR allows multiple segments in the air. The working of CodeOR is presented in Figure 9.11 where *(k-1)th, kth,* and *(k+1)th* segments are "in flight" simultaneously for the communication between source node *s* and destination node *t*. The segments of data stream from source *s* are broadcasted and received by *1-hop* downstream, neighboring nodes *a, d* and *i* for further transmission to their downstream neighbors. As the neighbors obtained a sufficient number of coded packets to decode the segment, the source *s* will start to transmit the next segment while the previous segment is "in flight".

9.3.1.4 Sending Window and Acknowledgment

In multi-hop communication, where the bandwidth of links is different, it needs to know the number of packets being forwarded by a node so that the recipient node does not get overwhelmed. The recipient node may drop a packet due to buffer overflow if its upstream neighbor sends packets at higher rate than the recipient's sending to its downstream neighbors. The sending window limits the number of segments buffered at the source that can be transmitted at any instant.

Since multiple packets are in the air concurrently, CodeOR implemented two acknowledgements. The first is end-to-end acknowledgement (E-ACK) sent by the destination to the source via the shortest path to notify that the segment of data packet has been received at the destination. The second is hop-to-hop acknowledgement (H-ACK) and is used by intermediate nodes to indicate to upstream nodes that the sufficient number of encoded packets have been received in a current segment, so move on to next segment. In the Figure 9.11, as *1-hop* neighbors *a, d* and *i* of source *s* obtain the sufficient coded packets of current segment, they will send the H-ACK packet to notify upstream node, i.e. *s*, to start transmitting next segment.

9.3.1.5 Performance

It has been observed that CodeOR achieves higher throughput gain than its predecessor MORE due to simultaneous segment transmission. As the number of hops increases between source and destination MORE becomes inefficient whereas CodeOR exploits all network resources to allow concurrent data transmission. Ideally, the sending window size should be equal to the delay-bandwidth product between source and destination. CodeOR increases as the window size increases until it reaches up to 6 for the long path whereas for the short path it is unchanged with window sizes.

9.3.2 SlideOR

MORE provides a stop-and-wait mechanism to reflect an extreme trade-off between the overhearing of duplicate packets at intermediate nodes and the deficiency of the path between source and destination if very few nodes forward the packet. It has an open challenge to move on to the next node when a source node stops transmitting the coded packets of a particular segment. Moreover, CodeOR partially addresses this open problem by transmitting multiple segments in a pipeline fashion, but it is exceptionally challenging in CodeOR to decide the optimal time to move the transmissions of new segment.

9.3.2.1 Overview

SlideOR addresses this problem by encoding source packets in overlapping sliding windows. Further, SlideOR is inspired by online network coding and, contrary to segmented network coding where the encoded packet of one segment is incapable of decoding the next segment, the coded packets of different overlapping sliding windows can be advantageous to each other in SlideOR protocol. The implementation is based on the following two critical issues:

1. How far, i.e. fast/slow to advance the sliding window.
2. The next-hop may re-encode the received packet with a different sliding window position, which may boost encoding-decoding complexity.

9.3.2.2 Encoding and Decoding Mechanism

The data stream at the transmitting node is divided into packets and allocated a sequence number as an identity of that packet. At any instant, when the source node wins the contention of transmission slot and the MAC layer allows for transmission, the source node randomly encodes some consecutive packets of output buffer which is referred to as sliding window size. Hence, the encoding algorithm produces the coded packet x as:

$$\mathrm{x} = \sum_{i=k}^{k+W} \alpha_i E_i$$

where k is the sequence number of next packet E and α is a randomly chosen coefficient from Galois field such as $GF\ (2^8)$.

The decoding of a received packet at any node is similar to solving a linear equation system as every coded packet is a linear equation of source packets at the previous hop. Every node needs to transmit the coding coefficient along with the coded packet for further decoding process.

9.3.2.3 *Sliding Window Process*

A feedback mechanism is executed after the successful reception of new coded packet. The assumption used in SlideOR is the reliable transmission of the ACK packet between communicating nodes. The destination node sends ACK to the source node via shortest path between source and destination that carries the sequence number of next expected packet. This sequence number is used to advance the window at the source station. Hence, it is straightforward that in sequential packet transmission, the source will enhance the window by one from $[k, k+W-1]$ to $[k+1, k+W]$ after receiving the ACK frame.

9.3.2.4 *Performance*

Sliding-window-based opportunistic routing is much simpler and substantially easier to implement. It reduces the complicacy of the scheduling mechanism of multiple segment transmission concurrently by just introducing the window advancement. The literature shows the throughput gain is higher than the MORE as well as CodeOR protocol when the window size is increased. SlideOR can transmit the coded packet concurrently during ACK propagation.

9.4 Research Aspects

In recent years, network coding is more popular in software-defined networks. Many industries like *hp, Cisco, D-link* etc. are incorporating network coding techniques to enhance the network throughput, and reduce packet delay and communication bandwidth. Network coding-based access point are available in the market for *Wi-Fi* signal distribution.

Software-Defined Networks (SDN) has the potential to virtualize the services like buffering, scheduling and routing over the internet. The literature (Hansen et al., 2015; Krigslund et al., 2015; Liu & Hua, 2014; Zhu et al., 2015) gives the view of network-coding-enabled SDN to promote the high speed 5G connectivity. Network coding techniques exploit the centralized control architecture of OpenFlow (SDN) to create fertile ground for a secure and more efficient network.

References

Ahlswede, R., Cai, N., Li, S. Y., & Yeung, R. W. (2000). Network information flow. *IEEE Transactions on Information Theory*, 46(4), pp. 1204–1216.

Biswas, S., & Morris, R. (2005). ExOR: Opportunistic multi-hop routing for wireless networks. *ACM SIGCOMM Computer Communication Review*, 35(4), pp. 133–144.

Chachulski, S., Jennings, M., Katti, S., & Katabi, D. (2007). *Trading structure for randomness in wireless opportunistic routing*, 37(4), pp. 169–180.

Chi, K., Jiang, X., & Horiguchi, S. (2009). A more efficient cope architecture for network coding in multihop wireless networks. *IEICE Transactions on Communications*, 92(3), pp. 766–775.

Huang, C. M., Lan, K. C., & Tsai, C. Z. (2008, March). A survey of opportunistic networks. In *22nd International Conference on Advanced Information Networking and Applications-Workshops, AINAW*, Okinawa, Japan, pp. 1672–1677.

IEEE Computer Society LAN MAN Standards Committee. (1997). Wireless LAN medium access control (MAC) and physical layer (PHY) specifications. *IEEE Standard 802.11-1997*, pp. 1–445.

Katabi, D., Katti, S., Hu, W., Rahul, H., & Medard, M. (2006, February). On practical network coding for wireless environments. In *International Zurich Seminar on Communications*, Zurich, Switzerland, pp. 84–85.

Katti, S., Katabi, D., Balakrishnan, H., & Medard, M. (2008, August). Symbol-level network coding for wireless mesh networks. In *ACM SIGCOMM Computer Communication Review*, Seattle, WA, 38 (4), pp. 401–412.

Katti, S., Rahul, H., Hu, W., Katabi, D., Médard, M., & Crowcroft, J. (2006, September). XORs in the air: Practical wireless network coding. In *ACM SIGCOMM Computer Communication Review*, Pisa, Italy, 36 (4), pp. 243–254.

Lin, Y., Li, B., & Liang, B. (2008, October). CodeOR: Opportunistic routing in wireless mesh networks with segmented network coding. In *IEEE International Conference on Network Protocols, ICNP*, Orlando, FL, pp. 13–22.

Lin, Y., Liang, B., & Li, B. (2010, March). SlideOR: Online opportunistic network coding in wireless mesh networks. In *INFOCOM, Proceedings IEEE*, San Diego, CA, pp. 1–5.

Liu, J., Goeckel, D., & Towsley, D. (2007, May). Bounds on the gain of network coding and broadcasting in wireless networks. In *INFOCOM, 26th IEEE International Conference on Computer Communications. IEEE*, Barcelona, Spain, pp. 724–732.

Pelusi, L., Passarella, A., & Conti, M. (2006). Opportunistic networking: Data forwarding in disconnected mobile ad hoc networks. *IEEE Communications Magazine*, 44(11), pp. 134–141.

Rodriguez Aranguren, S. (2013). *A Survey of Opportunistic Network techniques for Message Forwarding* (Master's thesis, Universitat Politècnica de Catalunya).

Shen, H., Bai, G., Zhao, L., & Tang, Z. (2012). An adaptive opportunistic network coding mechanism in wireless multimedia sensor networks. *International Journal of Distributed Sensor Networks*, 8(12), pp. 565–604.

Zhang, J., Chen, Y. P., & Marsic, I. (2008, March). Network coding via opportunistic forwarding in wireless mesh networks. In *Wireless Communications and Networking Conference, WCNC*, Las Vegas, NV, pp. 1775–1780.

Zhang, J., Chen, Y. P., & Marsic, I. (2010). MAC-layer proactive mixing for network coding in multi-hop wireless networks. *Computer Networks*, 54(2), pp. 196–207.

Zhao, F., & Médard, M. (2010, January). On analyzing and improving COPE performance. In *Information Theory and Applications Workshop (ITA)*, San Diego, CA, pp. 1–6.

References for Advanced Reading

Hansen, J., Lucani, D. E., Krigslund, J., Médard, M., & Fitzek, F. H. (2015). Network coded software defined networking: Enabling 5G transmission and storage networks. *IEEE Communications Magazine*, 53(9), pp. 100–107.

Krigslund, J., Hansen, J., Lucani, D. E., Fitzek, F. H., & Médard, M. (2015, May). Network coded software defined networking: Design and implementation. In *Proceedings of European Wireless; 21st European Wireless Conference*, Budapest, Hungary, pp. 1–6.

Liu, S., & Hua, B. (2014, September). NCoS: A framework for realizing network coding over software-defined network. In *IEEE 39th Conference on Local Computer Networks (LCN)*, Edmonton, Canada, pp. 474–477.

Zhu, D., Yang, X., Zhao, P., & Yu, W. (2015, August). Towards effective intra-flow network coding in software defined wireless mesh networks. In *24th International Conference on Computer Communication and Networks (ICCCN)*, Las Vegas, NV, pp. 1–8.

References and Further Reading

10

Taxonomy of Security Attacks in Opportunistic Networks

Gabriel de Biasi and Luiz F. M. Vieira

CONTENTS

10.1 Introduction ... 194
10.2 Classification of Attackers ... 195
  10.2.1 Insider versus Outsider Attackers ... 195
  10.2.2 Active versus Passive Attackers ... 195
  10.2.3 Malicious versus Rational Attackers ... 195
10.3 Security Attributes in Opportunistic Networks ... 195
  10.3.1 Authentication ... 195
  10.3.2 Nonrepudiation ... 195
  10.3.3 Availability ... 196
  10.3.4 Data Integrity ... 196
  10.3.5 Privacy ... 196
10.4 Authentication Schemes in Opportunistic Networks ... 196
  10.4.1 Signature-Based Schemes ... 196
  10.4.2 Verification-Based Schemes ... 196
  10.4.3 Cryptography-Based Schemes ... 196
10.5 Security Attacks ... 197
  10.5.1 Sybil Attack ... 197
  10.5.2 Black Hole and Selective Forwarding Attack ... 198
  10.5.3 Sinkhole Attack ... 199
  10.5.4 Wormhole Attack ... 200
  10.5.5 Jellyfish Attack ... 200
  10.5.6 GPS Spoofing Attack ... 201
  10.5.7 Impersonation Attack ... 202
  10.5.8 Bogus Information Attack ... 202
  10.5.9 Distributed Denial of Service ... 203
  10.5.10 Man-in-the-Middle Attack ... 204
  10.5.11 Rushing Attack ... 205
  10.5.12 Colluding Misrelay Attack ... 206
  10.5.13 Link Spoofing Attack ... 206
  10.5.14 Snare Attack ... 207
  10.5.15 Blackmail Attack ... 207
  10.5.16 Node Replication Attack ... 208
  10.5.17 Desynchronization Attack ... 209
10.6 Conclusion ... 209
References ... 210

10.1 Introduction

Although we can characterize traditional computer networks by allowing communication between a source and a destination using a direct routing path, other network architectures propose the use of algorithms that allow nodes be able to communicate even if there is no direct path between them. The opportunistic network belongs to a category of wireless networks that make use of various techniques for message transmission, such as direct transmission, flooding, prediction based, context based, and location based.

Normally, the nodes of opportunistic networks need to exchange messages with each other to obtain knowledge of their local neighborhood and be able to create a possible path to forward a message. From these concepts, some attackers can create new approaches to cause damage to the network or gain some benefit, such as generating false neighborhood messages to disrupt the connections or not forwarding packets, among other types of attacks.

Computer networks have always suffered from security attacks. In the context of opportunistic networks, attackers may abuse the method of packet forwarding to obtain important information from nodes or infiltrate the network as a legitimate node. These attacks can affect various network security threats, such as authentication, availability, privacy, and data integrity (Figure 10.1).

This chapter discusses several security attacks that exist on computer networks that might directly or indirectly affect all kinds of opportunistic networks. We classify the types of attacks according to the damage applied to the security attributes. We also show various research work proposing mechanisms to protect and mitigate these security issues.

We organized this chapter as follows: Section 10.2 details the classification of the entities that might act as an attacker on opportunistic networks and our possible goals. Section 10.3 presents the security attributes needed to deploy a secure opportunistic network. Section 10.4 shows the classification of authentication schemes that helps ensure the security attributes in many network contexts. Section 10.5 describes several security attacks, how they are executed, and the existing methods of mitigation and protection proposed in academia. Finally, Section 10.6 concludes the chapter, presenting the taxonomy of the attacks and their security attributes.

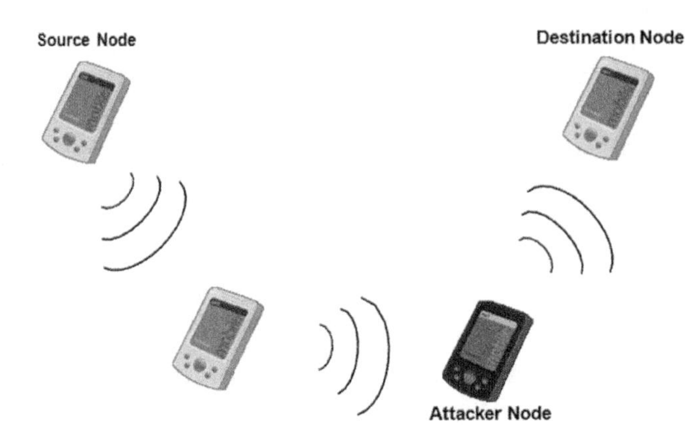

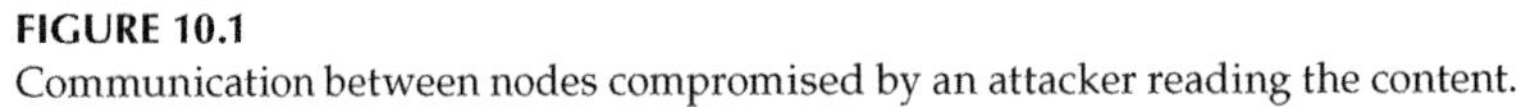

FIGURE 10.1
Communication between nodes compromised by an attacker reading the content.

10.2 Classification of Attackers

Several entities are potential attackers of an opportunistic network, such as coordinated groups, adversary companies, government agencies, or any individual who has a specific interest. The reasons for the attack may be monetary gain, political reasons, or even intellectual challenges.

We consider the entities that generate the security attacks as attacker nodes. According to their behavior and the type of attack they want to perform, we classify them differently. We organize the classification as follows.

10.2.1 Insider versus Outsider Attackers

Insiders are attacker nodes authenticated on the network, capable of exchanging messages normally without bringing attention to them. We consider the security attacks performed by these nodes very dangerous because the other nodes on the network trust them. However, outsiders are not able to conduct communication within the network because they are not properly connected, and their attacks are potentially less dangerous.

10.2.2 Active versus Passive Attackers

Active attackers perform actions within the network, acting directly on the communications, sending false messages or not forwarding messages, for instance. On the other hand, passive attackers are limited to just listening to the communication channel to get some kind of information from the nodes or just generating noise.

10.2.3 Malicious versus Rational Attackers

Malicious attackers aim to disrupt network connectivity even without any personal benefit or objective. However, rational attackers have their goals set before performing their security attacks.

10.3 Security Attributes in Opportunistic Networks

Opportunistic networks have important security requirements in order to provide safety communications. The following items present the main security attributes for a safe and reliable network (Raya and Hubaux, 2005).

10.3.1 Authentication

A node inside the network can only exchange messages with another legitimate node. The authentication of the nodes has an important role in the security of the network. In some applications, there is a key management infrastructure to ensure the authenticity.

10.3.2 Nonrepudiation

Nonrepudiation is a security mechanism, which means that whenever a message request is required, the sender and the receiver cannot deny it. The network needs to guarantee that all nodes follow this attribute.

10.3.3 Availability

The channels on the bandwidth reserved for the communications must be available for all nodes to use, even if it is under some security attack. New nodes may arrive with new security messages, and the communication channel must be prepared to receive them. In this case, the network must have an operation policy, for instance.

10.3.4 Data Integrity

Data integrity is the preservation and assurance of the accuracy and consistency of data throughout the lifecycle of messages on the network. This term is broad in scope and can have widely different meanings, depending on the context.

10.3.5 Privacy

This attribute is similar to data integrity. However, the problem relies on whether other nodes accessed the packet-sensitive data. The network must prevent unauthorized access to the content. Networks may apply some encryption approaches, like public-key cryptography, for this attribute.

10.4 Authentication Schemes in Opportunistic Networks

In this section, we introduce the types of authentication schemes used in various types of computer systems and that can be adapted to opportunistic networks. Besides the fact that there are many proposals for the detection and defense of specific attacks in mobile ad hoc networks (MANETs) and wireless sensor networks (WSNs), there are authentication schemes that provide both detection and defense of various types of security attacks at the same time, ensuring the security requirements discussed earlier. Manvi and Tangade (2017) classify some authentication schemes as follows.

10.4.1 Signature-Based Schemes

Using a digital signature in an opportunistic network guarantees the requirements of authentication, data integrity, and nonrepudiation. However, digital signature schemes do not guarantee the privacy attribute. We can classify these schemes as single-user signature schemes or group signature.

10.4.2 Verification-Based Schemes

Verification is an essential element for schemes that also work with signatures. There are proposals made by researchers to verify message authentication codes (MACs) cooperatively; in other words, the nodes belonging to the network authenticate a fixed number of messages, reducing the overhead and time required for the calculations. We can classify these schemes as batch verification schemes or cooperative message authentication schemes.

10.4.3 Cryptography-Based Schemes

Public-key cryptography, which is a well-known Internet communication protocol, can also be used to authenticate messages in opportunistic networks. However, it is necessary to consider the computational cost of calculating the results. The symmetric approach, consisting of only a private key between the two points, is simple to perform the calculations but does not guarantee nonrepudiation, and if someone discovers the private key, the network might become unsafe. These schemes can be categorized as asymmetric cryptography schemes, public-key infrastructure certificate schemes, elliptic curve cryptography schemes, asymmetric cryptography schemes, MAC schemes, and hash function schemes, among others.

10.5 Security Attacks

In this section, we present the main security threats existing in the literature. These attacks affect nearly all previously discussed security attributes. However, security attacks in opportunistic networks are not limited only to those listed below.

10.5.1 Sybil Attack

The Sybil attack happens when an attacker uses their own equipment to simulate false nodes around it using another identity, called Sybil nodes. Thus, the Sybil nodes perform another type of security attack using their identity while the attacker node is behaving normally. Depending on the network security level, these identities can be

Generated IDs: The network allows the insider nodes to generate valid identities to exchange messages and does not have strict control over who uses them.

Stolen IDs: The attacker must find a way to get real identities from other nodes in order to deploy a Sybil node capable of message exchange in the network.

Figure 10.2 shows a scenario during a Sybil attack. The attacker node B is generating two Sybil nodes, C and E, that can actually communicate with other legitimate nodes and perform other security attacks without compromising node B.

The distributed approach for vehicular ad hoc networks (VANETs) proposed by Grover et al. (2010) uses each roadside unit (RSU) to calculate and store different parameter values about the nodes passing by them, during a given observation period. After that, the

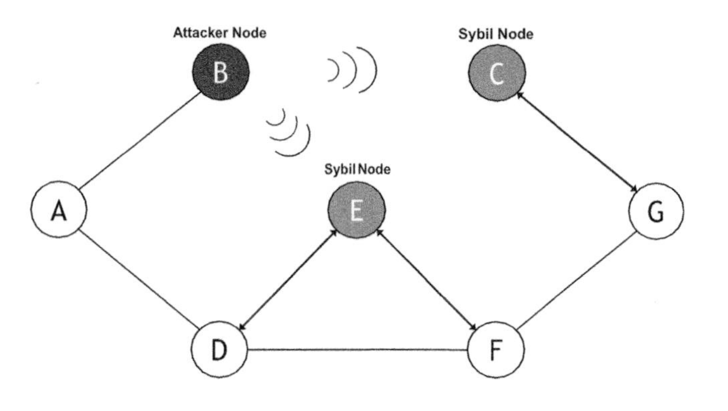

FIGURE 10.2
Sybil attack.

algorithm combines the values of the distance between the vehicle and the RSU, the signal strength, and the angle between the nodes and the RSU. If some nodes have same values for the parameters, the algorithm classifies these as Sybil nodes. However, they performed the simulations on a straight road with no curves, due to the calculations with the angles between the vehicles and the RSU.

In another approach, Grover et al. (2011) suggest that each node exchange groups of its neighboring nodes periodically and perform the intersection of these groups. If some nodes observe that they have similar neighbors for a significant duration of time, the algorithm classifies these similar neighbors as Sybil nodes. In this case, they evaluate the proposed approach on the realistic traffic scenario. The approach proposed by Kumar and Maheshwari (2014) uses an algorithm called priority batch verification algorithm (PBVA) to classify the requests obtained from multiple nodes and perform an immediate response to emergency nodes with less time delay. At the same time, the system also prevents Sybil attackers by restricting timestamps provided by RSUs.

10.5.2 Black Hole and Selective Forwarding Attack

An attacker node that drops any packets received characterizes the black hole attack. The attacker keeps broadcasting best routes for forwarding the packets to its direct neighbors to change the routes to it. However, the node drops all packets immediately, compromising network availability.

There are many variations of the execution of this security attack, divided into main three strategies:

Black hole attack: A single node performs the attack in the network, dropping all packets that arrived.

Cooperative black hole attack: Multiple nodes perform the attack at the same time. This is commonly the most used strategy.

Gray hole attack: Multiple nodes also perform the attack at the same time. However, the attacker chooses strategically which packets will be dropped.

Figure 10.3 illustrates the behavior of a cooperative black hole or gray hole if nodes B and E are connected, in some way, to switch to the attack mode. The nodes behind them do not receive the messages because the black hole nodes dropped them.

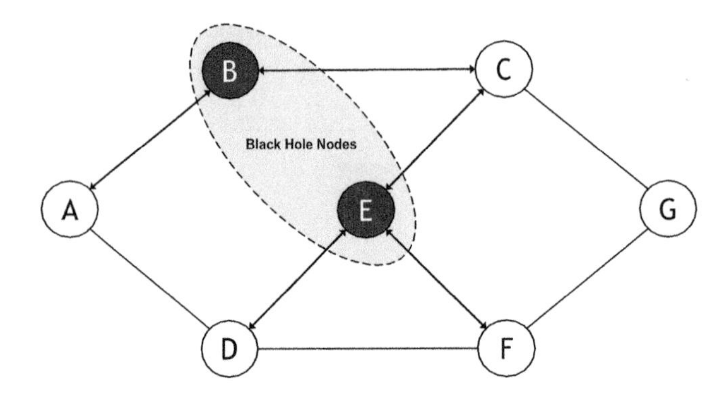

FIGURE 10.3
Cooperative black hole attack.

The approach presented by Alheeti et al. (2015a) suggests an intrusion detection system that uses artificial neural networks (ANNs) and fuzzed data to detect black hole attacks on the network. Their approach also uses the features extracted from the trace file as auditable data to detect the attack.

To avoid this attack, the approach proposed by Almutairi et al. (2014) suggests using a trusted table on each node, in order to evaluate the reliability of neighboring nodes during each step of the simulation. To evaluate their approach, they perform simulations using VANET Car Mobility Manager (VaCaMobil). However, this approach works only for a single black hole attack.

Based on an earlier approach, Alheeti et al. (2015b) used neural network concepts again to detect nodes that now are behaving as gray holes. The algorithm defined the output of the neural network in two possible results: normal or abnormal behavior. They used the Network Simulator 3 (NS-3) and Simulator of Urban MObility (SUMO) for the mobility models. The results obtained with this neural network were very promising with a low error rate.

10.5.3 Sinkhole Attack

The sinkhole attack is characterized by falsifying the quality of the route according to the routing algorithm to force the neighbors to forward their packets through it, and thus taking control of the packet flow of a certain network area. After achieving the desired packet flow control, the attacker node can start other types of network security attacks, such as selective forwarding (Section 10.5.2) or jellyfish (Section 10.5.5).

In Figure 10.4, we have an example of an opportunistic network using optimized link state routing protocol (OLSR) as the routing algorithm. The attacker node E sends HELLO messages stating that the quality of its route is the maximum possible (255), and because of that, its neighbors choose the attacker as their new multipoint relay (MPR).

In VANETs with position-based routing (PBR), Alsharif et al. (2011) proposed an algorithm that performs plausibility checks, that is, checks related to the distance traveled, current speed and density of the network, communication range, and current map location, in order to verify any abnormal behavior of vehicles. In addition, there are checks regarding the content of the message and the timestamp of the packet, and the communication channel is listened to to verify that the requested node actually forwarded the message. Called PBR-PC, the algorithm can completely mitigate the sinkhole attack in high-density

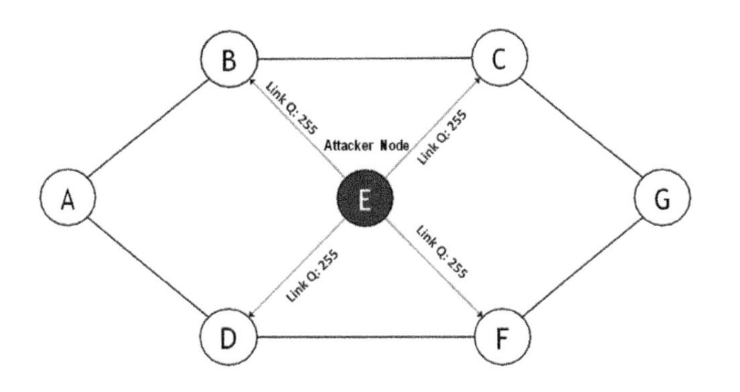

FIGURE 10.4
Sinkhole attack.

networks. In addition, they also simulated other attack approaches, such as routing loop, smart sinkhole, and wormhole (Section 10.5.4), and generated similar results.

10.5.4 Wormhole Attack

The wormhole attack is a variation of the sinkhole attack, although it uses two or more attackers to create packet tunnels to decrease the number of hops to reach distant nodes. This behavior causes neighbors to use the wormhole path to route their packets, allowing attackers to have control of the packet flow of a specific area of the network.

Hu et al. (2006) introduced a new wormhole detection protocol called TIK, which implements "leashes" on the network. Leashes are series of temporal rules to ensure the integrity of the network. For instance, a temporal leash characterizes a packet arriving at a destination earlier than expected. Using the leash information, the TIK protocol is able to authenticate communication between nodes and, at the same time, detect wormholes.

In Figure 10.5, node B wants to communicate with node F in an opportunistic network that uses the ad hoc on-demand distance vector (AODV) as the routing algorithm. When node B sends a RREQ message, the message passing through the wormhole first arrives at node F and the RREP message follows the same path, causing new packet routes to pass through the wormhole, and the attackers have control of part of the network.

10.5.5 Jellyfish Attack

The jellyfish attack on opportunistic networks is characterized by one or more attackers delaying and reordering the packets purposely, causing a decrease in quality of service on the network, especially in existing security services.

Chen et al. (2006) proposed an approach called throughput-feedback routing (TUF). First, the algorithm estimates and uses an approximate throughput value as the threshold to detect that a route has abnormal behavior. Even with the detection of false positives, the algorithm avoids attackers by creating new routes for these connections with low-throughput values, mitigating the effect of the attack. The results of the simulations show that in addition to mitigating jellyfish attacks, other availability attacks were avoided, such as rushing (Section 10.5.11) and denial of service (DoS) attacks (Section 10.5.9).

In Figure 10.6, node A sends a message informing something about the network situation. Upon receiving the message, the attacker node D purposely delays the forwarding for

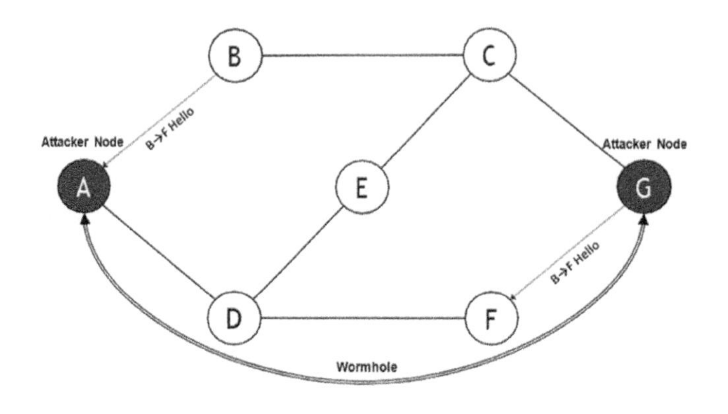

FIGURE 10.5
Wormhole attack.

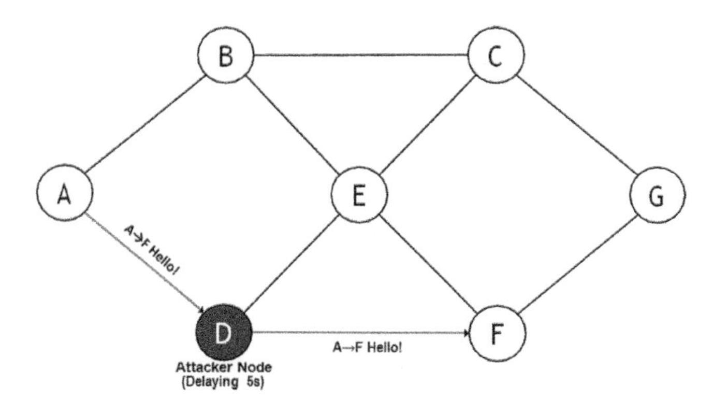

FIGURE 10.6
Jellyfish attack.

5 seconds. This delay may have already invalidated the message content, causing the node behind not to receive the correct information regarding the network situation.

10.5.6 GPS Spoofing Attack

GPS spoofing fools the GPS receivers present on nodes with false signals, and times differ from their actual physical locations and exact times (Larcom and Liu, 2013). Spoofing might be one of the most dangerous attacks in opportunistic networks because the set of mechanisms to detect this kind of attack must be very precise.

Figure 10.7 demonstrates the behavior of an attacker node on the network. The strong GPS signal generated by the antenna overlaps the satellite signal, allowing the attacker node E to send wrong locations to the legitimate nodes. Some nodes may still receive the signal from the satellites, but it implicates data integrity issues because they are receiving incoherent data at the same time.

The approach presented by Zhang et al. (2012) shows that an algorithm using a second receiver connected to a monopole antenna could help to detect fake signals of GPS and also detect the duration of the attack. The results demonstrated a quick detection compared with related work.

Using a base station equipped with multiple antennas, the approach proposed by Yan et al. (2015) utilizes channel observations to identify malicious nodes, also equipped with

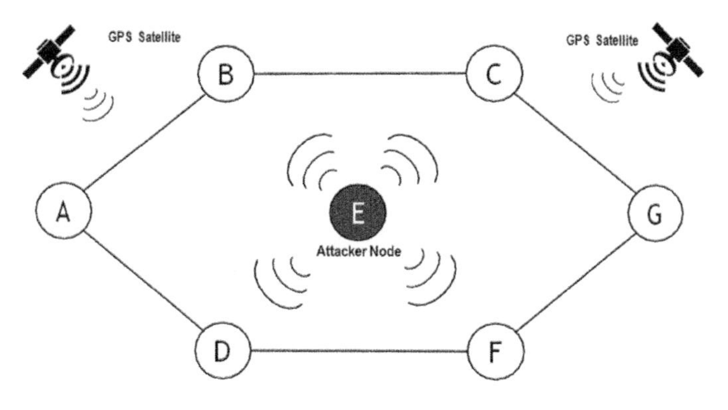

FIGURE 10.7
GPS spoofing attack.

multiple antennas, which spoof the GPS signal in their location. The analysis shows how the performance of a location spoofing detection system increases as the Rician K-factor of the channel between the base station and legitimate nodes increases.

Tippenhauer et al. (2011) identify from which locations and with what precision the attacker needs to generate its signals in order to spoof the receivers successfully. They also investigate the practical aspects of a satellite-lock takeover, when a victim node receives spoofed signals after first being locked from legitimate GPS signals.

10.5.7 Impersonation Attack

Whenever an attacker node on the network broadcasts security messages for their advantage or steals IDs from genuine nodes, this is an impersonation attack. It is important to have a trusted authentication system on the network to avoid this kind of attack.

The proposed scheme for VANETs made by Chhatwal and Sharma (2015) floods the beacon packets into the network to discover the presence of neighboring nodes and their accurate position. They manage the authentication scheme through VANET content fragile watermarking, which is a technique for hiding the information, as it prevents illegal manipulation of the content. They evaluate their proposal using simulations and analyzing the beaconing overhead, routing stretch, and node load.

An approach by Prathima et al. (2015) uses RSUs to take responsibility for checking for message integrity and authenticating the users who reduce the burden of individual nodes from authenticating each other. However, the approach does not guarantee the total detection of attackers with other identities. Using SUMO, they evaluate their scheme simulating various scenarios.

Ying and Nayak (2014) proposed an approach using an efficient authentication protocol (EAF), which employs smart cards based on the password of nodes and uses the dynamic login; the authors provide the anonymity of authentication. Figure 10.8 shows how the attacker node B would launch the attack. Sending messages to other nodes, the attacker pretends to be another node to get sensitive information.

10.5.8 Bogus Information Attack

The bogus information attack occurs when an attacker broadcasts multiple messages on the network to cause confusion in the system or damage other network nodes. The most viable solution is to check the occurrence of malicious data transmitted on the network

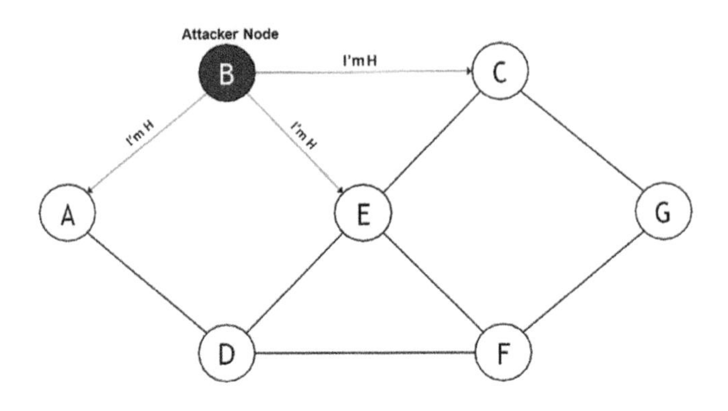

FIGURE 10.8
Impersonation attack.

to detect which node is performing the attack. Therefore, many approaches propose the use of MAC attached to the packets to validate their content, and that the attacker cannot generate false data without being authenticated in the network.

Figure 10.9 shows the behavior of an attacker node E performing a bogus information attack. According to its choices, it may send specific messages to the other nodes to obtain some benefit or just to cause damage.

The approach presented by Golle et al. (2014) suggests that every node present in the network must search for possible explanations for the data it has collected based on the fact that malicious nodes may be present. However, there is a cost involved in analyzing and characterizing the packet data, and there are privacy issues while having access to the packet data.

10.5.9 Distributed Denial of Service

The DoS attack in distributed form is very similar to attacks on websites nowadays, but in the case of opportunistic networks, the victim is a specific node that receives a huge amount of false data.

Pathre et al. (2013) proposed an approach that works similarly to the methods for the detection of bogus information attacks. Verify if the nodes of the network are receiving many packets with false information, in this case with the assistance of RSUs. With the effective detection of the attackers, they are marked for not participating more in the routing algorithm through a broadcast message done by the RSUs, losing authorization of belonging to the network.

A closer approach to practical application, Biswas et al. (2012) made a deeper analysis of the consequences of a distributed denial of service (DDoS) attack on the MAC layer of IEEE 802.11p (Han et al., 2012). Through simulations, they analyze the synchronization-based DDoS attack by a small group of attackers and present different mitigation techniques to avoid this attack.

The dissertation of Naik (2012) proposed a new offensive measure for the detection, mitigation, and prevention of this attack. It showed an authentication scheme using SYN and ACK messages to detect a DDoS attack based on the exchange messages between the nodes. The algorithm detects the DoS attack by modifying the characteristics of SYN messages during a TCP connection and analyzing when there is a flood of SYN messages and the storage structure is populated. To perform simulations, he used OMNeT++ for network simulation and SUMO for traffic simulation.

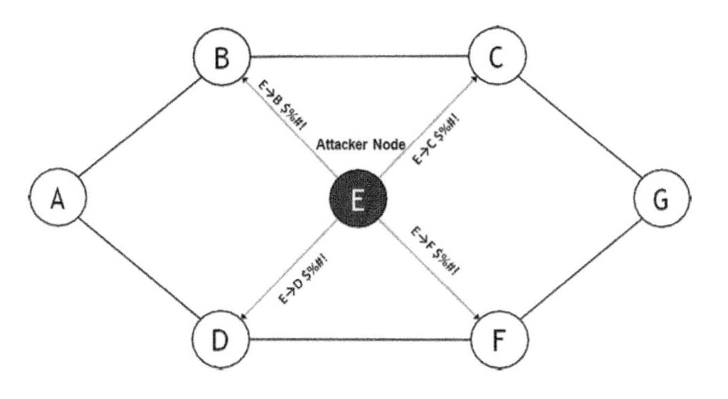

FIGURE 10.9
Bogus information attack.

Figure 10.10 shows an example of a DDoS attack. Several nodes belonging to the same attacker unite to perform the distributed attack on the victim node E. Upon receiving numerous security messages, the victim may make wrong decisions when dealing with these messages.

10.5.10 Man-in-the-Middle Attack

The man-in-the-middle attack happens when an attacker is able to change the data in a packet that is going through without corrupting the packet. With this capability, the attacker may create false scenarios inside the network without exposing their identity. Usually, the attacker is not aiming to harm the network directly, but is trying to get some benefit from the communication.

Figure 10.11 shows a simple scenario of an attack. An attacker node E receives a security message that makes claims about the network situation, but it changes the contents of the message by sending wrong information.

The approach presented by Ravi (2014) suggests that to protect the privacy of nodes and avoid this attack, one solution is to hide the identities of the nodes with the use of pseudonym keys, and whenever a node needs to communicate, it uses its pseudonym key to communicate with other nodes. They used Network Simulator 2 (NS-2) to perform the simulations with the proposed protocol.

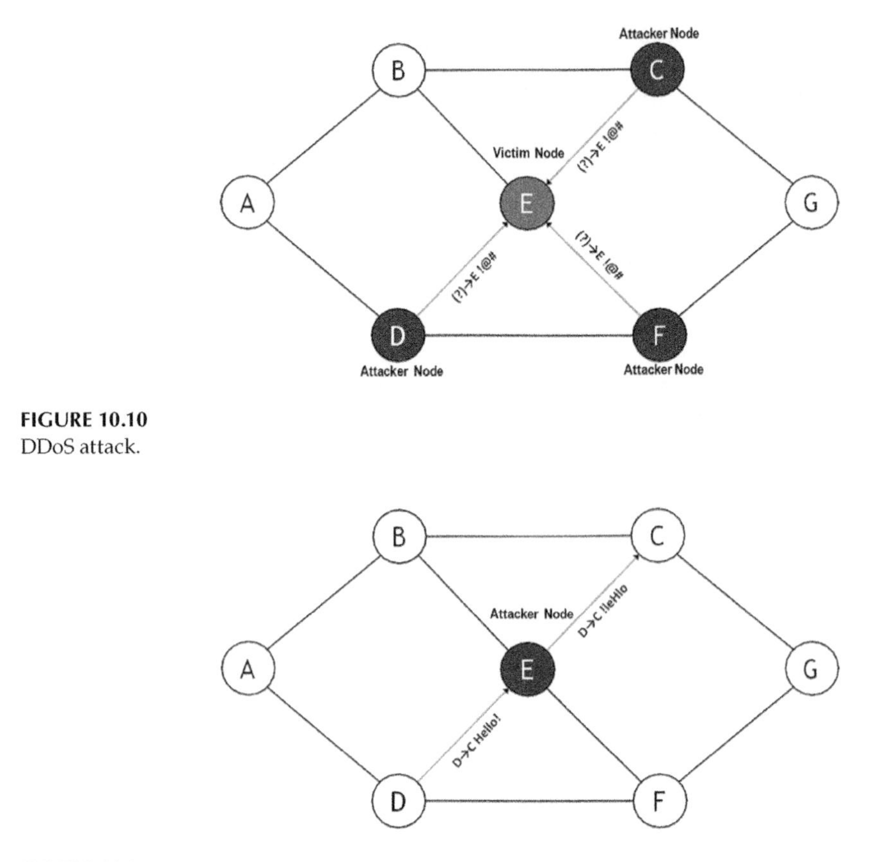

FIGURE 10.10
DDoS attack.

FIGURE 10.11
Man-in-the-middle attack.

The proposed scheme for vehicular networks by Zhang et al. (2008) is based on the premise that the RSUs are responsible for verifying the authenticity of the messages sent from vehicles and for notifying the results for every node while protecting the identity of users by preventing attackers from intercepting messages without being noticed.

Mejri et al. (2016) propose an approach for secure communication between vehicle platoons, using a private key shared with group members previously and a variation of the Diffie-Hellman algorithm to mitigate the attack of man in the middle. To evaluate their approach, they used NS-3 in conjunction with SUMO. The simulations varied the number of nodes belonging to the group and their speeds.

10.5.11 Rushing Attack

In on-demand routing protocol networks, a source node must flood nearby neighbors to find a valid route to the destination. Attackers can take advantage of this behavior and start their flooding discovery to get a valid route for this connection and so be able to initiate other types of attacks.

In Figure 10.12, node A wants to send a message to node G and starts flooding for path finding. When the attacker node B receives the message, it initiates its flood to get the packet route between the nodes and can execute other attacks. Upon receiving the legitimate message, node G drops the message because it has already received from the attacker.

An approach presented by Zapata and Asokan (2002) shows the SAODVs, which are two security mechanisms for the authentication of message exchanges of the routing algorithm AODV. The first one implements a digital signature to authenticate the nonmutable fields of the messages and uses hash chains to handle the hop count of the messages. These hash chains ensure that every node on the network is able to verify if an attacker has not changed the hop count.

Rushing attack prevention (RAP) is a defense method presented by Hu et al. (2003) that claims to protect the network nodes against rushing attacks. Several defense mechanisms are used together, such as secure neighbor detection, secure route delegation, and randomized "route request" forwarding.

The discovery of neighbors happens with a message handshake, in order to verify if the neighboring node is within a transmission range for communication. To validate the RAP, they perform the simulations to verify the packet delivery ratio by varying the attack models employed.

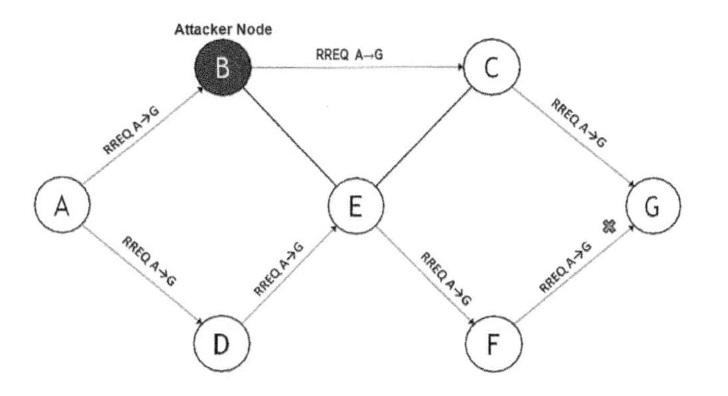

FIGURE 10.12
Rushing attack.

10.5.12 Colluding Misrelay Attack

In networks using OLSR as a routing algorithm, two attackers collude to disrupt the connection of a victim node. The first attacker informs the victim node whose path is closest to the communication, making it their new MPR. However, the second node can perform modifications in the packet, such as dropping them, modifying their content, and sending new TC messages to disrupt the communication.

A study by Kannhavong et al. (2007) shows the impact of a colluding misrelay attack on a mobile node network using OLSR as the routing algorithm. For the simulations, they used NS-2 in an environment with attackers sending false advertisements of two-hop neighbors while the second attacker drops TC messages. There has been a significant drop in the delivery ratio of messages in the environment without any mitigation method against this attack, which makes it a very important issue for future work.

In Figure 10.13, nodes A and G are initiating a connection, but the attackers are within the route chosen by the algorithm and can perform modifications and drop packets from the connection.

10.5.13 Link Spoofing Attack

In networks with OLSR as the routing algorithm, the HELLO and TC messages allow communication between the nodes, divulging to their near nodes their neighborhood state information. However, an attacker can generate HELLO and TC messages with false data, creating a different connection scenario and disrupting current communications, creating a link spoofing attack.

In Figure 10.14, the attacker node F informs by a TC message to nodes D, E, and G that node A is an active neighbor, but it is incorrect information and aims to disrupt the communications destined to node A.

An approach to mitigating the link spoofing attack proposed by Jeon et al. (2012) is to perform some modifications to the OLSR algorithm. The first modification is the range of HELLO messages, allowing them to be sent to the two-hop neighbors as well. All nodes have a trust table with their two-hop neighbors to manage the "trust flag" between these nodes for possible new MPRs. With these mechanisms, the method is able to mitigate attacks, such as adding nonexistent nodes and deleting node neighbor information.

However, we have another approach that does not imply making modifications to the OLSR algorithm, working based only on logs provided by the nodes according to the

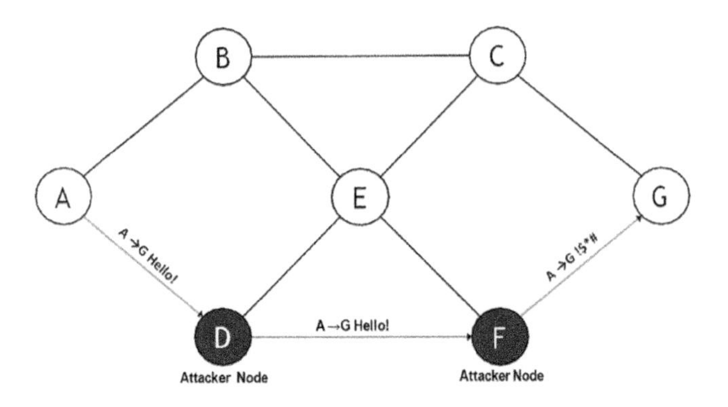

FIGURE 10.13
Colluding misrelay attack.

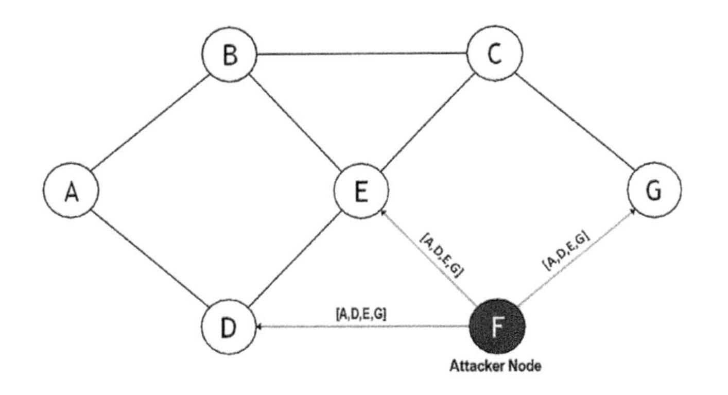

FIGURE 10.14
Link spoofing attack.

behavior of the network. Alattar et al. (2012) created a signature for a link spoofing attack so that the generated logs could detect it. A detection algorithm is presented with a series of rules for checking, using a cooperative trust system, and confidence levels. They validated the results using network simulations where attackers start the link spoofing attack. At the end of the rounds, the attackers obtained very low confidence values and the algorithm was able to detect them.

10.5.14 Snare Attack

The snare attack is a very specific attack where a group of nodes may contain important information coming from a node considered very important on the network (VIN). Attackers attempt to obtain the security information from any node of this network to become insiders to obtain the routing information to the VIN.

To combat this specific type of attack, a protocol called ASRPAKE was presented by Lin et al. (2007), providing anonymity for the route between the source and destination nodes and integrating a key exchange mechanism for the routing algorithm. In addition, they applied the DECOY mechanism to detect compromised nodes, while the VIN is not affected during network operation. For the anonymous route protocol, they use a ring signature scheme based on the elliptic curves cryptosystem (ECC).

In this scheme, any of the members can check if one of the members of the group has signed a given message, but this scheme has serious consequences on overhead issues. The DECOY engine allows the VIN to choose a number of nodes as a decoy, and they all share a cryptographic key with the VIN. Whenever the VIN receives a route request, VIN randomly chooses one of its decoys to respond to the request, protecting the node from possible security attacks.

In Figure 10.15, a compromised node A is used by the attackers as an insider to find the routing path to the VIN, making the entire network compromised.

10.5.15 Blackmail Attack

In opportunistic networks implementing trusted tables, it is possible for a group of attackers present on the network to inform neighboring nodes that a particular node is an intruder on the network in order to change the trust table of neighboring neighbors and thus deny connectivity to the victim. This security attack is known as a blackmail attack.

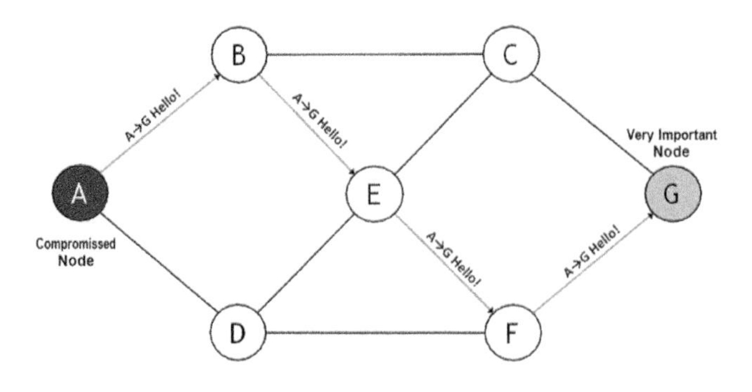

FIGURE 10.15
Snare attack.

Perrig et al. (2002) propose a series of new protocols for the security of sensors networks, including the simple NDEF exchange protocol, used for data confidentiality, authentication, integrity, and freshness, and μTESLA, providing authenticated broadcast.

In Figure 10.16, two attackers send to their neighbors one message stating that the victim node is in their blacklists, even though they did not have any communication before. This behavior transforms into a cascading effect that denies the network service to node E.

10.5.16 Node Replication Attack

In authenticated opportunistic networks, there is the possibility of a compromised node and all its security data, such as cryptographic keys and identity, being lost. Attackers might create clone nodes in the network using this data, and the clones can normally communicate even on an authenticated network.

In Figure 10.17, node A was compromised and its security data leaked. Attackers with this information can create clones of node A and communicate with the other nodes within the authenticated network.

Because it is a very common attack on WSNs, many works have implemented mitigation methods for this scenario. Parno et al. (2005) present a centralized proposal where the base station verifies the location of all the nodes of the network to find clones. After that, the nodes use several distributed methods using broadcast, deterministic multicast, and random multicast to send advertisement messages.

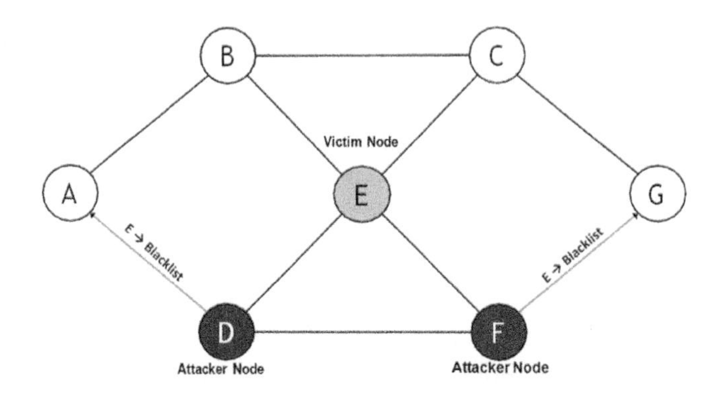

FIGURE 10.16
Blackmail attack.

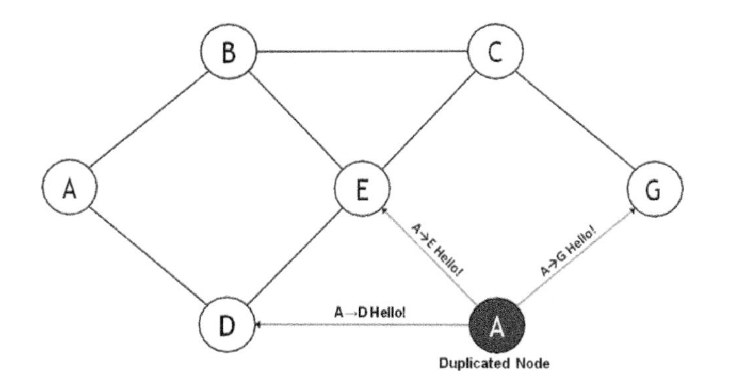

FIGURE 10.17
Node replication attack.

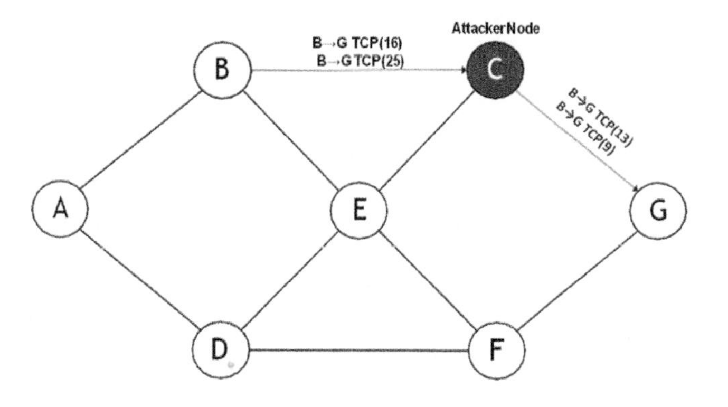

FIGURE 10.18
Desynchronization attack.

10.5.17 Desynchronization Attack

In a desynchronization attack, the attacker is on the route of communication between the nodes and performs modifications in the transport layer, forging the sequence number of the packets and changing control flags. Trying to retransmit their messages, the nodes cannot exchange useful messages and have their connection completely interrupted.

In Figure 10.18, nodes B and G attempt to create a connection between them, although the attacker node C keeps changing the sequence number of the packets, disrupting the communication.

10.6 Conclusion

In this chapter, we presented a taxonomy of security attacks in opportunistic networks. We presented the security attributes, and several attacks have also been discussed with approaches to prevent them. In Table 10.1, the security attacks are presented according to the security attribute that is affected.

Several research works are growing on this subject. Possible future work will be the inclusion of environment context analysis to help infer and detect attackers of any type,

TABLE 10.1

Taxonomy of Security Attacks and Their Attributes

Security Attribute	Security Attack
Authentication	Impersonation Sybil Snare Blackmail Node replication
Availability	Black hole Sinkhole Wormhole Jellyfish DDoS Rushing Link spoofing
Data integrity	Bogus information Man in the middle GPS spoofing Colluding misrelay Desynchronization

choosing the appropriate metrics where the context-aware methods might work better in different scenarios.

Researchers are suggesting other types of architectures to avoid certain security attacks, such as applying software-defined network concepts, centralizing the control of data flow but resulting in an increase of data overhead through the network.

References

Alattar, M., Sailhan, F., and Bourgeois, J. (2012, June). Trust-enabled link spoofing detection in MANET. In *2012 32nd International Conference on Distributed Computing Systems Workshops*. Macau, pp. 237–244. doi: 10.1109/ICDCSW.2012.27.

Alheeti, K. M. A., Gruebler, A., and McDonald-Maier, K. D. (2015a). An intrusion detection system against black hole attacks on the communication network of self-driving cars. In *2015 Sixth International Conference on Emerging Security Technologies (EST)*. Braunschweig, pp. 86–91. doi: 10.1109/EST.2015.10.

Alheeti, K. M. A., Gruebler, A., and McDonald-Maier, K. D. (2015b). On the detection of grey hole and rushing attacks in self-driving vehicular networks. In *2015 7th Computer Science and Electronic Engineering Conference (CEEC)*. Colchester, pp. 231–236. doi: 10.1109/CEEC.2015.7332730.

Almutairi, H., Chelloug, S., Alqarni, H., Aljaber, R., Alshehri, A., and Alotaish, D. (2014). A new black hole detection scheme for VANETs. In *Proceedings of the 6th International Conference on Management of Emergent Digital Ecosystems*. ACM, New York, NY, Article 23, 6 pages. doi: 10.1145/2668260.2668262.

Alsharif, N., Wasef, A., and Shen, X. (2011, June). Mitigating the effects of position-based routing attacks in vehicular ad hoc networks. In *2011 IEEE International Conference on Communications (ICC)*. Kyoto, pp. 1–5. doi: 10.1109/icc.2011.5962855.

Biswas, S., Mii, J., and Mii, V. (2012). DDoS attack on wave-enabled VANET through synchronization. In *Global Communications Conference (GLOBECOM), 2012 IEEE*. Anaheim, CA, pp. 1079–1084. doi: 10.1109/GLOCOM.2012.6503256.

Chen, R., Snow, M., Park, J. M., Refaei, M. T., and Eltoweissy, M. (2006, Nov). Nis02-3: Defense against routing disruption attacks in mobile ad hoc networks. In *IEEE GLOBECOM 2006*. San Francisco, CA, pp. 1–5. doi: 10.1109/GLOCOM.2006.269.

Chhatwal, S. S. and Sharma, M. (2015). Detection of impersonation attack in VANETs using buck filter and VANET content fragile watermarking (VCFW). In *2015 International Conference on Computer Communication and Informatics (ICCCI)*. Coimbatore, pp. 1–5. doi: 10.1109/ICCCI.2015.7218093.

Golle, P., Greene, D., and Staddon, J. (2014). Detecting and correcting malicious data in VANETs. In *Proceedings of the 1st ACM International Workshop on Vehicular Ad Hoc Networks*. ACM, New York, NY, 29–37. doi: 10.1145/1023875.1023881.

Grover, J., Gaur, M. S., and Laxmi, V. (2010). A novel defense mechanism against Sybil attacks in VANET. In *Proceedings of the 3rd International Conference on Security of Information and Networks*. ACM, New York, NY, 249–255. doi: 10.1145/1854099.1854150.

Grover, J., Gaur, M. S., Laxmi, V., and Prajapati, N. K. (2011). A Sybil attack detection approach using neighboring vehicles in VANET. In *Proceedings of the 4th International Conference on Security of Information and Networks*. ACM, New York, NY, 151–158. doi: 10.1145/2070425.2070450.

Han, C., Dianati, M., Tafazolli, R., Kernchen, R., and Shen, X. (2012). Analytical study of the IEEE 802.11p MAC sublayer in vehicular networks. *IEEE Transactions on Intelligent Transportation Systems*, Vol. 13, No. 2, pp. 873–886. doi: 10.1109/TITS.2012.2183366.

Hu, Y.-C., Perrig, A., and Johnson, D. B. (2003). Rushing attacks and defense in wireless ad hoc network routing protocols. In *Proceedings of the 2nd ACM Workshop on Wireless Security*. ACM, New York, NY, pp. 30–40. doi: 10.1145/941311.941317.

Hu, Y.-C., Perrig, A., and Johnson, D. B. (2006, February). Wormhole attacks in wireless networks. *IEEE Journal on Selected Areas in Communications*, Vol. 24, No. 2, pp. 370–380. doi: 10.1109/JSAC.2005.861394.

Jeon, Y., Kim, T. H., Kim, Y., and Kim, J. (2012, October). LT-OLSR: Attack-tolerant OLSR against link spoofing. In *37th Annual IEEE Conference on Local Computer Networks*. Clearwater, FL, pp. 216–219. doi: 10.1109/LCN.2012.6423612.

Kannhavong, B., Nakayama, H., and Kato, N. (2007). A study of colluding misrelay attacks in OLSR MANET. In *Tohoku-Section Joint Convention Record of Institutes of Electrical and Information Engineers* Japan, p. 28. Retrieved from http://ci.nii.ac.jp/naid/130005444492/en/. doi: 10.11528/tsjc.2007.028.

Kumar, P. V. and Maheshwari, M. (2014). Prevention of Sybil attack and priority batch verification in VANETs. In *2014 International Conference on Information Communication and Embedded Systems (ICICES)*. Chennai, pp. 1–5. doi: 10.1109/ICICES.2014.7033926.

Larcom, J. A. and Liu, H. (2013). Modeling and characterization of GPS spoofing. In *2013 IEEE International Conference on Technologies for Homeland Security (HST)*. Waltham, MA, pp. 729–734. doi: 10.1109/THS.2013.6699094.

Lin, X., Lu, R., Huafei, Z., Ho, P. H., Shen, X., and Cao, Z. (2007, June). ASRPAKE: An anonymous secure routing protocol with authenticated key exchange for wireless ad hoc networks. In *2007 IEEE International Conference on Communications*. Glasgow, pp. 1247–1253. doi: 10.1109/ICC.2007.211.

Manvi, S. S. and Tangade, S. (2017). A survey on authentication schemes in VANETs for secured communication. *Vehicular Communications*, Vol. 9, pp. 19–30. doi: 10.1016/j.vehcom.2017.02.001.

Mejri, M. N., Achir, N., and Hamdi, M. (2016, January). A new group Diffie-Hellman key generation proposal for secure VANET communications. In *2016 13th IEEE Annual Consumer Communications Networking Conference (CCNC)*. Las Vegas, NV, pp. 992–995. doi: 10.1109/CCNC.2016.7444925.

Naik, M. (2012). Early detection and prevention of DDoS attack in VANET. (Master's Dissertation, National Institute of Technology, Odisha, India) Retrieved from http://ethesis.nitrkl.ac.in/6817/1/Early_Naik_2015.pdf.

Parno, B., Perrig, A., and Gligor, V. (2005, May). Distributed detection of node replication attacks in sensor networks. In *2005 IEEE Symposium on Security and Privacy (Euro S&P'05)*. pp. 49–63. doi: 10.1109/SP.2005.8.

Pathre, A., Agrawal, C., and Jain, A. (2013). A novel defense scheme against DDoS attack in VANET. In *2013 Tenth International Conference on Wireless and Optical Communications Networks (WOCN)*. Bhopal, pp. 1–5. doi: 10.1109/WOCN.2013.6616194.

Perrig, A., Szewczyk, R., Tygar, J., Wen, V., and Culler, D. E. (2002, September). SPINS: Security protocols for sensor networks. *Wireless Networks*, Vol. 8, No. 5, pp. 521–534. doi: 10.1023/A:1016598314198.

Prathima, P., Rajendiran, K., Ranjani, G. S., Kurian, P., and Swarupa, S. (2015). Simple and flexible authentication framework for vehicular ad hoc networks. In *2015 International Conference on Communications and Signal Processing (ICCSP)*. Melmaruvathur, pp. 1176–1180. doi: 10.1109/ICCSP.2015.7322690.

Ravi, R. J. (2014). Privacy protection against man-in-the-middle attacks in vehicular ad hoc networks. *International Journal of Advanced Research in Electrical, Electronics and Instrumentation Engineering*, Vol. 3, No. 4, pp. 9035–9042.

Raya, M. and Hubaux, J.-P. (2005). The security of vehicular ad hoc networks. In *Proceedings of the 3rd ACM Workshop on Security of Ad Hoc and Sensor Networks*, ACM, New York, NY, pp. 11–21. doi: 10.1145/1102219.1102223.

Tippenhauer, N. O., Pöpper, C., Rasmussen, K. B., and Capkun, S. (2011). On the requirements for successful GPS spoofing attacks. In *Proceedings of the 18th ACM Conference on Computer and Communications Security*. ACM, New York, NY, pp. 75–86. doi: 10.1145/2046707.2046719.

Yan, S., Malaney, R., Nevat, I., and Peters, G. W. (2015). Location spoofing detection for VANETs by a single base station in Rician fading channels. In *2015 IEEE 81st Vehicular Technology Conference (VTC Spring)*. Glasgow, pp. 1–6. doi: 10.1109/VTCSpring.2015.7145917.

Ying, B. and Nayak, A. (2014). Efficient authentication protocol for secure vehicular communications. In *2014 IEEE 79th Vehicular Technology Conference (VTC Spring)*. Seoul, pp. 1–5. doi: 10.1109/VTCSpring.2014.7022900.

Zapata, M. G. and Asokan, N. (2002). Securing ad hoc routing protocols. In *Proceedings of the 1st ACM Workshop on Wireless Security*. ACM, New York, NY, USA, pp. 1–10. doi: 10.1145/570681.570682.

Zhang, C., Lin, X., Lu, R., and Ho, P. H. (2008). Raise: An efficient RSU-aided message authentication scheme in vehicular communication networks. In *2008 IEEE International Conference on Communications*. Beijing, pp. 1451–1457. doi: 10.1109/ICC.2008.281.

Zhang, Z., Trinkle, M., Qian, L., and Li, H. (2012). Quickest detection of GPS spoofing attack. In *2012 IEEE Military Communications Conference*. Orlando, FL, pp. 1–6. doi: 10.1109/MILCOM.2012.6415722.

11

Pervasive Trust Foundation for Security and Privacy in Opportunistic Resource Utilization Networks

Ahmed Al-Gburi, Abduljaleel Al-Hasnawi, Raed Salih, and Leszek Lilien

CONTENTS

11.1 Introduction 213
11.2 Background Information on Oppnets 214
    11.2.1 Oppnet Structure and Operations 214
    11.2.2 Oppnet Helpers and Their Categories 216
    11.2.3 Oppnet Security and Privacy Challenges 217
11.3 Background Information on Pervasive Trust Foundation (PTF) 217
11.4 Literature Review 218
    11.4.1 Concepts Related to Oppnets 218
    11.4.2 Pervasive Trust Foundation for Security and Privacy 219
    11.4.3 Other Related Work 219
11.5 Proposed Solution: PTF-Based Oppnet Architecture (POA) 220
    11.5.1 POA Structure 221
    11.5.2 POA Operations 222
11.6 Conclusions 225
Key Terminology and Definitions 225
References 226
Bibliography for Advanced/Further Reading 227

11.1 Introduction

Coexisting heterogeneous computing systems too often are unable to collaborate, wasting resources of underutilized systems while encountering resource limits in busy systems. This is particularly visible when an overburdened system is a neighbor of an underutilized system, and the latter is unable to "help" the former.

To deal with this situation, the paradigm and technology of *opportunistic resource utilization networks* (oppnets) was introduced by Lilien et al. (2006a). The distinctive feature of oppnets is their use of "helpers" that assist in reaching the goals of oppnets. With helpers, oppnets expand in an opportunistic and ad hoc manner by acquiring more needed resources of any kind; these are resources from, among others, underutilized neighbors.

Other kinds of opportunistic networks known in the literature are a proper subset of oppnets, since opportunism in them is usually limited to a subset of oppnet capabilities (such as only opportunistic communication capabilities or opportunistic message forwarding). The following characteristics taken together distinguish oppnets from other opportunistic and collaborative systems: (1) support for the *helper* paradigm—oppnet growth and expansion based on the idea of finding helpers that can assist in reaching the goals of the oppnets; (2) *opportunistic* use of all kinds of computing *resources*—not limited, for example, to communication resources; (3) *ad hoc* operation—for most of the oppnet lifetime (as described below, preceded by predefined operation—as a seed oppnet—at the beginning of the oppnet lifetime); (4) *universality* of the helper nodes—regardless of the system or device make or function, and through any communication media, protocols, and so forth; and (5) *lack of* third-party *mediators*—since interactions among oppnets and their helpers can take place without third parties.

Security and privacy, the major challenges in any distributed or pervasive system, are critical in oppnets due to their highly heterogeneous nature. oppnets can be viewed as middleware for generalized ad hoc networks, so they inherit all security and privacy issues of ad hoc networks. In addition, oppnets introduce their own specific problems. We believe that *trust* should underlie all security and privacy controls (solutions)—if utilized effectively and efficiently. We use the paradigm of the *pervasive trust foundation* (*PTF*) (Lilien et al., 2009, 2010a), which was proposed, among others, for pervasive computing systems and next-generation networks (such as the Internet of Things).

We outline a new architecture for oppnets called *PTF-based oppnet architecture* (*POA*). POA integrates PTF into the oppnet structure and operations, placing PTF under the control of oppnet's supervisor, called *decentralized oppnet controller* (DOC). PTF provides to oppnet all security and privacy services during the entire oppnet lifetime. Among others, PTF enforces secure and privacy-preserving handling of data transmitted between oppnet entities and verifies that a helper joining the oppnet is not malicious.

In this chapter, we propose using trust (in the form of the PTF paradigm) as a basis for assuring security and privacy in oppnets. The chapter is organized as follows. Section 11.2 provides background information on oppnets, including a discussion of the security and privacy challenges in oppnets. Section 11.3 overviews PTF. Section 11.4 discusses related work, including solutions related to oppnets and using trust as a security and privacy foundation. Section 11.5 describes the structure and operations of POA. Section 11.6 concludes the chapter.

11.2 Background Information on Oppnets

Opportunistic resource utilization networks (oppnets) are a paradigm for a special kind of networks that grows by discovering *foreign entities* (systems or devices that are not a part of the current oppnet) that possess desired resources and integrating them into oppnets as *helpers* (Lilien et al., 2006a). This expansion by finding helpers in an opportunistic and ad hoc way is the most distinctive feature of oppnets.

A *potential helper* is a foreign entity that is oppnet-enabled, that is, has enough communication capabilities to be contacted by an oppnet searching for helpers. Once invited by the oppnet, it can join the oppnet as an *(actual) helper* and, due to the possession of certain resources, help the oppnet in its job (Lilien et al., 2006a, 2007).

11.2.1 Oppnet Structure and Operations

This section shows the basic oppnet structure and its activities and operations.

Oppnet Structure: oppnet consists of three major types of components or nodes (Lilien et al., 2007, 2010b; Kamal et al., 2008), as shown in Figure 11.1.

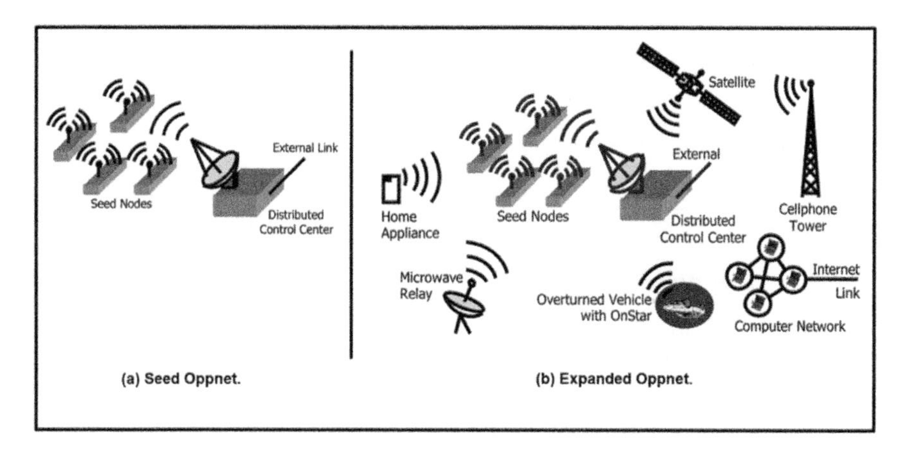

FIGURE 11.1
Oppnet growth from (a) a seed oppnet into (b) an expanded oppnet. (From Lilien, L., et al., Opportunistic networks for emergency applications and their standard implementation framework, in *IEEE International Performance, Computing, and Communications Conference (IPCCC)*, New Orleans, LA, 2007, 588–593.)

1. *Seed nodes*: A set of oppnet nodes constituting the *seed oppnet*, employed together at the time of the initial oppnet deployment. It might be very small—in the extreme, a single node.
2. *Control center (CC) nodes*: A subset of oppnet nodes together implementing a distributed CC, responsible for managing all interactions and processing, including oppnet growth and shrinking (e.g., when removing malicious nodes from an oppnet).
3. *Helper nodes*: Nodes that were originally foreign but became integrated into an oppnet to assist it in reaching its desired goals. For example, the *expanded oppnet* in Figure 11.1b admitted the following helpers (counterclockwise from bottom left): (a) a computer network, wired to the Internet; (b) a cellphone infrastructure (represented by the cellphone tower), contacted via the oppnet's cellphone peripheral; (c) a satellite, contacted via a direct satellite link; (d) a home area network, contacted via an intelligent appliance (e.g., a refrigerator) with a wireless link; (e) a microwave network, contacted via a microwave relay; and (f) body area networks (BANs) of the occupants of an overturned car, contacted via OnStar (Lilien et al., 2007).

Basic Oppnet Operations: The CC manages growth of a seed oppnet (Figure 11.1a) into an expanded oppnet (Figure 11.1b). The process involves looking for useful foreign entities and admitting them to help in the realization of the oppnet goals. An entity that accepts an invitation to help makes an obligation to keep on helping as long as the oppnet needs it, and becomes a helper. When a helper completes its tasks, it is released from its obligation by the oppnet and is free to return to its original duties (Lilien et al., 2010b).

The basic sequence of oppnet operations is summarized in Figure 11.2 (Lilien et al., 2010b). A distributed CC (a.k.a decentralized command center) supervises the operations of its oppnet throughout the entire oppnet's life. If an oppnet needs more resources to achieve its goals, the cycle of oppnet growth (including discovering, evaluating, admitting, and integrating candidate helpers) is repeated. Once the goals of an oppnet are achieved, the oppnet assists the helpers in restoring the state they were in before joining the oppnet and releases them (Lilien et al., 2006a, 2007, 2010b).

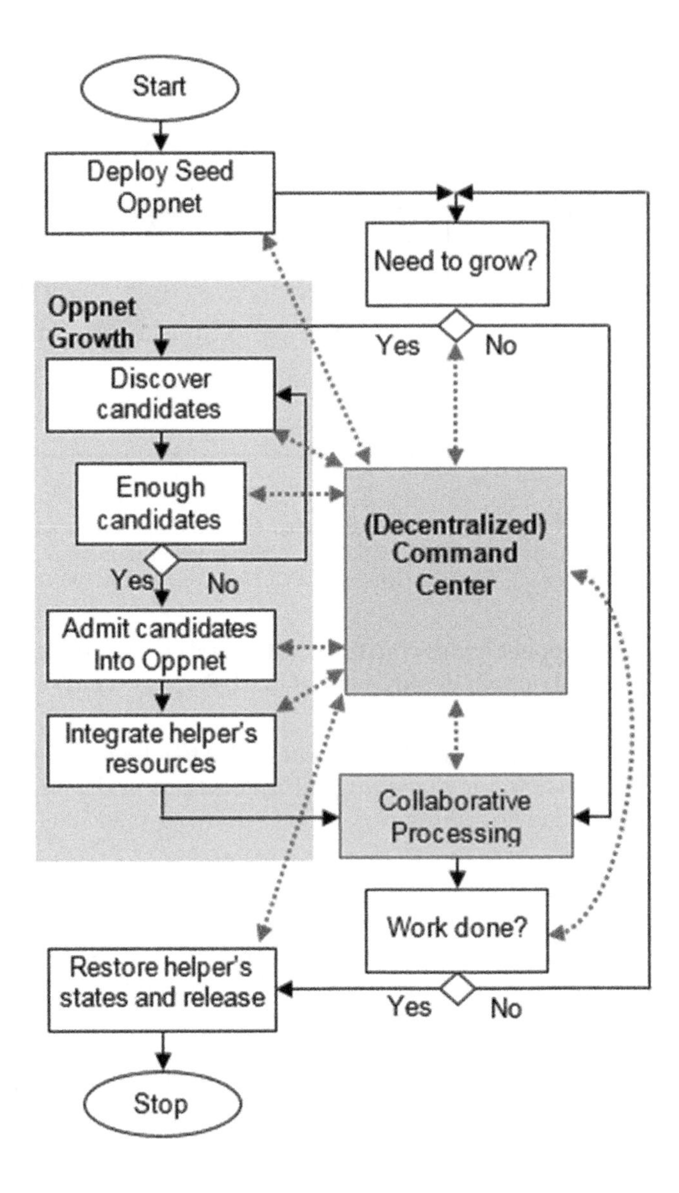

FIGURE 11.2
Basic oppnet operations (cf. Lilien et al., 2010b).

11.2.2 Oppnet Helpers and Their Categories

An entity that is sufficiently capable of downloading and installing a minimal subset of oppnet middleware modules necessary to contact oppnets becomes *oppnet-enabled*. Any oppnet-enabled entity is a potential helper (Lilien et al., 2010b). A helper can possess diverse capabilities and "skills" potentially useful for the oppnets, such as specialized software, communication, computing, storing, sensing, or actuating capabilities. Oppnet helpers generally can be classified into three main categories (Lilien et al., 2010b; Alduailij and Lilien, 2015):

1. *Potential helpers*: A set of oppnet-enabled devices within the reach of an oppnet, which may serve it as future helpers.

2. *Candidate helpers*: A subset of potential helpers that meet the needs of an oppnet duty, which are asked or ordered to join an oppnet (in ordinary or critical situations, respectively).
3. *Actual helpers*: A subset of the candidate helpers selected and admitted by an oppnet. Actual helpers (usually referred to simply as helpers) include two subcategories (Lilien et al., 2006a, 2007; Alduailij and Lilien, 2015):
   a. *Regular helpers*: Can discover, invite, and admit other helpers; they provide significant resources to the oppnet that admitted them.
   b. *Lightweight helpers* (a.k.a *lites*): Have very limited resources, including very restricted communication capabilities (usually due to their weak native hardware/software capabilities).

11.2.3 Oppnet Security and Privacy Challenges

Providing controls for security and privacy challenges is crucial for oppnets, oppnet-enabled applications, and their users. In fact, Lilien et al. (2006b, 2010b) have gone as far as referring to the security and privacy challenges as the "make it or break it" issue. oppnets, like most other nontrivial technologies, can be malevolent—deployed to harm humans, their artifacts, and their technical infrastructures.

Oppnets can be viewed as middleware for generalization of ad hoc networks or systems, and, therefore, they inherit all security and privacy challenges of ad hoc networks. Lilien et al. (2006a) presented six security and privacy challenges that must be investigated to ensure security and privacy for oppnets and their sensitive data. These challenges are: (1) increasing trust and secure routing, (2) helper privacy and oppnet privacy, (3) protecting data privacy, (4) ensuring data integrity, (5) identifying the most dangerous attacks and sketching solutions, and (6) intrusion detection.

We concentrate on protecting the security and privacy of data in oppnets. We consider various kinds of threats and attacks, including both *outsider attacks*, from the outside of an oppnet, and *insider attacks*, from within an oppnet. Attacks include denial of service (DoS), eavesdropping, information leakage, ID spoofing, man in the middle (MITM) attacks, black hole and packet dropping (gray hole) attacks, and wormholes (Lilien et al., 2006a, 2006b). For instance, MITM may occur when a node sends a help request to rescuers and a malicious node alters this help message.

An oppnet should also prevent a malicious entity from joining as a helper. In addition, any foreign or oppnet entity may either maliciously or inadvertently reveal or destroy sensitive data. Since motivations of attackers cannot be known for sure, even inadvertent disclosures of sensitive data—for example, not malicious but just selfish (Buttyan and Hubaux, 2007)—must be treated as an attack.

11.3 Background Information on Pervasive Trust Foundation (PTF)

The PTF paradigm (Lilien et al., 2009, 2010a) uses trust as the basis for security and privacy in computing systems. PTF involves two major principles. First, trust should be used as a foundation for security and privacy (as well as many other services not considered in this chapter, such as collaboration or routing). Second, trust should be considered as pervasive in computing systems, and must be considered either *explicitly,* in large or open systems (such as oppnets), or *implicitly,* in small or closed systems.

A PTF module processes trust information for its oppnet. The trust value that Entity O (e.g., oppnet's CC) currently has for Entity H (e.g., an actual helper) determines the set of actions that Entity H is authorized by Entity O to perform. Entity O calculates the trust value for Entity H based on Entity O's direct experience with Entity H and reputation information (indirect experience) obtained by Entity O about Entity H from other entities.

The trust level for an oppnet entity must be reevaluated with an appropriate frequency since it changes dynamically over time.

11.4 Literature Review

This section discusses concepts related to oppnets, work that considers trust a foundation for security and privacy, and other related work.

11.4.1 Concepts Related to Oppnets

Pelusi et al. (2006) and Conti et al. (2010) consider opportunistic networks as one of the most interesting evolutions of mobile ad hoc networks (MANETs), characterized by self-organizing networking and the store-carry-and-forward paradigm. In opportunistic networks, mobile nodes are enabled to communicate with each other even if a route connecting them never exists. Specifically, node mobility is exploited as an opportunity to deliver data among disconnected parts of a network. Therefore, routes are established dynamically. Messages are routed between a sender and destinations, and any potential intermediate node can be used opportunistically as the next hop, provided it is likely to bring the message closer to the final destination. The authors indicate that node mobility plays a crucial role as it permits the bridging of disconnected clouds of nodes, and ultimately enables end-to-end communications despite connectivity weakness. They envision employing opportunistic networks in the next-generation Internet.

Alajeely et al. (2016) define opportunistic networks (referring to them as "OppNets") as an evolution of MANETs in which many wireless nodes communicate opportunistically with each other by means of store-carry-and-forward. A variety of wireless technologies can be used to transfer data, including WiFi, WiMAX, and Bluetooth. No end-to-end connection is established between the source and the destination nodes before sending messages; the nodes usually have high mobility, low density, limited power, and short radio ranges. Due to these features, OppNets face serious security and privacy issues, such as challenges for node authentication and access control; data confidentiality, integrity and availability; and routing, security, privacy, and trust management.

Overall, we believe that oppnets—as defined earlier (Lilien et al., 2006a)—are significantly different from opportunistic networks—as defined above by other researchers, such as Pelusi et al. (2006), Conti et al. (2010), Alajeely et al. (2016). Most importantly, oppnets enable not only opportunistic communications but also opportunistic use of all kinds of resources, services, or capabilities (including hardware, software, and human skills) possessed by foreign entities that happen to be within the oppnet's reach. Another distinctive feature is that an oppnet starts as a relatively small predefined network (a seed oppnet) that grows into an expanded oppnet.

11.4.2 Pervasive Trust Foundation for Security and Privacy

Bhargava et al. (2004) introduce the concept of pervasive trust. They recognize that using trust in computing systems is already common. Trust is and should be pervasive in many interactions between people, organizations, animals, and even artifacts. Hence, *pervasive trust* should also be considered in computing environments. The authors argue that trust can support security and privacy, facilitating solutions.

Lilien et al. (2009, 2010a) propose the PTF paradigm for next-generation networks, including the Future Internet. They claim that trust can also be used to support many other functions, for example, routing, data aggregation, time synchronization, and even stimulating cooperation in autonomous wireless networks. The authors claim that: (1) security without trust (not based on trust) is more difficult to achieve than security with trust (based on trust); (2) confusing narrowly defined trust with broadly defined trust leads to denying the need for a pervasive trust foundation; (3) ignoring trust leads to high risks; (4) trust can be used implicitly only after making a conscious decision that there is a sufficient trust level; and (5) using broadly defined trust is necessary for comprehensive and consistent consideration of trust by any security service serving any network layer.

11.4.3 Other Related Work

Network Function Virtualization (NFV): The NFV paradigm separates software implementation of network functions from the underlying hardware by leveraging virtualization technologies and programmable hardware. Networks using NFV improve the flexibility of network services by provisioning and reducing the time to market for new services (Han et al., 2015).

NFV opens the door to over-the-top (OTT) services, in which networks may appear substantially different to their users from their underlying infrastructure. For example, a network provider can provide a single global network view with consistent security policies, even though it does not have a presence in all countries; it is done by virtualizing local networks and clouds for those countries. Incorporation of computing and storage resources in the guts of the network itself enables rethinking what constitutes a network interface, namely, going beyond a "bit pipe" (Joshi and Benson, 2016); this results in considering flexible service chains, store-and-forward content distribution services, network-based Big Data analytics, and sensory data aggregation services (Joshi and Benson, 2016).

We believe that the solutions outlined in this chapter can be extended to NFVs. As an example, the idea of oppnet-enabled entities could be used in NFVs.

Software-Defined Networks (SDN): A SDN is a network virtualization paradigm that separates network software from the underlying hardware. The separation of the control logic from data forwarders can be realized by means of a well-defined programming interface between the switches and the SDN controller. With SDN, network switches become simple forwarding devices and the control logic is implemented by a centralized controller (Kreutz et al., 2015). The controller exercises direct supervision over the states of data-forwarding elements via a well-defined application programming interface (API). An example of such an API is OpenFlow (McKeown et al., 2008); depending on the rules installed by a controller on an OpenFlow switch, the switch can be instructed to behave like a router, switch, or firewall, or perform other roles (e.g., be a load balancer or traffic shaper).

We believe that oppnets can take advantage of the SDN technology to implement their middleware, called *Oppnet Virtual Machine* (OVM) (Alduailij and Lilien, 2015). On the other hand, the so-called "primitives" used in OVM can benefit the SDN technology (e.g., by developing an SDN analogy to making entities oppnet-enabled).

11.5 Proposed Solution: PTF-Based Oppnet Architecture (POA)

The proposed security and privacy architecture for oppnets, called PTF-based oppnet architecture (POA), is based on the state diagram for a single oppnet helper (Al-Gburi, 2017), shown in Figure 11.3.

POA integrates PTF into the oppnet structure and operations, placing it under the control of oppnet's supervisor, previously called the decentralized CC and now renamed decentralized oppnet controller or DOC (Al-Gburi, 2017). PTF underlies all security and privacy services in its oppnet performed during its entire lifetime. Among others, PTF-based services ensure secure data transmissions among oppnet entities and verify that a helper joining an oppnet is not malicious.

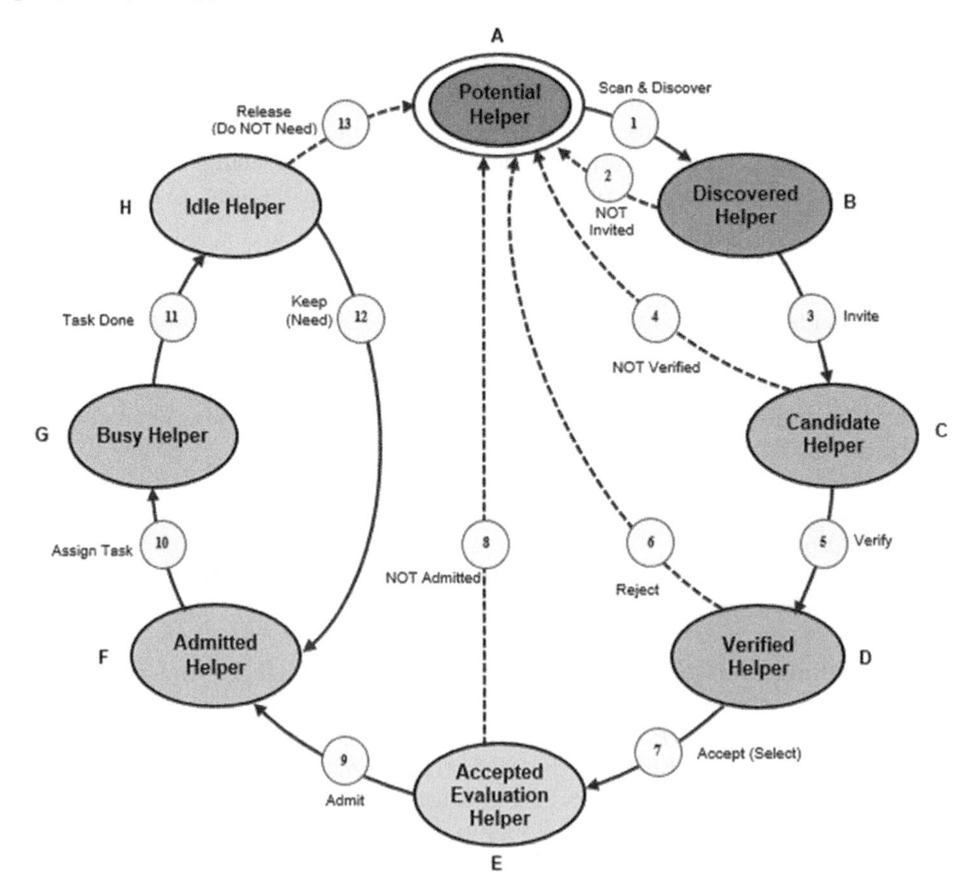

FIGURE 11.3
The state diagram for an oppnet helper. (From Al-Gburi, A., Protecting interaction security and data privacy in oppnets using pervasive trust foundation and active data bundles, PhD proposal, Department of Computer Science, Western Michigan University, Kalamazoo, November 2017.)

11.5.1 POA Structure

The POA structure for a seed oppnet and for an expanded oppnet is shown in Figure 11.4.* The PTF is integrated into the oppnet structure and operations to ensure security and privacy-preserving oppnet's interactions (Al-Gburi, 2017). The DOC and PTF are briefly described below.

Decentralized Oppnet Controller (DOC): It is a decentralized software that controls all oppnet's interactions and activities. DOC has the following beneficial properties for oppnets:

1. Controls all oppnet operations.
2. Makes decisions to grow its oppnet by admitting other new nodes, or to shrink its oppnet by releasing or removing nodes (expelling malicious nodes).
3. Monitors collaborative processing.
4. Supervises and monitors PTF processing.
5. Detects suspicious behaviors of oppnet entities based on the feedback from collaborative and PTF processing. DOC constitutes the second line of defense in detecting insider attacks.

Pervasive Trust Foundation (PTF): The PTF implements the trust foundation for security and privacy in oppnets. It is envisioned as a basis for all interactions among oppnet entities. Whenever an entity interacts with other entities, trust is considered either explicitly (in large-scale or open oppnet applications) or implicitly (in small-scale and closed oppnet applications).

Trust processing in PTF requires a decentralized approach to process (calculate, update, and store) trust values for a multitude of heterogeneous oppnet or oppnet-enabled entities. In addition, PTF adds the following benefits to oppnets:

1. Supports security and privacy services that constitute the first line of defense against insider/outsider attacks.
2. Supports security and privacy services that perform helper verification (including authentication, authorization, and checking for a sufficient trust level) to ensure that a helper joining an oppnet is not malicious.

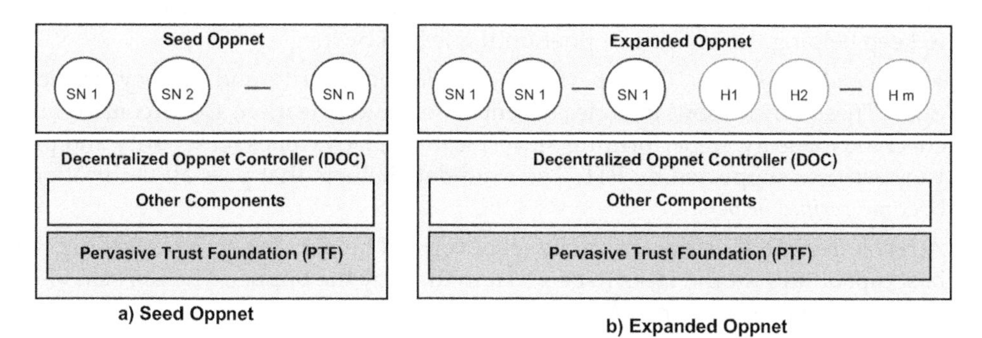

FIGURE 11.4
Basic POA structure: (a) seed oppnet and (b) expanded oppnet. (cf. Al-Gburi, 2017).

* Some components (identified as "Other Components" in Figure 11.4) are not shown in detail due to not being critical for understanding the POA structure.

3. Supports secure interactions among oppnet entities during a continuous expansion.
4. Supports security and privacy services that ensure secure and privacy-preserving data transmissions among oppnet entities.
5. Provides support for boosting DOC performance.

11.5.2 POA Operations

The basic POA operations are shown in Figure 11.5. The solid-line arrows indicate control flow between architecture blocks, and the broken-line arrows indicate data flow. Additionally, the uppercase letters (A–H) indicate the current helper state (as identified in Figure 11.3) for helpers being processed. For example, before entering the *Discover Candidate Helpers* block, a helper is in State A (*Potential Helper* in Figure 11.3); when a helper leaves this block, it is in State B (*Discovered Helper*).

DOC controls every single block in Figure 11.5, but control interactions by DOC are not shown to avoid overcrowding this figure with lines and arrows. The four major activities controlled by DOC are (1) oppnet growth, (2) collaborative processing, (3) PTF processing, and (4) terminating oppnet work. These activities are described in turn.

Oppnet Growth: When an oppnet needs to grow into a bigger one, it initiates the oppnet growth process, which ensures security by admitting helpers only after verifying them by security services (e.g., authentication) supported by PTF.

The growth process (cf. the *oppnet Growth* block in Figure 11.5) includes the following major steps:

1. *Discover candidate helpers*: This step performs two major functions. First, it scans within its range for potential helpers. Second, it identifies a subset of potential helpers that it wishes to invite. These helpers become *candidate helpers.*
2. *Invite candidate helpers*: Oppnet invites (or orders in some situations, e.g., life-or-death circumstances) selected candidate helpers to join oppnet to help it in a specific job. An invited candidate helper replies to this invitation with either *Accept* or *Reject*. A candidate helper that replies *Accept* expresses an obligation to keep on helping the inviting oppnet as long as the oppnet needs it. (An invited candidate helper can reply *Reject* for any reason; a candidate helper ordered to join is obliged to keep helping the inviting oppnet until released by it.)
3. *Verify candidate helpers*: DOC performs a series of security and privacy verifications. These verifications include authentication, authorization, access control, and checking integrity. In performing all verifications, DOC relies on security and privacy services supported by PTF. The candidate helpers that pass all verifications become *verified helpers.*
4. *Accept helpers*: DOC selects from the set of verified helpers the ones possessing the best capabilities for the tasks to be given to them by the oppnet. The selected verified helpers become *accepted helpers.*
5. *Enough evaluated helpers*? DOC decides whether its oppnet has a sufficient number of helpers (including just accepted helpers and the ones accepted in earlier growth iterations). Note that in Figure 11.3 there is no state corresponding to the *Enough Evaluated Helpers?* block. The reason is that this block is related to a set of helpers, while the helper state diagram is for a single helper.

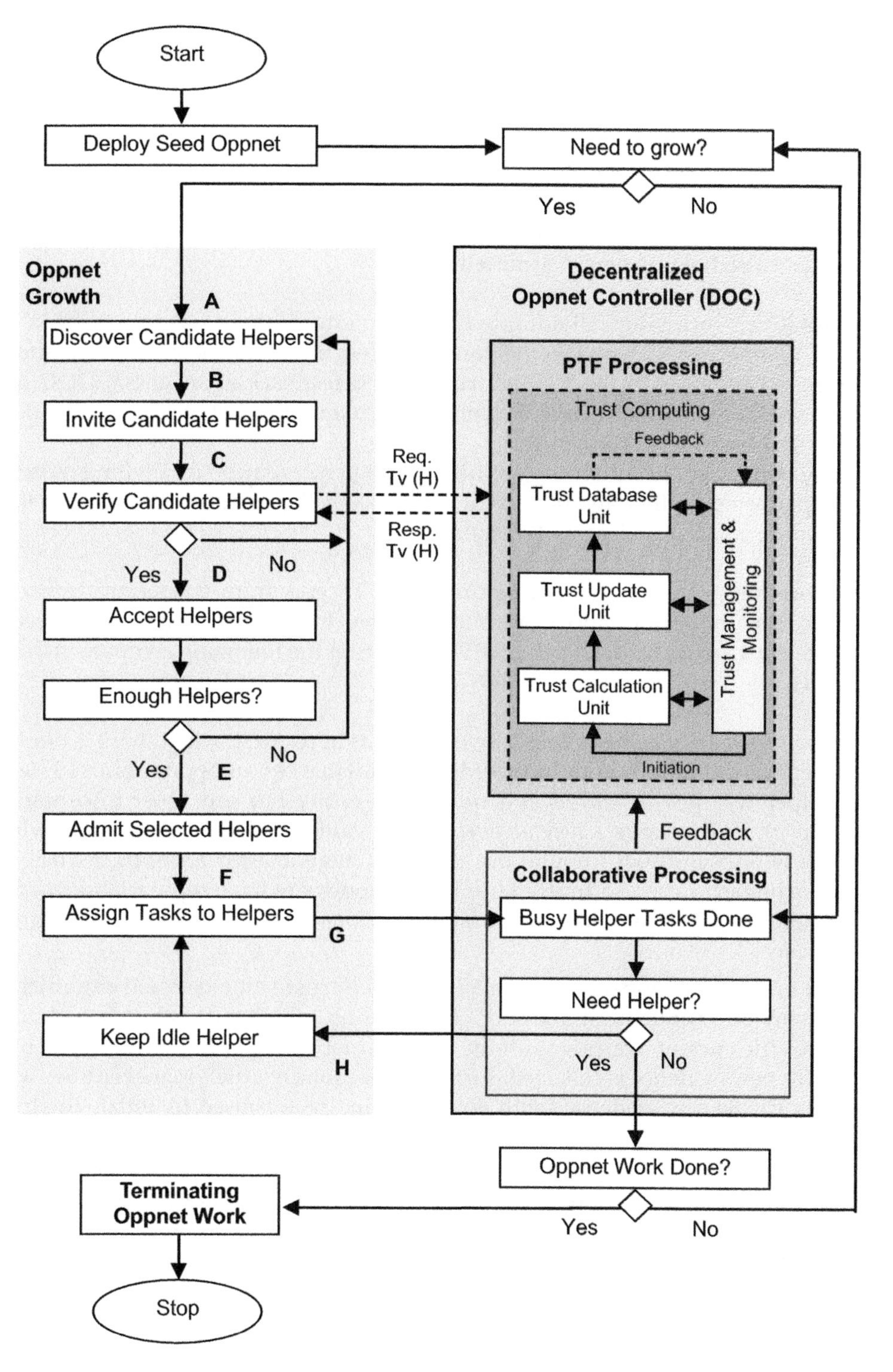

FIGURE 11.5
Basic POA operations. (From Al-Gburi, A., Protecting interaction security and data privacy in oppnets using pervasive trust foundation and active data bundles, PhD proposal, Department of Computer Science, Western Michigan University, Kalamazoo, November 2017.)

6. *Admit selected helpers*: All accepted helpers are admitted into oppnet, becoming *admitted helpers*.
7. *Assign tasks to helpers*: All newly admitted helpers receive their tasks, in this way helping their oppnet (in the process integrating their resources with other resources of their oppnet). They become *busy helpers*.
8. *Keep idle helper*: A helper completing its tasks becomes an *idle helper*. It can be released by DOC or—if it might be needed for executing other tasks—returned to the pool also holding newly admitted helpers.

Collaborative Processing: All admitted helpers perform their assigned tasks in a collaborative manner. DOC manages helpers' statuses, monitors their behavior and performance, and provides to the PTF its trust-related feedback about them. DOC makes decisions whether to release a given helper or return it to the pool of admitted helpers (to use them for more tasks in the future).

If during oppnet job execution any oppnet node expresses a need for inviting more helpers, DOC performs another iteration of the *Oppnet Growth* procedure. When oppnet completes its entire job, DOC releases all helpers.

PTF Processing: As indicated in Figure 11.5, PTF gets from cooperative processing a feedback about the performance of admitted helpers (busy and idle). It uses this feedback to calculate trust values for each helper. PTF consists of the following components, shown in Figure 11.5:

1. *(Initial) trust calculation*: We have two subcases here. First, when there is neither direct nor indirect (reputation) evidence available for an oppnet Entity E (e.g., a helper was just admitted and no oppnet entity has any trust information about it), PTF assigns a *default initial* trust value for Entity E. Second, when there is only indirect (reputation) evidence available for an oppnet Entity E (e.g., a helper was just admitted but DOC happens to have reputation information about it), PTF calculates an *initial* trust value for Entity E based on indirect evidence.
2. *Trust update*: Any oppnet entity must possess a trust value before it can interact with any other entity. If it does not, it receives an initial trust value as explained above. The current (starting with the initial) trust value for Entity E evolves over time as new evidence is collected during interactions of other oppnet entities with Entity E. The new evidence (both direct and indirect) is used to update the trust value for Entity E.
3. *Trust database*: Trust database stores all trust values for all oppnet entities. It also stores at least one previous trust value for each entity (if useful for analysis, it might even store the whole history of trust changes). We propose using a relational database with tuples, including attributes such as entity ID, its trust value, and the identification of tasks that affected the current trust value.

Terminating Oppnet Work: When the whole oppnet job is completed, DOC releases all helpers. Instead of purging from the trust database all information about released helpers, DOC might keep some or all of it for a possible future use (in case the same entity is a potential helper for future runs of the oppnet).

11.6 Conclusions

Opportunistic resource utilization networks face many security and privacy challenges that need to be addressed. In our view, the common key to improving both security and privacy in oppnets is using trust.

We exploit the *Pervasive Trust Foundation* (PTF) paradigm, in which trust is the basis for security and privacy. We describe a new architecture for oppnets called *PTF-based oppnet architecture* (POA), which is an expanded version of the original oppnet architecture enhanced with PTF. POA integrates PTF into the oppnet structure and operations, so PTF can serve oppnet for all security and privacy services performed during the entire oppnet lifetime.

We expect that POA and its future extensions will significantly decrease the security and privacy risks in oppnets. In particular, POA can significantly decrease the security and privacy risks associated with interactions of entities within oppnet, as well as interactions of oppnets with potential helpers.

We are currently working on the evaluation of the POA solution via simulation experiments implemented in Java and SimJava.

Key Terminology and Definitions

Opportunistic resource utilization networks (oppnets): Oppnets are a generalization of ad hoc networks, and grow by discovering and integrating as their *helpers* foreign devices or systems that possess desired resources. The basic idea is that helpers can improve or speed up the applications run by an oppnet. The range of helper-related oppnet actions includes the following: scan and discover candidate helpers, invite candidate helpers, verify candidate helpers, accept the best verified helpers, admit selected accepted helpers, assign tasks to admitted helpers, and release idle helpers when no longer needed.

Helper: This is an entity that can assist an oppnet in its activities and accepts an invitation to help the oppnet. Before it joins an oppnet, a future helper is a *candidate helper;* a candidate helper can possess diverse capabilities and skills potentially useful for the oppnet (such as specialized software, communication, computing, storing, sensing, or actuating capabilities). A helper, selected from among (potentially a broad set of) candidate helpers and integrated into an oppnet, becomes an *actual helper* (or, using a more precise categorization, an *admitted helper*).

Oppnet security and privacy: Security in oppnets involves ensuring secure interactions among oppnet entities, which are communicating to collaborate on achieving specific tasks. This requires many verification processes, such as authentication, authorization, and checking integrity. Privacy in oppnets protects sensitive information of the interacting entities and ensures that this information stays confidential; that is, it is not disclosed to unauthorized parties.

Pervasive trust foundation: A paradigm that involves two major principles. First, trust should be used as a foundation for security and privacy (as well as many other services not considered in this chapter, such as collaboration or routing). Second, trust should be considered pervasive in computing systems, and must be considered either *explicitly,* in large or open systems (such as oppnets), or *implicitly,* in small or closed systems.

References

Alajeely, M., Doss, R., and Ahmad, A. A. (2016, May). Security and trust in opportunistic networks—A survey. *IETE Technical Review*, 33(3), 256–268.

Alduailij, M. A. and Lilien, L. T. (2015, June). A collaborative healthcare application based on opportunistic resource utilization networks with OVM primitives. In *IEEE International Conference on Collaboration Technologies and Systems (CTS)*, Atlanta, GA, 426–433. https://doi.org/10.1109/CTS.2015.7210461.

Al-Gburi, A. (2017, November). Protecting interaction security and data privacy in oppnets using pervasive trust foundation and active data bundles. PhD proposal, Department of Computer Science, Western Michigan University, Kalamazoo.

Bhargava, B. and Lilien, L. (1981, May). Feature analysis of selected database recovery techniques. In *National Computer Conference, ACM*, Chicago, IL, 543–554.

Bhargava, B., Lilien, L., Rosenthal, A., and Winslett, M. (2004, September/October). Pervasive trust. *IEEE Intelligent Systems*, 19(5), 74–77.

Buttyan, L. and Hubaux, J. P. (2007). *Security and Cooperation in Wireless Networks: Thwarting Malicious and Selfish Behavior in the Age of Ubiquitous Computing*. Cambridge University Press, Cambridge.

Conti, M., Giordano, S., May, M., and Passarella, A. (2010, September). From opportunistic networks to opportunistic computing. *IEEE Communications*, 48(9), 126–139. https://doi.org/10.1109/MCOM.2010.5560597.

Han, B., Gopalakrishnan, V., Ji, L., and Lee, S. (2015, February). Network function virtualization: Challenges and opportunities for innovations. *IEEE Communications*, 53(2), 90–97.

Joshi, K. and Benson, T. (2016, December). Network function virtualization. *IEEE Internet Computing*, 20(6), 7–9. https://doi.org/10.1109/MIC.2016.112.

Kamal, Z. H., Lilien, L., Gupta, A., Yang Z., and Batsa, M. (2008). New UMA paradigm: Class 2 opportunistic networks. In Y. Zhang et al. (Eds.), *Unlicensed Mobile Access Technology: Protocols, Architectures, Security, Standards and Applications*. Auerbach Publications, Taylor & Francis Group/CRC Press, Boca Raton, FL, 349–392.

Kreutz, D., Ramos, F. M., Verissimo, P. E., Rothenberg, C. E., Azodolmolky, S., and Uhlig, S. (2015, January). Software-defined networking: A comprehensive survey. *Proceedings of the IEEE*, 103(1), 14–76.

Lilien, L., Al-Alawneh, A., and Ben Othmane, L. (2009, November). Some thoughts on the pervasive trust foundation for the future Internet architecture. Position Presentation at CERIAS Security Seminar, Center for Education and Research in Information Assurance and Security, Purdue University, West Lafayette, IN.

Lilien, L., Al-Alawneh, A., and Ben Othmane, L. (2010a, September). The pervasive trust foundation for security in next generation networks. In *ACM New Security Paradigms Workshop (NSPW)*, Concord, MA, 129–142.

Lilien, L., Gupta, A., Kamal, Z. E. H., and Yang, Z. (2010b, March). Opportunistic resource utilization networks—A new paradigm for specialized ad hoc networks. *Computers and Electrical Engineering*, 36(2), 328–340. https://doi.org/10.1016/j.compeleceng.2009.03.010.

Lilien, L., Gupta, A., and Yang, Z. (2007, April). Opportunistic networks for emergency applications and their standard implementation framework. In *IEEE International Performance, Computing, and Communications Conference (IPCCC)*, New Orleans, LA, 588–593.

Lilien, L., Kamal, Z. H., Bhuse, V., and Gupta, A. (2006a, March). Opportunistic networks: The concept and research challenges in privacy and security. In *International Workshop on Research Challenges in Security and Privacy for Mobile* and *Wireless Networks (WSPWN)*, Miami, FL, 134–147.

Lilien, L., Kamal, Z. H. and Gupta, A. (2006b, September). Opportunistic networks: Research challenges in specializing the P2P paradigm. In *3rd International Workshop on P2P Data Management, Security and Trust (PDMST)*, Krakow, Poland, 722–726.

McKeown, N., Anderson, T., Balakrishnan, H., Parulkar, G., Peterson, L., Rexford, J., and Turner, J. (2008, March). OpenFlow: Enabling innovation in campus networks. *ACM SIGCOMM Computer Communication Review*, 38(2), 69–74.

Pelusi, L., Passarella, A., and Conti, M. (2006, November). Opportunistic networking: Data forwarding in disconnected mobile ad hoc networks. *IEEE Communications*, 44(11), 134–141. https://doi.org/10.1109/MCOM.2006.248176.

Bibliography for Advanced/Further Reading

Lilien, L., Kamal, Z. H., Bhuse, V., and Gupta, A. (2007). The concept of opportunistic networks and their research challenges in privacy and security. In K. Makki et al. (Eds.), *Mobile and Wireless Network Security and Privacy*, Springer US, Norwell, MA, 85–117.

Lilien, L., Kamal, Z. H., Gupta, A., Woungang, I., and Tamez, E. B. (2011). Quality of service in an opportunistic capability utilization network. In M. Denko et al. (Eds.), *Mobile Opportunistic Networks: Architectures, Protocols and Applications*, Auerbach Publications, Taylor & Francis Group, Boca Raton, FL, 173–204.

12

Future Networks Inspired by Opportunistic Networks

Anshul Verma, Mahatim Singh, K. K. Pattanaik, and B. K. Singh

CONTENTS

12.1 Introduction ... 229
12.2 Bio-Inspired Next-Generation Networks ... 230
12.3 Mobile Ubiquitous LAN Extensions ... 231
12.4 Shared Wireless Infostation Model ... 233
12.5 Vehicular Delay-Tolerant Networks ... 234
12.6 Social Networking for Pervasive Adaptation ... 235
12.7 Service Platform for Social-Aware Mobile and Pervasive Computing ... 236
12.8 ZebraNet ... 238
12.9 FreeNet ... 239
12.10 DakNet ... 241
12.11 Haggle ... 243
12.12 Conclusion ... 244
References ... 244

12.1 Introduction

An opportunistic network is an evolution of the classic mobile ad hoc network (MANET). A MANET is characterized by infrastructure-less, autonomous, and mobile nodes. Nodes communicate directly when they are within communication range of each other. Every node can play two roles: end nodes (source node or destination node) and intermediate node (relay node or router). In a MANET, end-to-end connection between the source and destination is necessary for eventual transmission of any message. However, when nodes are highly mobile, the connection opportunity between nodes may become intermittent. As a result, traditional MANET routing protocols are not able to perform eventual transmission due to lack of an end-to-end path between the source and destination (Borgia et al., 2005). Therefore, several properties of traditional MANETs, such as the disconnection of nodes, mobility of users, network partitions, and link instability, are treated as drawbacks. This makes the communication between nodes significantly more difficult. The aim of the opportunistic network is to exploit such intermittent connectivity to provide communication between nodes and to design applications for such an environment.

Opportunistic networks are formed out of portable mobile devices carried by people without the assumption of any preexisting network infrastructure (Pelusi et al., 2006; Verma and Srivastava, 2011). They provide connection opportunities between mobile devices by exploiting their mobility while removing the physical end-to-end connection requirement. Disconnections, partitions, mobility, and so forth, are treated as features instead of drawbacks (Conti and Kumar, 2010; Verma et al., 2013). The eventual transmission is achieved by using the store-carry-forward approach, in which intermediate nodes are used to store messages during the nonavailability of a forwarding opportunity toward the destination, and any future contact opportunity with other mobile devices is exploited to bring the messages closest to the destination. In the literature, opportunistic networks are also known as intermittently connected mobile networks (ICMNs).

An opportunistic network is similar to a delay-tolerant network (DTN) (Fall, 2003), which is defined by the DTN IRTF Research Group (https://irtf.org/concluded/dtnrg). A DTN consists of a disconnected group of networks, and nodes within each network are well connected through predefined topology, but communication between independent networks is established through a DTN overlay. In general, the point (node) in each network that connects with DTN overlay is already known. In order to achieve end-to-end connectivity, the DTN overlay exploits occasional communication opportunities among the networks, which might be either scheduled over time or unscheduled. No prior knowledge is assumed about the possible points of disconnections in opportunistic networks, and the existence of separate well-connected subnetworks is not assumed. Therefore, despite the various similarities between DTNs and opportunistic networks, the routing approaches between both are quite different.

Indeed, despite the fact that opportunistic network research is still in its early stages, the opportunistic networking concept is nowadays exploited in a number of concrete applications. Various applications have been developed using opportunistic networks to analyze human mobility and social interaction patterns and, on the basis of that, to build efficient message routing models that incur minimum message delays (Jain et al., 2004; Verma and Pattanaik, 2017). Opportunistic networks are also applied to interdisciplinary projects focusing on wildlife monitoring to perform novel studies of animal migrations and interspecies interactions. The use of opportunistic networks to bring Internet connectivity to rural areas is also a trend. Opportunistic networks represent an easy-to-deploy and extremely cheap solution to bring Internet connectivity to rural areas (Boldrini et al., 2008; Conti and Giordano, 2007a, 2007b).

This chapter describes various existing real-life applications and case studies of opportunistic networks. Mainly objectives, motivations, limitations, and implementation approaches are discussed in depth about each application. The rest of the chapter is organized as follows. Sections 12.2 through 12.11 extensively describe real-life applications of opportunistic networks. Finally, the chapter is concluded in Section 12.12.

12.2 Bio-Inspired Next-Generation Networks

The new emerging pervasive environments are producing a huge amount of information. This is not manageable by the current Internet protocol because it was developed almost 40 years ago and was not planned to work in these emerging pervasive environments. The communication rules imposed by this protocol on the low-cast devices, such as sensors or tags, directly restrict the fundamental goals of such environments. Because pervasive and ubiquitous computing are expected to be a key enabler for all intelligent systems in the near future, there is a need to develop a totally different approach to provide communication in these environments that can fulfill their requirements. Working principles of biological systems can be exploited to completely redefine the working principles, such as structure, control, function, and communication, of the emerging pervasive environments. On the basis of this phenomenon, biological genetics and their evolution mechanisms are combined with pervasive computing concepts and lead to the autonomous and autonomously self-adaptive systems, such as service-oriented communication systems. Bio-inspired next-generation networks (BIONETS) works on the same phenomenon and considers an omnipresent low-cost pervasive environment as a randomly self-organizing by-product of a collection of self-optimizing services (Carreras et al., 2004).

In BIONETS, a service is defined by chromosomes and many services are associated with living organisms. Due to the presence of chromosomes, a service evolves and adapts to the environment constantly and autonomously. Chromosomes consist of genes that are the smallest data unit of service and define all the functional characteristics of an organism and service. Similar to the biological systems, a complete life cycle can be defined for the organisms and services. The life cycle starts from the birth of an organism, goes through the reproduction, and ends with the death. All these stages are clearly described in Carreras et al. (2004). Reproduction and evolution processes are influenced and inherited from the natural phenomenon.

The fitness of the organism's genetic information is measured by determining its suitability with the surrounding environment and the exchange of the genetic information with the environment. Therefore, information is locally exchanged between mating organisms on demand instead of end-to-end communication. The organisms are evaluated for natural selection by the corresponding environment on the basis of their fitness. As a consequence, best organisms (services) come out.

In this way, BIONETS describes a novel idea to exchange information in ubiquitous networks. The main aim is exchanging information on demand between movable users and disseminating information instead of forwarding data packets. Genetics models are used to evaluate and select a population of services suitable for a set of user nodes. The suitability of these services leads their growth on the user nodes; otherwise, they are declined. Along with the growth of populations, services are also evaluated by using mutations and selection of the fittest (Altman et al., 2010).

In order to justify the correctness of the developed model in a realistic scenario, a parking lot scenario is considered in which a city is divided into blocks and each gene represents the occupancy status (free or occupied) of the parking spot. The service continuously suggests the nearest available parking slot to the user and updates the suggestion according to the user's movement. Initial simulation results have proven the correctness of the model; however, the model should be extended to accommodate more realistic user behaviors and dynamic scenarios (Carreras et al., 2007).

12.3 Mobile Ubiquitous LAN Extensions

The Internet is the main tool of communication, and it is adapting to the ubiquitous nature day by day, but its arbitrary installation is limited by costly and tedious infrastructure. A network model and a data acquisition system have been proposed for the sporadic networks with an aim to enhance the networking competence. Previously, some similar network protocols have been developed, but those are more concerned with the higher-frequency networks that interact randomly between the nodes. In this proposed network model, the nodes are placed highly distant to interact directly and only wish to communicate unidirectionally with a single server. To exhibit the utility of the proposed model for other research areas where small sensors are dispersed over a wide sporadic area and for a case study, a traffic analysis mechanism is developed to help with traffic administration and city planning (Shah et al., 2003).

In the specified wide sporadic network area, the quantity and position of the motes are taken randomly and Berkeley Mica-Motes is used to collect the data and use it in required systems. The modest quantity and low cost of the motes allow them to be dispersed and move everywhere with ease, and also to be settled at the proper position and then buried.

In the implementation, Intel personal servers are serially connected with a single mote, known as a mobile ubiquitous LAN extension (MULE). These MULEs are linked with a host that moves through the region where motes are placed, and as per the system specification, the host does not necessitate any proceeding for the system to function. This implies that personal servers are free to take random paths. The hosts form the network framework with no knowledge and hence act as uninformed agents, as their quantity is not stated and can be changed. A host can be anything—a vehicle, a human, or anything that can carry a personal server.

The personal server is installed on the vehicles (specially four-wheelers and above) in the proposed model for data acquisition from the motes dispersed on the streets, and to aggregate the gathered data, the central server is installed at vehicle stations. Like most data acquisition mechanisms, our model is up and operational within weeks.

When a MULE bypasses or returns to a central server, it clears out its data to the data server with the help of the Bluetooth radio protocol. The main control catalogs the data and mentions the lacking items. The MULE is then offloaded with acknowledgments, generated by it, for delivery to every mote it can potentially assist (Gao et al., 2007).

Since mote data is devised to be enduring, it is not allowed to clear its data unless it has no free space or it receives a confirmation for the data it has sent already. To make the communication more flexible, the mote field can be divided into different groups, and when a MULE acquires its data, the data server assigns a group number to the MULE. Based on its group assignment entry, the MULE will attempt to interact only with motes present in the assignment.

As far as the security and efficiency of the model are concerned, some mechanism is introduced. For the data integrity, encryption is performed on the collected data. In the encryption, a major part of the computation will be performed at the personal server so that motes have to perform little work to conserve power. To prevent unauthorized access, an authorization protocol is applied to the sporadic ad hoc network, the personal server, and the motes; this protocol limits unauthorized access and hence increase the mote battery life.

Limited space and power is the major concern of the model; the motes have less memory and power life than personal servers, whereas personal servers have less memory and power life than the main servers. For this reason, major parts of the mote's work are pushed to the personal server and major parts of the personal server's work are pushed to the main server as well. The data server has no such limitation of space and power.

MULEs, along with low implementation and configuration of the model, will introduce a new system for researchers, government and private organizations, and others who require the ability to collect data from a sporadic distributed network in a cheap and dependable way. The main aim of the model is to automate the data gathering to the maximum limit (Kazemeyni et al., 2012).

12.4 Shared Wireless Infostation Model

Capacity and delay are two major concerns of wireless ad hoc networks. Eventually, a trade-off occurs between these two for certain analysis, particularly focusing on two extremes: either minimize the delay or maximize the capacity. However, during these extremes, the same trade-off is instantiated by different strategies to a numerous extent. Infostation network methodology is such an example that facilitates increasing the capacity of a mobile network at the cost of delay, and it provides a geographical intermittent coverage at an excessive rate. In an extension to the infostation approach, a network model is defined, known as shared wireless infostation model (SWIM), with an integration of ad hoc networking (Small and Haas, 2003). The SWIM model with a moderate hike in storage need enables further enhancement of the capacity–delay trade-off.

An ad hoc network setup has been defined in a biological information collection system where nodes are represented by whales tagged with a radio device to show that SWIM can be applied to deal with practical problems. With the introduction of multitiered operation, SWIM can be enhanced and realized by representing seabirds as mobile data acquisition nodes. An analytical formula was developed for distribution of end-to-end delays that helps in calculating the storage requirements. The biologic system consists of random delays. The system may consider that old data is as useful as the latest data since the animals' behavior remained unchanged over the years and data is time dependent on any timescale. The derived analytical formula can be used to numerically evaluate the system model on certain measures, which helps to analyze the impact of these measures on system performance. It also authorizes close inspection of the capacity–delay trade-off.

A research group at WINLAB2 first proposed the infostation architecture (http://www.winlab.rutgers.edu/docs/focus/Infostations.html). The architecture includes low-power base stations and should be adapted in such applications where substantial delays can be tolerated. Infostation offers very high data speed to users in a small geographical region, as it has a firm radio signal quality. Infostation has the absence of continuous coverage; hence, to supply data at high speed, it has to bear the cost of intermittent connections. High delays occur often when a node wants to transmit data that is out of the range of the infostation for an excess amount of time and should always directly transmit to an infostation. The infostation exploits the mobility of a node to trade connectivity for capacity, and when a node returns to the range of the infostation, it transmits with a high data rate (Small and Haas, 2003).

As discussed above, SWIM extends the capacity–delay trade-off, which enables information to reach the infostation by replicating and defusing (sharing) itself in the network. Information takes the service of mobile nodes as physical carriers to propagate, which somehow overcomes the limitation of the infostation model, where a node should be present in the neighborhood of an infostation to communicate. When a node carrying the information arrives at any of the infostations, the information is offloaded to the network and removed from the node. The whole methodology of SWIM replicates the concept of disease expansion, as harmful bacteria of the disease move from an infected body (a node in the network) to a normal body by sharing (information transmission), and when the infected carrier reaches any healing center (infostation), harmful bacteria are removed (information offload). The carrier stores the identity of the packets offloaded to the infestation, and that information will help in discarding all such packets in the future; it resembles the event of developing immunity to a disease.

One of the main advantages of SWIM, the delay until one of the replicas reaches the infostation, is fairly reduced by distributing the packets throughout the mobile nodes. However, this distribution drains the network capacity comparatively fast. This implies that a capacity–delay trade-off still persists. A solution has been developed to control this trade-off by controlling the parameters of the distribution, for example, limiting the probability of packet transmission between two consecutive nodes (as the probability of infection in the disease expansion model), the transmission area of each node (infection gap), or the quantity and circulation of infostations (quantity and position of healing centers). SWIM also enables analysis of the bulk of storage needed to understand an appropriate occurrence of this trade-off (Small and Haas, 2003).

12.5 Vehicular Delay-Tolerant Networks

The promising applications of the vehicular ad hoc network have been receiving more attention in recent years because of some of the network's special features compared with traditional networks. The range of its applications covers road safety, traffic management, and driving assistance, as well as providing connectivity in isolated regions and building a network for rescuer teams in the case of disaster. Its features include that it is sporadic and intermittent, no end-to-end connectivity is required, and it has random delays, high latency, and excessive erroneous and irregular data rates.

A bundle protocol related to the architecture and communication of DTNs has been proposed by the DTN Research Group (https://irtf.org/concluded/dtnrg). DTN is the basis of the vehicular delay-tolerant network (VDTN), which is a breakthrough-based approach to enable vehicular communication that aims to overcome the limitations of lengthy delays and random networks (Isento et al., 2013). In the circumstance of high round-trip delays, all the needed information of a transaction is "bundled" together, which reduces the round-trip exchange quantity. A bundle is basically a large-sized packet of aggregated datagrams under the network layer adapting an IP approach, and it consists of an originating timestamp, convenient life signifier, service assignment class, and length signifier. A bundle layer in a DTN is a message-intended overlay layer, below the application layer, that combines the application data units into one or more data units known as bundles. To maintain transmission of the data between disconnected parts of the DTN-based network, the substantial movement of the vehicles and the opportunistic contacts between them are utilized.

The store-carry-forward approach of DTNs is used by VDTN, in which bundles are forwarded to reach the target asynchronously by hopping over the moving vehicles equipped with short-range Wi-Fi devices. The main issues of VDTNs, related to their layered architecture, aggregation and de-aggregation strategies, routing protocols, fragmentation, scheduling and dropping methodologies, and performance assessment tools, are extensively described in Isento et al. (2013). The bundle layer uses store-carry-forward bundle-switching methodology instead of traditional end-to-end communication methodology. It accumulates and forwards bundles between nodes. Following a path that is eventually directed to the target node, messages are forwarded from one node to another node by hopping.

The applications and methodologies of DTNs are used in an underwater network that enables communication for contamination tracking, navigation assistance, and disaster prediction, and it has been used by biologists to track and monitor the behavior of wildlife species over an extensive region (Ahmed et al., 2015).

The VDTN's consideration of two planes for out-of-band signaling, that is, the control plane and the data plane, is used for connection setup between two successive nodes and the transfer of bundles, respectively. In the proposed VDTN, three kinds of nodes are defined: terminal nodes, mobile nodes, and relay nodes. Terminal nodes enable connection to the user, mobile nodes lift bundles between nodes (fixed and mobile), and relay nodes have the responsibility of enhancing network connectivity and analogous bundle delivery, and they are located at crossroads as fixed nodes.

To evaluate the performance of the proposed VDTN, a laboratory test bed (VDTN@Lab) and a real test bed (developed by FIAT, Brazil) are used. To provide and enhance the bundle delivery ratio and delivery wait, different routing protocols have been defined. Buffers at VDTN nodes are managed by scheduling and dropping techniques, while fragmentation reduces the retransmission of the same bundle, which upgrades the overall network performance. A simple network management protocol–based administration is defined to assist the monitoring of VDTN nodes and trace the inconsistency (Isento et al., 2013).

12.6 Social Networking for Pervasive Adaptation

With continuous technology improvement, the world is witnessing rapid growth in small mobile devices with capacity for storage, processing, and information broadcasting in our daily life. Though this offers a medium to produce, gather, and transmit new insights and information in the distinct facet of human life, it also raises certain challenges for its accomplishment. It requires advancement in flexibility, adaptation, and dynamicity as the existing end-to-end network methodology is irrelevant. The Social Networking for Pervasive Adaptation (SOCIALNETS) project (http://www.social-nets.eu/) of EU Future and Emerging Technologies (FET) desires to beat these challenges and assists humans in acclimating and illustrating agility beyond other species, by understanding, representing, and utilizing crucial attributes (Allen et al., 2008). It is expected that highly efficient pervasive communication and content arrangement can be facilitated by establishing the competence for devices to socially network.

Communication is essential for information and knowledge exchange in human society, but the existing network design and artificially developed content arrangement methods limit human communication to strictly adopt engineering paradigms to assist the transmission. The inclusion of such engineering paradigms in the wireless network might be the main reason impeding the evolution of technology like MANETs or sensor networks. It can be observed and justified that recent human-centric techniques are inspired by basic features of human nature and self-organization. Projects like SOCIALNETS are required for human centricity.

The basis of the SOCIALNETS project is to adopt the framework of human affairs and cooperation to present a solution to the concerns of the pervasive network, and this is ideal for conditions and circumstances and constantly persists in human culture. As the name suggests, it is closely related to the end users of the technology and combines the design and concepts of sociology, complex systems, computer science, and network engineering. Human social association covers varied disciplines, with the growth of natural analogies used to assist self-organization, like the autonomic nervous system and biological conventions (Allen et al., 2008).

The SOCIALNETS encompasses both online social networks and opportunistic wireless networks. The project investigates in what way the social network can be implemented for data delivery and gathering with security and assurance. It is a comprehensive assumption that implementation of the basic features of human nature, of extremely efficient and dynamic mutual affairs, along with the social network will lead to an immensely efficient, progressive, reliable, versatile, and human-centric pervasive communication system. The project has the following major findings:

- Understanding the fundamentals of human behavior through social aspects and physical mechanisms for circulation, which include the movement of information in intranetworks and internetworks through processes like gossip and biological transmission.
- Evolution of a flexible social framework for the transmission of data in unicast and broadcast schemes. It includes community discovery and defining protocols to allow data transmission through human social organization using opportunistic networking methodology.
- Behavioral content handling methods, such as (1) the social arrangement of commodities with human social design to obtain efficiency for access, (2) a social plan to encourage coordinated conduct and mutual interchange, and (3) social techniques for rational pushing.
- Development of generalized social protection to deal with random faith and coordination using trust bonding within a decentralized framework. It enables us to acquire confidence in data storage functionalities, secrecy in data access and exchange, and coordination among decentralized nodes.
- Complete architectural design to demonstrate how different components of the SOCIALNETS can be integrated to assist each other through a social network layer.
- Determination of a potential imminent application and a way to implement it through the wireless framework. Concerned applications could be social network blogging without the availability of the Internet, opportunistic networking, and enabling of communication in developing areas.

The aim of SOCIALNETS is to exploit the key features of human social interaction to develop new methodologies for communication and information distribution in large, heterogeneous, and reliable pervasive networks. The successful implementation of this concept demands collaboration and cooperation among the diverse fields of sociology, complex systems, network engineering, and computer science (Allen et al., 2008).

12.7 Service Platform for Social-Aware Mobile and Pervasive Computing

Empowering the smart devices of a mobile user with the ability to search and use the available resources of the surroundings conveniently leads them to construct and access a rich set of services, and such services can enhance the serviceability of a mobile phone beyond limits. Particularly, a distributed task execution will be needed for using the different resources, like raw computational power, social networking connections, or readings of sensors across a set of different devices. To provide distributed task execution in opportunistic pervasive networks, the Social-Aware Mobile and Pervasive Computing (SCAMPI) framework has been proposed (Pitkänen et al., 2012). This architecture supports effective opportunistic communication between diversified sensors, individual communication devices, and resources submerged in the surroundings by implementing influencing human social behaviors. SCAMPI facilitates the composition of a wide range of applications that can utilize a bundle of service components present in the neighborhood by abstracting resources as service ingredients following a service-oriented scheme.

The motive of the SCAMPI project was to empower the end user of smart devices to use their own resources and opportunistically utilize the resources available in their surroundings in a secure and reliable way. The project exploits all the available resources of the surroundings that could be fixed equipment, such as a Wi-Fi router, CCTV, and resources available on other devices, opportunistically and on-demand. SCAMPI investigates new direction aspects analogous to traditional service-oriented schemes, and its applications facilitate a personal and isolated device with an additional wide range of functionalities. Services in SCAMPI abstract the resources of personal devices with the resources of the surroundings and deliberately granted by the users. Its networking framework is considered highly unstable in comparison with the traditional Internet model, and hence requires a complete redesign of the technical solutions (Pitkänen et al., 2012).

The features of service-oriented human pervasive networks combine the features of both conventional pervasive networks and opportunistic networks. It is assumed that in pervasive networks mobile users are dispersed in a region with the majority of devices capable of sharing networking and computing resources. An opportunistic network is considered highly dynamic with unstable topology, and hence to maintain continuity in networking and computing tasks, it is necessary to use opportunities to establish a connection with other devices. This technique integrates access to the infrastructure in the complete service-centric design. Infrastructure access that is considered a service, such as granting support for accessing traditional Internet infrastructure, will be seen as an opportunistic service; such access can be composed of attaining a required target, for example, delivering a segment of content to a user outside of the physical neighborhood.

Pervasive and opportunistic networking methodologies are the base of the SCAMPI project. To understand this, one should examine the meaning and deployment of a service-centric platform for mobile and pervasive networking and shift from absolute networking issues. Accordingly, the project focuses on the collaborative mechanism to provide shared services using the available software and hardware resources of the user's device, instead of filling pure networking concerns. The technical approach will be focused on human factors. One side of the technical solution established inside the project examines the creative example of social-centric services provided by SCAMPI explanations, while the other side focuses on knowledge about the social nature of the users, which will be utilized as root contextual information. SCAMPI uses a wide variety of resources, whether physical resources, such as processing units, digital cameras, Wi-Fi routers, and sensors, or soft resources, such as programs and files, as serving items (Pitkänen et al., 2012).

SCAMPI can provide services over an opportunistic network that make it different from others. SCAMPI is an autonomous and infrastructure-less system that forms a network using its mobile devices automatically. Hence, its implementation for any service appears as one of the complex and challenging tasks. Without any fundamental infrastructure and infrastructure assistance, SCAMPI is capable of adopting pure opportunistic networks.

The main goal of SCAMPI is to focus on an infrastructure-less computing environment that is useful for users in the wide area of applications and can also be useful in extreme circumstances, such as disaster and emergency situations.

Several research outcomes support the SCAMPI framework as an approach to opportunistic computing, and it can provide a computing platform in one the toughest environments that have been discussed above. The whole analysis explains how the association of analytical modeling, low-level techniques, and artificial learning by experiments in social networks helps to build a service-rich infrastructure for mobile users. It is considered that such architecture can provide various emerging social- and context-aware applications that will enable adaptable application development with up-to-date smart devices (Pitkänen et al., 2012).

12.8 ZebraNet

Wireless communication and mobile computing are essential techniques for implementing several new computing applications on mobile devices. The essential areas of the wireless sensor networks are applications involving the independent utilization of computing, sensing, and wireless networking devices for both commercial and scientific means. The ZebraNet system is a mobile sensor network based on wireless peer-to-peer networking methodology, proposed basically for wildlife tracking and monitoring purposes (Juang et al., 2002). In the ZebraNet system, an electronic tracking collar (node) consists of a GPS chip, nonvolatile flash memory, wireless radio transceiver, and tiny user-programmable CPU, attached to the animal roaming across a broad forest area. These collars behave as a peer-to-peer network and forward the stored information to the base station. Usually, such wildlife regions do not have the facility of cellular networks or broadcast services; hence, ad hoc or peer-to-peer networking is the only way to provide communication. Since the researchers themselves are moving and base stations are also not stable, additional challenges are raised despite there being several ad hoc protocols available. The main design objectives of ZebraNet are to minimize the utilization of energy, storage, and other resources, preserving the reliability of the system with the high success rate of "data homing." It is quite obvious that the developed domain-oriented protocol and its energy trade-off will have common adaptability in other wireless and sensor-based applications (Liu et al., 2004).

A ZebraNet system with 30 nodes is implemented in Mpala Research Centre of central Kenya to investigate the behavior of zebras in their natural habitat. The ZebraNet project was basically proposed to work in remote locations, such as wildlife sanctuaries, where it is assumed that no fundamental infrastructure of networking and communication is available. As a consequence, ZebraNet uses peer-to-peer communication to transmit the data. The researchers gather the collected data from the animals roaming within their communication range with the help of handheld mobile devices. In literature, a wide discussion of ad hoc sensor networks is available, whereas very little has been discussed regarding mobile sensor networks with a movable base station, and even less on implementing real systems (Zhang et al., 2004). The ZebraNet project offers some exclusive contributions:

- This project considers itself to be the first to exercise protocols for a mobile sensor network with mobile base stations. It assumes that the base stations will be available occasionally only when the researchers are moving for data-gathering purposes, and researchers will upload the data while passing through the region.

- The ultimate goals of this project are to track the mobility patterns of zebras, which are quite unknown, and to determine why, when, and how zebras commence long-term migrations, which is also a critical biological question. It is intended to bootstrap the mobility pattern with recent, little well-processed biological data to develop the initial protocols, which later can be processed and adapted as the first deployed system that enables learners to study in detail the movements of zebras, especially their long-term migrations.
- The ZebraNet data acquisition system follows the same communication model as the other sensor networks in which data is collectively sent toward the base stations. The developed protocol of ZebraNet is optimized for this data acquisition communication scheme and to avoid a high degree of latency in this application domain.
- With the help of the real system, energy measurement and energy trade-off will be measured for the operational ZebraNet prototype hardware.

As discussed earlier, ZebraNet uses the peer-to-peer networking scheme to deliver information to researchers' mobile base stations without the presence of any cellular or telecommunication networks. After a successful experiment, the system will analyze some points, like the effect of the battery, weight limits on energy, storage for the system, and its protocols, for further improvements (Juang et al., 2002).

12.9 FreeNet

FreeNet is an application based on the peer-to-peer networking strategy, and it allows data to be displayed, replicated, and accessed while preserving the identities of both the producer and consumer (Clarke et al., 2001). It functions as a network of similar nodes that pool their storage area together to keep the data items, and it assists the path request to the most probable physical location of the data. It does not have broadcast search or common file indexing facilities; hence, the files are accessed in a location-free mode. The requested files are dynamically replicated in the neighborhood of the applicant and removed when and where they are not required. It is very difficult to find the actual source and destination of a file passing through the network, and a node administrator cannot easily determine or control the files of its own node.

Networked computer systems are widely being adapted to store and transmit information. However, this provides less security to users and stores data files in some predefined or fixed location, creating a prime breakdown point. A system should implement high-security and -reliability measures to satisfy the demand of an individual to preserve their privacy, and also protect the prime breakdown point from complete data loss or flooding attack. FreeNet is a distributed data storage and retrieval system developed to focus on privacy and availability issues. The system runs around many personal computers as a location-independent distributed file system, and grants data input, storage, and access in an unidentified way. Its major design goals are (Clarke et al., 2001)

- Anonymity for both producers and consumers of information
- Deniability for storers of information

- Resistance to attempts by third parties to deny access to information
- Efficient dynamic storage and routing of information
- Decentralization of all network functions

The system is developed to react adaptively to usage sequence, transparent movement, replication, and the clearing of files needed to enable efficient services without relying on broadcast searches and integrated file indexing schemes. Permanent file storage cannot be ensured even though a large number of nodes with enough memory space will join in the near future. The system runs at the application layer with the assumption of a secured transport layer without relying on it. The system requires anonymity only for FreeNet file operations, not for general network utilization.

FreeNet is in its developing stage as a freeware software project, and its introductory implementation can be downloaded from its official website (www.freenetproject.org). It is extending day by day and was originally developed at the University of Edinburgh.

Nodes in FreeNet interact with each other to store and access the data files, whose names are defined by location-independent keys. Every node has a specific local storage dedicated to the network for reading and writing purposes, and it also has a dynamic routing table containing the addresses of other nodes and their keys. It is assumed that the majority of the users of the system will run the nodes to increase the security measures, as well as the total storage capacity of the network (Clarke, 1999).

This system can be considered a cooperative distributed file system, having location independence and transparent slow replication. Like distributed.net, which allows general users to share unutilized CPU cycles on their computing device, FreeNet allows them to share unutilized disk spaces; however, distributed.net preserves this facility for itself. The requests for keys are passed from node to node through a sequence of proxy networks, where each node locally decides the next forwarder. Routes vary according to the requested key. The routing algorithms for data storage and access are proposed to adaptively adapt the routes over time for efficient system performance using only local knowledge. This is highly required because a node only has access to its immediate upstream and downstream nodes in the proxy sequence, to manage privacy.

Analogous to IP's time-to-live (TTL), every request in FreeNet is assigned hop-to-live bounds, which decrement at every node to avoid infinite chaining. It also has a unique pseudorandom identifier to avoid the loops caused by duplicate requests; the preceding node simply rejects it when it occurs and forwards it to a different node. This process runs until the request either is successfully processed or exceeds its hop-to-live bound, according to the output; the result is forwarded back to the origin node through the same sequence of chain. There is no hierarchy among the nodes; hence, a central breakdown point persists. If a new node wants to join the network, it first discovers the address of one or more nodes and then starts sending the request to them (Clarke, 1999).

The FreeNet networks offer an effective and anonymous information storage and access mechanism, with the help of cooperative nodes dispersed over many processing devices. In addition to the effective routing algorithms, it maintains the secrecy and availability of information while preserving high scalability. The initial implementation of a test version has shown good results, with a huge amount of downloaded and circulated copies. Due to the property of the anonymous system, it is not possible to count how many are using it and how well its mechanisms perform, but factual results are very satisfactory (Clarke et al., 2001).

12.10 DakNet

Despite the fact that establishment of communication has become very convenient in different parts of the world, residents of remote regions still have to cover a long distance to connect with others or get required documents, which residents of developed areas accomplish in a fraction of seconds. With an inappropriate assumption, the government established a copper-wired telephone connection in the villages, which later appeared to be more costly than the broadband wireless Internet connection with the advent of wireless technology. A wireless technology–based ad hoc network known as DakNet is used to enable digital connectivity with the help of the existing transportation and communication infrastructure. DakNet is derived from the Hindi word *Dak*, a word related the postal service, and integrates physical transportation and wireless data delivery to expand the Internet connectivity that an uplink, post office, or a cyber cafe provides (Pentland et al., 2004).

Real-time applications require huge capital investment, and therefore to compensate the receiver cost, it needs a high level of user adoption. The average urban populations are unable to afford personal communication equipment such as cell phones and computers, and to bear its cost, they need to share the communication framework. In such areas establishing a non-real-time framework and applications like voice messaging, e-mail, and bulletin boards, MMS and similar could be more convenient than the real-time telephonic or cellular services. Studies support the fact that the present infrastructure requires more manageability and underlying interactivity than does non-real-time connectivity for efficient rural information and communication technology (ICT) services (Hasson et al., 2003). Distressed with poor transportation, immoral pricing, and corruption, villagers are eager and ready to pay for digital facilities. A cost-effective communication technology like asynchronous or store and forward can be used without sacrificing the delivery of valuable user services, and can also build a local information repository that society members can add to and monitor with a capability to enable services like e-mail and voice messages.

DakNet, a wireless-based ad hoc network, was developed in the MIT Media Lab to enable asynchronous digital connectivity, proving that a combination of wireless and asynchronous technology can be presented as a roadmap to universal broadband connectivity (Hasson et al., 2003). In some of remote regions of India and Cambodia, DakNet has been implemented successfully with a significantly lower cost than the traditional landline infrastructure. It enables villagers to use an affordable Internet service.

As discussed above, DakNet uses the currently available facility of communication and transportation to enable digital connectivity in regions where it does not exist. In an approach to cost-effectiveness and power saving, DakNet transmits data over the short point-to-point connection between kiosks and portable storage devices, known as mobile access points (MAPs). A MAP is installed and powered by a vehicle with a generator; it delivers the data among kiosks and personal communication equipment (like an intranet) or hub connected through the Internet (for no real-time Internet access). Data stored in MAPs is transferred automatically with inexpensive Wi-Fi radio transceivers for every point-to-point connection at high bandwidth.

DakNet has two basic operational steps (Pentland et al., 2004):

- When a MAP-installed vehicle reaches inside the territory of a rural Wi-Fi-enabled kiosk, it uploads and downloads a significant amount of data.
- When it reaches any Internet access point or the hub, it automatically synchronizes all the collected data from the rural kiosks.

With every MAP-installed vehicle, these steps are performed repeatedly, which results in an economical wireless and continuous communication infrastructure.

It can be sufficient to enable regular information service with a single passing vehicle through rural areas with high connection quality. Though DakNet does not facilitate on-demand data delivery, it can transfer a huge amount of data (typically 20 MB) in each direction, and it can provide high data throughput in comparison to low-bandwidth technologies like telephone modem. The villages can utilize the same infrastructure to enjoy on-demand information access with an additional installation of wireless towers and antennas that will be entirely transparent to them, as they do not require new skills to use. With these additional installations, some new and more sophisticated services, like voice-over IP, which permits general real-time telephony, can be enabled. Asynchronous wireless broadband communication provides a foundation and migration path to a constantly on, broadband framework, and end-user utilities (Pentland et al., 2004).

Villagers in India and northern Cambodia are using electronic services like e-mail and voicemail with the help of DakNet connections. An e-governance project in India, Bhoomi, deployed the concept of DakNet as a tool for affordable rural connectivity, and it was also installed in remote regions of Cambodia in September 2003 for 15 solar-enabled rural schools, telemedicine centers, and a governor's office.

India's Bhoomi initiative: It was pioneered by Karnataka's state government to digitalize the land records and was deployed successfully at every district headquarters across the state to perfectly replace the paper-based record system (Chawla and Bhatnagar, 2004). Bhoomi's database is available to the villages of Daddaballapur up to 40 km away from Bhoomi's district headquarters, Taluka. In the deployment, a government bus equipped with DakNet MAP is used to deliver land record requests from every village kiosk to the Taluka servers. These servers handle the requests and produce the land records; further, these records are transported to every village kiosk by bus, where they are printed after giving a nominal charge to the kiosk manager. The bus travels three round-trips per day by passing through the hub and each village, and enables villagers to access Bhoomi's database. It provides several benefits to the villagers in terms of cost and ease of access.

E-mail facility to Cambodian schools: Almost 225 rural schools in Cambodia are managed by cambodiaschools.com with the intention to facilitate Internet access by enabling asynchronous connectivity to a backbone or hub (equipped with a satellite antenna in the provincial capital of Ban Lung), which consists of 256 Kbps links (Chea et al., 2009). This project is funded by some individual donors and the World Bank. Unlike the Bhoomi project, here the MAPs are installed on a Honda motorcycle or ox cart (few locations) because of the sophisticated territory in northern Cambodia. The outcome of this project has also been very satisfying, as has provided Internet facilities to the students of Cambodia for the first time.

Encouraged by the overwhelming response and low-cost implementation of DakNet, researchers are extending its reach to other countries, including Nigeria, Jordan, and the remaining part of Cambodia. Researchers are also trying to provide a key solution that will enable users to implement DakNet by themselves. A project like this should enable the first possibility of digital connectivity to the residents of isolated areas, as the study shows that increasing connectivity boosts economic growth. The main goal of such a project is to replace wired Internet connectivity with wireless techniques, but it can only be implemented after certain bureaucratic assessments, which include user satisfaction, economic growth, and system reliability. Governments can provide Internet facility to poor people of the world's remote areas only after successfully competing with the discussed bureaucratic assessments.

12.11 Haggle

Haggle is the recent networking approach to provide communication in a sporadic connectivity atmosphere (Nordström et al., 2012). In sporadic connectivity, there is no fixed network connectivity, whether it is Bluetooth, 802.11 WLAN, Ethernet, or others. It is intended for a radical exit from the current TCP/IP suite by utilizing application layer forwarding rather than the network layer. Haggle uses context awareness for message transmission between pervasive mobile devices even in sporadic connectivity. Opportunistic forwarding, authentication, privacy, reliability, and encouragement to cooperate are the major properties of Haggle. Some researchers have deployed this communication model to examine it based on a large-scale (around 500–800 nodes) experimentation, including issues like security, reliability, and information aging, and have communicated with sociologists to analyze its significance to Internet users. Haggle code can be downloaded from https://code.google.com/archive/p/haggle/.

Haggle has empowered a new group of applications with an excess of spatial locality and referred to this domain of networking as communities, and also enabled specific support for community forming and administration. Communities within Haggle make it easy to conclude and control the transmission of information (Scott et al., 2006). Haggle can be used as a technique for ad hoc content sharing with an aim to leverage the search for context-aware circulation. In pursuance of the importance to the users, Haggle prioritizes the contents and permits content that is high in demand to transmit more quickly than content that is not, even when the quantity of desired candidates is small; that is, interested candidates have a strong desire even though the content is not so popular. Unlike Haggle, earlier mobile content-sharing techniques were focused on maximizing the delivery ration and minimizing the delays regardless of importance to the users, but Haggle can dishearten the transmission of less desired contents so that energy, storage, and bandwidth can be saved and the dissemination of junk can be minimized (Odiyo, 2010).

PhotoShare enables users to share images captured with a smartphone's camera. MobiClique (Pietiläinen et al., 2009) and Opportunistic Twitter (Ristanovic et al., 2012) allow use of Facebook and Twitter in an ad hoc manner. MailProxy facilitates e-mailing with no infrastructure. Haggle-ETT (Martín-Campillo et al., 2010) grants an electronic triage tag for calamity regions; these are some prominent applications constructed on the top of Haggle. Haggle enables these applications to apparently share and transmit the contents over ad hoc networks and current LANs without the support of any mediate service provider. Neighboring devices share their interests and push the common contents in order of priority, and the received contents are stored locally until a new transmitting occasion appears. This implies that contents move according to their relative priority, without the requirement of end-host identity or address.

In comparison with related systems like DTN (Fall, 2003), PodNet (Lenders et al., 2007), and existing pub/sub systems (Eugster et al., 2003), one of the major contributions of Haggle is the implementation of push-based search dissemination as the fundamental technique for content sharing. Other contributions are the advanced approach to use forwarding algorithms to estimate content representatives, accounting for the correct abstractions and basics, and how it can be combined together to build a system. Haggle enables smartphone users to share contents according to importance and prioritize them when node contacts are bounded by time.

12.12 Conclusion

Communication and computing mechanisms of opportunistic networks provide an emerging computing paradigm for pervasive and intermittent networks, having tremendous opportunities in the future. By seeing the current growth and utilization of ubiquitous devices in our daily life, it seems that opportunistic networking will become an important part of our daily life in the near future. Therefore, despite having too many existing applications on opportunistic networking, there is a need to develop more efficient solutions to the many unaddressed issues still remaining in this fascinating computing domain. To make this paradigm more popular, secure, reliable, and fault-tolerant software architectures are needed that are compatible with the highly dynamic, distributed, and intermittent networks. Along with all these properties, architectures should also have immense flexibility, scalability, adoptability, and user friendliness for successful acceptance by users. Opportunistic computing can be used to address the issues of many challenging application areas. The main critical application areas of opportunistic networking include disaster and crisis management systems, info-mobility services and intelligent transportation systems, pervasive healthcare systems, and social- and context-aware computing systems. This chapter presented many suitable existing frameworks, case studies, and experiments of the prominent application areas of the opportunistic networking domain. It also focused on several challenges and issues that an application designer faces while developing applications for opportunistic networks. Each existing application is explored by looking more in depth at some of the critical issues, including security, legacy system support, and user interfaces. The chapter presented an in-depth study of 10 existing applications that have shown different aspects of opportunistic networking.

References

Ahmed, S. H., Kang, H., and Kim, D. (2015, January). Vehicular delay tolerant network (VDTN): Routing perspectives. In *Consumer Communications and Networking Conference (CCNC), 2015 12th Annual IEEE* (pp. 898–903).

Allen, S. M., Conti, M., Crowcroft, J., Dunbar, R., Mendes, J. F., Molva, R., et al. (2008, October). Social networking for pervasive adaptation. In *Second IEEE International Conference on Self-Adaptive and Self-Organizing Systems Workshops, 2008 (SASOW 2008)*, Venice, Italy (pp. 49–54).

Altman, E., Dini, P., Miorandi, D., and Schreckling, D. (2010). Paradigms for biologically-inspired autonomic networks and services. *The BIONETS Project eBook*.

Boldrini, C., Conti, M., and Passarella, A. (2008). Autonomic behaviour of opportunistic network routing. *International Journal of Autonomous and Adaptive Communications Systems*, 1(1), 122–147.

Borgia, E., Conti, M., Delmastro, F., and Pelusi, L. (2005). Lessons from an ad hoc network test-bed: Middleware and routing issues. *Ad Hoc & Sensor Wireless Networks*, 1(1–2), 125–157.

Carreras, I., Chlamtac, I., De Pellegrini, F., and Miorandi, D. (2007). Bionets: Bio-inspired networking for pervasive communication environments. *IEEE Transactions on Vehicular Technology*, 56(1), 218–229.

Carreras, I., Chlamtac, I., Woesner, H., and Kiraly, C. (2004, October). BIONETS: BIO-inspired NExt generaTion networks. In *Workshop on Autonomic Communication* (pp. 245–252). Springer, Berlin.

Chawla, R. and Bhatnagar, S. (2004, May). Online delivery of land titles to rural farmers in Karnataka, India. In *Scaling Up Poverty Reduction: A Global Learning Process and Conference*, Shanghai (pp. 25–27).

Chea, S., Luo, M. M., and Bui, T. X. (2009, January). If you build it, they will use: Usage motivations and unintended effects of the Internet village Motoman project in rural Cambodia. In *42nd Hawaii International Conference on System Sciences, 2009 (HICSS'09)* (pp. 1–8).

Clarke, I. (1999). *A Distributed Decentralised Information Storage and Retrieval System* (doctoral dissertation, master's thesis, University of Edinburgh).

Clarke, I., Sandberg, O., Wiley, B., and Hong, T. W. (2001). Freenet: A distributed anonymous information storage and retrieval system. In *Designing Privacy Enhancing Technologies* (pp. 46–66). Springer, Berlin.

Conti, M. and Giordano, S. (2007a). Multihop ad hoc networking: The reality. *IEEE Communications Magazine*, 45(4), 88–95.

Conti, M. and Giordano, S. (2007b). Multihop ad hoc networking: The theory. *IEEE Communications Magazine*, 45(4), 78–86.

Conti, M. and Kumar, M. (2010). Opportunities in opportunistic computing. *Computer*, 43(1), 42–50.

Eugster, P. T., Felber, P. A., Guerraoui, R., and Kermarrec, A. M. (2003). The many faces of publish/subscribe. *ACM Computing Surveys (CSUR)*, 35(2), 114–131.

Fall, K. (2003, August). A delay-tolerant network architecture for challenged internets. In *Proceedings of the 2003 Conference on Applications, Technologies, Architectures, and Protocols for Computer Communications*, Karlsruhe, Germany (pp. 27–34).

Gao, S., Niu, Y., Huo, H., and Zhang, H. (2007, December). An energy efficient communication protocol based on data equilibrium in mobile wireless sensor network. In *International Conference on Mobile Ad-Hoc and Sensor Networks* (pp. 433–444). Springer, Berlin.

Hasson, A. A., Fletcher, R., and Pentland, A. (2003). DakNet: A road to universal broadband connectivity. In *Wireless Internet UN ICT Conference Case Study* (pp. 1–9).

Isento, J. N., Rodrigues, J. J., Dias, J. A., Paula, M. C., and Vinel, A. (2013). Vehicular delay-tolerant networks? A novel solution for vehicular communications. *IEEE Intelligent Transportation Systems Magazine*, 5(4), 10–19.

Jain, S., Fall, K., and Patra, R. (2004). Routing in a delay tolerant network. *ACM SIGCOMM Computer Communication Review*, 34(4), 145–158.

Juang, P., Oki, H., Wang, Y., Martonosi, M., Peh, L. S., and Rubenstein, D. (2002). Energy-efficient computing for wildlife tracking: Design tradeoffs and early experiences with ZebraNet. *ACM SIGARCH Computer Architecture News*, 30(5), 96–107.

Kazemeyni, F., Johnsen, E. B., Owe, O., and Balasingham, I. (2012, June). MULE-based wireless sensor networks: Probabilistic modeling and quantitative analysis. In *International Conference on Integrated Formal Methods* (pp. 143–157). Springer, Berlin.

Lenders, V., Karlsson, G., and May, M. (2007, June). Wireless ad hoc podcasting. In *4th Annual IEEE Communications Society Conference on Sensor, Mesh and Ad Hoc Communications and Networks, 2007 (SECON'07)*, San Diego, CA (pp. 273–283).

Liu, T., Sadler, C. M., Zhang, P., and Martonosi, M. (2004, June). Implementing software on resource-constrained mobile sensors: Experiences with Impala and ZebraNet. In *Proceedings of the 2nd International Conference on Mobile Systems, Applications, and Services*, Boston, MA (pp. 256–269).

Martín-Campillo, A., Crowcroft, J., Yoneki, E., Martí, R., and Martínez-García, C. (2010, February). Using Haggle to create an electronic triage tag. In *Proceedings of the Second International Workshop on Mobile Opportunistic Networking*, Pisa, Italy (pp. 167–170).

Nordström, E., Gunningberg, P., and Rohner, C. (2012). Haggle: Relevance-aware content sharing for mobile devices using search. http://user.it.uu.se/~erikn/papers/haggle-arch.pdf.

Odiyo, B. (2010). An investigative study of the INFANT-Haggle. http://uu.diva-portal.org/smash/get/diva2:297974/FULLTEXT01.pdf.

Pelusi, L., Passarella, A., and Conti, M. (2006). Opportunistic networking: Data forwarding in disconnected mobile ad hoc networks. *IEEE Communications Magazine*, 44(11), 134–141.

Pentland, A., Fletcher, R., and Hasson, A. (2004). Daknet: Rethinking connectivity in developing nations. *Computer*, 37(1), 78–83.

Pietiläinen, A. K., Oliver, E., LeBrun, J., Varghese, G., and Diot, C. (2009, August). MobiClique: Middleware for mobile social networking. In *Proceedings of the 2nd ACM Workshop on Online Social Networks*, Barcelona, Spain (pp. 49–54).

Pitkänen, M., Kärkkäinen, T., Ott, J., Conti, M., Passarella, A., Giordano, S., et al. (2012, August). SCAMPI: Service platform for social aware mobile and pervasive computing. In *Proceedings of the First Edition of the MCC Workshop on Mobile Cloud Computing*, Helsinki, Finland (pp. 7–12).

Ristanovic, N., Theodorakopoulos, G., and Le Boudec, J. Y. (2012, March). Traps and pitfalls of using contact traces in performance studies of opportunistic networks. In *INFOCOM, 2012 Proceedings IEEE*, Orlando, FL (pp. 1377–1385).

Scott, J., Crowcroft, J., Hui, P., and Diot, C. (2006, January). Haggle: A networking architecture designed around mobile users. In *WONS 2006: Third Annual Conference on Wireless On-Demand Network Systems and Services*, France (pp. 78–86).

Shah, R., Roy, S., and Jain, S. (2003). Data mules: Modeling a three-tier architecture for sparse sensor networks. *Elsevier Ad Hoc Networks Journal: Sensor Networks Applications & Protocols*, 1(2), 30–41.

Small, T. and Haas, Z. J. (2003, June). The shared wireless infostation model: A new ad hoc networking paradigm (or where there is a whale, there is a way). In *Proceedings of the 4th ACM International Symposium on Mobile Ad Hoc Networking & Computing*, Annapolis, MD (pp. 233–244).

Verma, A. and Pattanaik, K. K. (2017). Routing protocols in opportunistic networks. In *Opportunistic Networking: Vehicular, D2D and Cognitive Radio Networks*, (pp. 125–166). CRC Press, Boca Raton, FL.

Verma, A., Pattanaik, K. K., and Ingavale, A. (2013). Context-based routing protocols for oppnets. In *Routing in Opportunistic Networks* (pp. 69–97). Springer, New York.

Verma, A. and Srivastava, A. (2011). Integrated routing protocol for opportunistic networks. *International Journal of Advanced Computer Science and Applications*, 2(3), 85–92.

Zhang, P., Sadler, C. M., Lyon, S. A., and Martonosi, M. (2004, November). Hardware design experiences in ZebraNet. In *Proceedings of the 2nd International Conference on Embedded Networked Sensor Systems*, Baltimore, MD (pp. 227–238).

13

Time and Data-Driven Triggering to Emulate Cross-Layer Feedback in Opportunistic Networks

Rintu Nath

CONTENTS

13.1 Introduction ... 247
13.2 Introduction to Emulation in an Opportunistic Network ... 248
 13.2.1 Network Emulation Basics ... 248
 13.2.2 Simulation, Emulation, and Real World Experiments ... 248
 13.2.3 Advantages of Emulation in an Opportunistic Network ... 249
 13.2.4 Challenges ... 249
13.3 Emulation Design Aspects for an Opportunistic Network ... 250
 13.3.1 Pervasive Computing ... 250
 13.3.2 Delay-Tolerant Network (DTN) Architecture ... 250
 13.3.3 Bundle Protocols for Overlay Network ... 251
 13.3.4 Security: Authentication and Privacy ... 252
13.4 Emulator Testbeds ... 252
 13.4.1 Overview ... 252
 13.4.2 Quality Observation and Mobility Experiment Tools (QOMET) ... 252
 13.4.3 QOMB Testbed ... 254
13.5 Time and Data-Driven Triggering to Emulate Cross-Layer Feedback in Opportunistic Networks ... 254
 13.5.1 Pattern Generation ... 255
 13.5.2 Pattern and Scenario Files ... 255
 13.5.3 KauNet Triggering ... 255
13.6 Research Directions ... 256
Key Terminology & Definitions ... 256
References ... 257

13.1 Introduction

The performance of a network can be evaluated by simulation, emulation, and live experiments. However, experiments in an emulated environment provide several advantages in the case of opportunistic networks. Multi-hop mobile communication often suffers from connectivity breaks and low node density. Traditional network communication protocols may not be able to guarantee end-to-end connectivity in such cases. Delay-tolerant networking (DTN) is a communication networking paradigm that ensures reliable node-to-node communication where end-to-end connections may not be guaranteed and intermittent data come at different intervals. Simulators in such cases may provide results that are far from reality. Real-time experiments are ideal for assessing system performance before actual deployment. However, real-time systems have time and financial constraints. Moreover, systems that are designed to be deployed during disaster management or emergency conditions cannot be evaluated before the event. Network emulation is ideal in such cases. A number of emulators are available to evaluate protocols designed for opportunistic networks.

In this book chapter, I will deliberate upon time-driven triggering to emulate opportunistic networks, emulation design aspects, and different emulation abstraction techniques. Research directions in cross-layer information exchange emulation for an opportunistic network are discussed.

13.2 Introduction to Emulation in an Opportunistic Network

13.2.1 Network Emulation Basics

The rapidly growing distributed computing architecture and functionalities need adequate testing to meet quality of service (QoS) requirements (Esmailpour, 2016). Emulation platforms are ideally suited in such cases as it enables us to use the real implementation of protocols and applications without deploying the actual network and devices.

A network emulator is expected to measure the throughput and responsiveness of a network. It may also be used for troubleshooting for a real network. Network emulators are able to mimic a client/server configuration without the need for a router or even live traffic (Jin et al., 2014). Emulators are available as hardware or software solutions.

13.2.2 Simulation, Emulation, and Real World Experiments

Both practically and semantically, a network emulator is different from a network simulator. An emulator is used to test the performance of a real network. In a network emulator, end-systems such as a node can be attached to the emulator. A network emulator mirrors the network, which connects end-systems (Hahn et al., 2011).

A network simulator, on the other hand, is a mathematical modeling that creates a virtual network. A simulator is programmed to mimic an actual network under different load configurations with variable node and latency. OPNET, NetSim, and ns (open source) are examples of a few simulators.

The Common Open Research Emulator (CORE) is a network emulator for hybrid topologies (Ahrenholz, 2010). In real time, CORE builds a representation of a computer network. It was developed by a network technology research group, a part of the Boeing Research and Technology division. The emulator can be connected to physical networks and routers (Figure 13.1). It provides an environment for running real applications and protocols. CORE is used for the evaluation of networking scenarios and protocol testing. Scalability of a network may also be tested using CORE.

The backend of CORE is the CORE daemon, which manages emulation sessions. The graphical user interface (GUI) and the daemon communicate using a socket-based application programming interface known as the CORE API. The nodes and networks come together via interfaces installed on nodes. The emulator allows the running of different

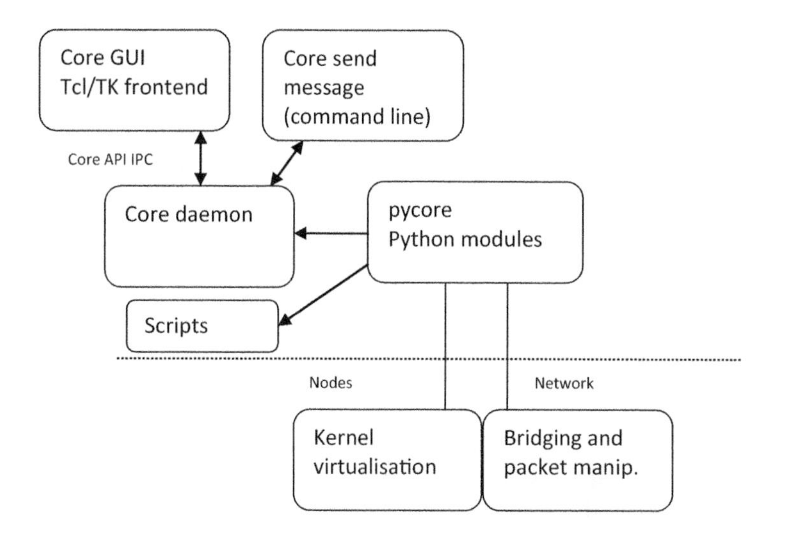

FIGURE 13.1
CORE architecture.

network components through socket-based API. The CORE daemon can also manage sessions of virtual nodes and networks. A CORE node is more lightweight than a full virtual machine. Tens of nodes can be created on a standard computer. Networking stacks can be re-used in CORE framework. The source code is available and easy to understand and modify.

CORE emulates networks using the virtual network stacks of operating systems. In order to emulate, CORE needs to run on commodity computers. CORE itself is not an instantiation of hardware or a testbed.

ns-3 is a network simulator used to simulate networking and routing protocols. It is a discrete-event-based simulator and is able to test routing and queuing algorithms and run multicast protocols and IP protocols, such as the User Datagram Protocol (UDP) or the Transmission Control Protocol (TCP), over wired and wireless networks. The ns-3 simulation core supports both IP- and non-IP-based networks (Figure 13.2).

ns-3 is built as a system of software libraries that work together. User programs can be written that links with these libraries. To deploy ns3, the target system should build the libraries first, then build the user program.

13.2.3 Advantages of Emulation in an Opportunistic Network

A large-scale opportunistic network requires extensive evaluation before actual deployment. Testbeds usually include tests over large networks that have multiple switches and nodes. Hence, network emulation can be a cost-effective solution before real deployment. A real client/server application service can be deployed in an emulated network. Scaling up is possible by increasing the number of clients on a single machine.

13.2.4 Challenges

Computing nodes in an opportunistic network are often mobile devices that have limited resources. One of the scenarios in configuring emulators is to consider nodes outside the emulator machine. UDP tunnels are created to transport Ethernet frames between the

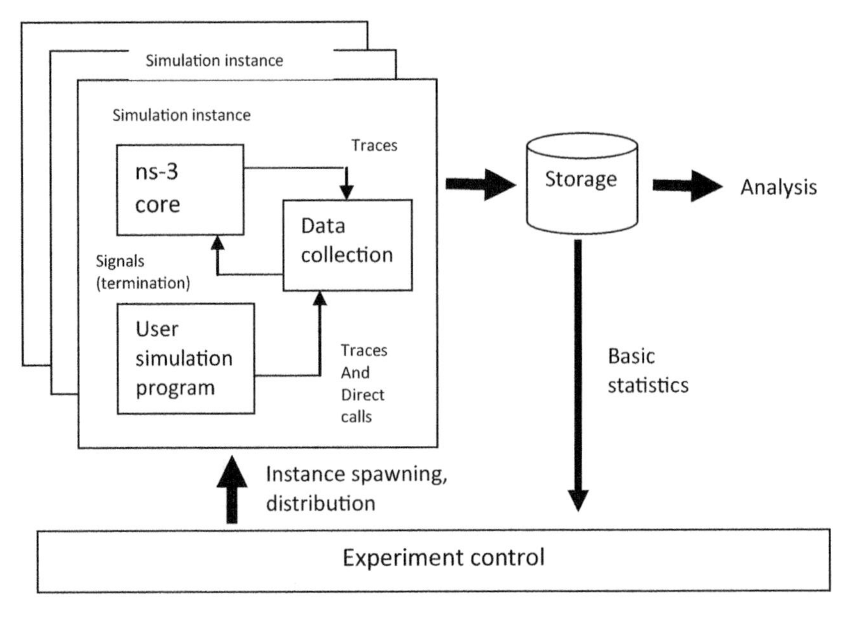

FIGURE 13.2
ns-3 architecture.

emulator and each application. Therefore, the distribution is transparent to the emulator. Such an environment may be hard to configure. Moreover, network scenarios may vary from simple topologies to complicated designs. Thus, configuration should be automated as much as possible.

13.3 Emulation Design Aspects for an Opportunistic Network

13.3.1 Pervasive Computing

Pervasive computing is the evolution of distributed computing where computing devices are integrated for information processing. It is also referred to as ubiquitous computing. It enables computation and exchange of information in a distributed platform whenever and wherever needed. Computing devices interact and communicate with each other over wireless networks.

In wireless networks, mobile devices are equipped to directly share resources without relying on any centralized servers or infrastructure support. The unavailability of a universal trust mechanism, the dynamic population of participating nodes, and the limited resources of computing nodes are some of the challenges of pervasive computing.

13.3.2 Delay-Tolerant Network (DTN) Architecture

A mobile ad hoc network like terrestrial mobile networks, sensor networks, or satellite communication often experience high latency, disconnection, and long queuing times. Other requirements like interoperability considerations and low power of computing devices make the task of reliable communication even more challenging. Delay-tolerant networking provides a network architecture that is independent of messaging service. DTN operates above the transport layer for various network architectures and provides store and forward functionality for dissimilar and special networks (Li et al., 2015). DTN is required to store messages in non-volatile memory when reliable delivery is required and provides name-mapping via globally unique tuples. The structure of a DTN scheduler is given in Figure 13.3.

A message transferred to a DTN node, which has large amounts of non-volatile storage, is classified as persistent. It can hold the message until the next communication opportunity. In DTN, an end-to-end routing path cannot be assumed. Instead, routes are classified as a cascade of time-dependent contacts called communication opportunities, characterized by start and end times, latency, direction, and capacity.

13.3.3 Bundle Protocols for Overlay Network

In DTN, a bundle is defined as a data unit. Multiple instances of the same bundle may exist in different parts of a network. However, different instances may have different representations. One of the instances may be in the local memory of a bundle node, whereas another instance may be in the transit. A bundle payload, or simply "payload", is the data to be transmitted to the bundle destination. If entire data is being carried, it is called nominal payload. Payload can be fragmented, for example the first *N* bytes of an *M* byte length data, where $N<M$. The destination address is provided as one of the parameters during bundle transmission. The node in a DTN that send or receives bundles is called a bundle node. A bundle node may be a process running on a general-purpose computer. However, a general bundle node has three components: (i) a bundle protocol agent; (ii) convergence layer adapters; and (iii) an application agent.

The bundle protocol agent of a node is a component that executes the procedures of the bundle protocol. A convergence layer adapter sends and receives bundles on behalf of the bundle protocol agent. The application agent of a node is the node component that utilizes the bundle protocol (BP) services (Figure 13.4).

BP is useful in networks that have intermittent connectivity, high probability of bit errors, and variable delay. BP is implemented at the application layer forming a store and

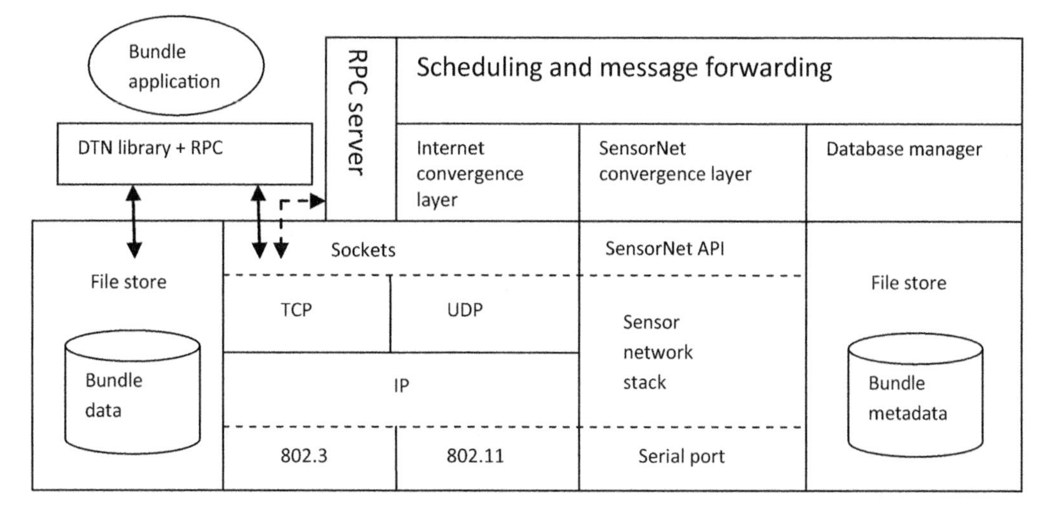

FIGURE 13.3
Structure of DTN scheduler. Multiple convergence layer, one per protocol stack provide a common interface to message scheduler.

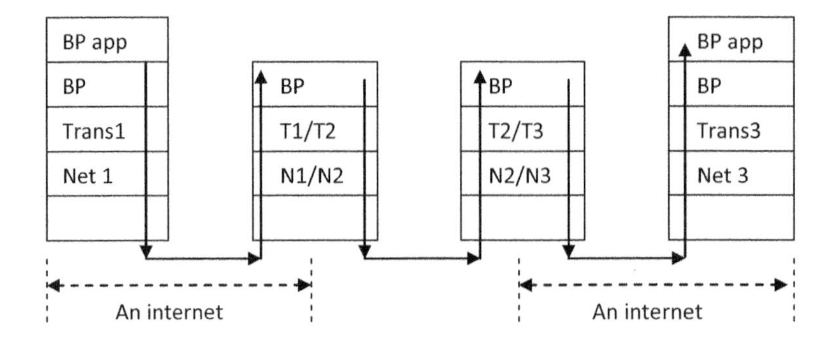

FIGURE 13.4
Bundle protocol sits at application layer.

forward network. BP is characterized by its ability to take advantage of opportunistic networks by late binding of overlay network.

13.3.4 Security: Authentication and Privacy

An opportunistic network uses open networks to transmit data. Intermediate nodes may be malicious (attack the proper network operations without considering their own gains) or selfish (intend to maximize their own gains while minimizing their contributions to it). In DTNs, bundles may traverse over underlying heterogeneous networks. The modification of messages (or bundles) in transit for malicious purposes is a big security threat. Due to the resource-scarcity characteristics of DTNs, unauthorized access and use of DTN resources can be a serious concern. Lack of end-to-end connectivity not only brings challenges to routing but also makes the existing security solutions unsuitable.

The Bundle Security Protocol Specification defines three types of security blocks that may be included in a bundle: i) the Bundle Authentication Block (BAB); ii) the Payload Integrity Block (PIB); and iii) The Payload Confidentiality Block (PCB).

13.4 Emulator Testbeds

13.4.1 Overview

Emulation provides fully controlled and reproducible environments where applications can be co-located in a local lab to develop and debug (Beuran et al., 2012). It is easy to reproduce an emulated network. It is cost-effective and scalable.

A testbed is a framework for testing protocols in the real world. The testbed consists of four elements: i) components (hardware and software); ii) monitoring architecture; iii) database; and iv) GUI for storing and analyzing results.

13.4.2 Quality Observation and Mobility Experiment Tools (QOMET)

It is possible to create a scalable wired-network infrastructure to perform wireless network emulation. QOMET is a wireless local area network (WLAN) emulator with two-stage scenario-driven emulation. In the first stage, the user provides a scenario representing a real world. The output of the first stage is a description of the network states at successive moments of time. Based on this, wireless environment conditions are produced in the second stage.

The scenario-driven architecture has two stages. In the first stage, from real-world scenario representation network quality degradation (ΔQ) description is created which corresponds to the real-world events (Figure 13.5). The change in network service quality between two points is called quality degradation and is denoted as ΔQ. In the second stage, emulator configuration is done using ΔQ description calculated in the first stage.

It is essential to calculate signal attenuation due to the distance for emulator configuration. Attenuation is modeled as large-scale path loss and small-scale fading component. If γ is the path loss exponent of the channel, the large-scale path loss over distance d is expressed as d^{γ}. Fading represents the deviation of signal from estimated attenuation (Figure 13.6).

Received power at a distance d from a reference point is expressed as:

$$U_R(d) = U_R(0) - 2.3\log\left(d^{\gamma}\right) - X_{\sigma}$$

where:

$U_R(0)$ is the power of a reference point
d^{γ} is the path loss
X_{σ} is Gaussian distributed random variable representing fading component

Frame error rate (FER) is a function of received power $U_R(d)$ and is represented as:

$$FER_1 = FER_s\, e^{\gamma(S-U_R)},$$

where:

FER_s is adapter specific frame error rate
FER_1 is equal to FER_s when $U_R(d)$ is equal to threshold value S

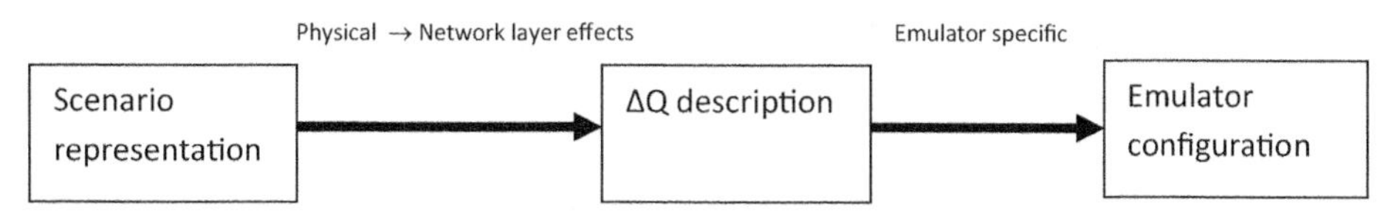

FIGURE 13.5
Two-stage scenario-driven approach to WLAN emulation.

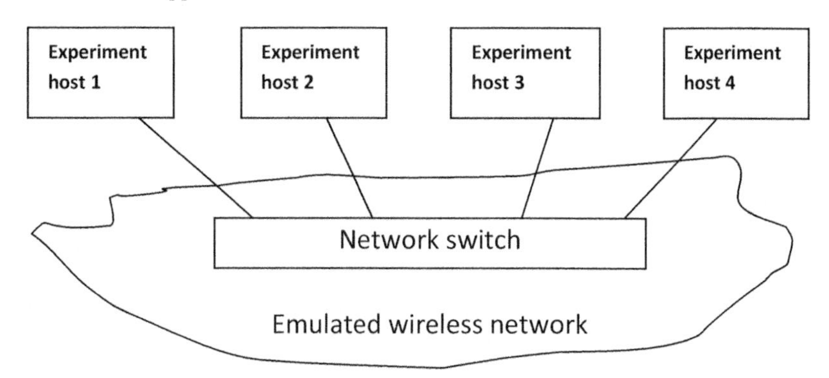

FIGURE 13.6
Distributed emulation setup.

There is another issue with the emulator – it is the ability to discern signal from noise. Bit error rate (BER) can be represented as a function of signal-to-noise ratio (SNR) and received power $U_R(d)$.

The next step is to introduce delay D. Delay is calculated as the weighted average of the delays due to retransmission. $U_R(d)$, FER, BER, and D are the emulator parameters for the data link layer. The next step is to compute network layer parameters. The packet loss rate (PLR) and perceived bandwidth (BW) are the two parameters need to be modeled. PLR is a function of FER.

13.4.3 QOMB Testbed

QOMB is a testbed designed for the evaluation of wireless network systems, protocols, and applications. The testbed uses the wireless network emulation set of tools, Quality Observation and Mobility Experiment Tools (QOMET). Corresponding to a given scenario, the testbed reproduces real-time wireless network conditions (Beuran, 2012).

Depending on traffic conditions, QOMB testbed supports multi-hop routing protocols such as Optimized Link State Routing Protocol (OLSR), and dynamic communication condition computation. Real implementation of routing protocols can also be carried out.

Corresponding to real-world events, QOMET uses a scenario representation to compute the point-to-point network quality degradation (ΔQ). ΔQ description is applied in real time using a dynamic library routine in the kernel module. Thus, the user-defined scenario of wireless conditions is recreated on the wired network testbed. Several processing steps are done in QOMB in order to realize traffic conditions and support multi-hop routing. Using the QOMB testbed, it is possible to implement routine protocols in different network scenarios.

13.5 Time and Data-Driven Triggering to Emulate Cross-Layer Feedback in Opportunistic Networks

Opportunistic networking protocols, apart from the exchange of packets, need to react to the dynamics of the underlying network. In order to support IP-level emulation of applications and protocols that react to lower layer events, emulation triggers may be implemented. Emulation triggers can be synchronized with other emulation effects and can emulate arbitrary cross-layer feedback.

The chapter focuses on the KauNet emulation system. The system offers emulation-based experiments with a greater control and flexibility. It is possible to reproduce the origin of patterns through collected traces and analytical expressions.

KauNet is designed based on a well-known emulator, Dummynet. Dummynet is a live network emulation tool, originally designed for testing networking protocols. It is also used for bandwidth management and the implementation of various scheduling algorithms. Dummynet has been in use for several years and provides a stable codebase. Dummynet can emulate packet loss, incorporate delay, and apply bandwidth restrictions. KauNet extends Dummynet by its ability to introduce bit errors. KauNet allows bit errors, packet losses, delay, and bandwidth changes on a per-packet or per-millisecond basis.

KauNet uses patterns that describe the desired changes. At the beginning of emulation, patterns are inserted into the kernel of KauNet. It allows exact control when conditions change, thus ensuring fine-grained control. Command line tools are used to incorporate dynamic events. Common usages of KauNet emulator are the verification of the transport layer protocol, performance evaluations of the transport layer and the application layer, and the emulation of hand-over scenarios.

Two modes of operation of KauNet are time-driven and data-driven. In the time-driven mode, the index advances every millisecond irrespective of any data is transferred or not. In case of data-driven mode, the index advances for each packet. The maximum resolution of a trigger pattern is, therefore, one millisecond or one packet. However, in case of a bit error pattern, the index advances according to the number of bits.

13.5.1 Pattern Generation

Initially, patterns are created and saved in kernel space. During emulation, emulator behavior is controlled by these patterns. It is possible to specify different flows to be sent to different pipes. This allows multiple connections between multiple hosts to be emulated using many different patterns. The *ipfw* command can be used to create multiple rules. A new pattern may be specified which will be used after the completion of the current pattern. The default setting is wrap-around and starts from the beginning after the completion of one pattern cycle.

In addition to the kernel-based pattern management, a command line tool *patt_gen* is also available to create and manage patterns. It is also capable of importing uncompressed pattern descriptions from simple text files. These text files can be generated by off-line simulators or trace collection equipment. In addition to command line tool, a GUI-based tool *pg_gui* is also available.

13.5.2 Pattern and Scenario Files

For each aspect, like packet losses, bit errors, delay, variable bandwidth, etc., a separate pattern file is required. Experiments require the simultaneous variation of more than one aspect – hence multiple patterns need to be managed at emulation runtime. The pattern files are stored and imported into the kernel using the *patt_gen* utility.

To simplify the management of pattern files, a scenario file format is defined. A scenario file consists of several pattern files with an additional header. Scenario files are also created using the *patt_gen* utility. The scenario file header includes a scenario ID (SID) and a textual description of the scenario. The emulator is configured to accept SIDs that have the correct checksum digit.

13.5.3 KauNet Triggering

The information-passing functionality is called triggers in KauNet. Triggers convey position control information during runtime. Trigger patterns can be either data- or time-driven. KauNet triggering is more prevalent in scenarios involving opportunistic networking. Cross-layer information like intermittent connectivity, inherent in an opportunistic network, can be emulated by triggering (Pérennou et al., 2011).

Consider a scenario with intermittent connectivity and availability of bandwidth during the period of connection changes. In such a case, the bandwidth variations that occur during the connectivity period can be modeled. The bandwidth variation model is combined with trigger patterns that generate the upward flowing connectivity information. The bandwidth and trigger patterns are synchronized with each other to form a consistent emulation scenario.

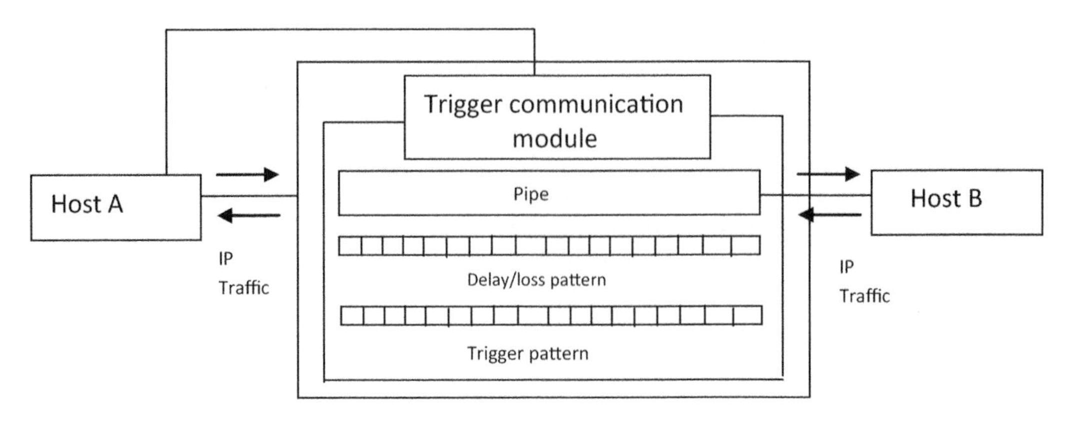

FIGURE 13.7
Time-driven emulation triggering.

Figure 13.7 shows a general view of an emulation setup using triggers. In this setup, two hosts are connected using a KauNet-enabled host. The KauNet host emulates the conditions of the particular link or network. Emulation is carried out for bandwidth variability, delay, bit error, and packet loss. The trigger pattern is located at the KauNet host. Trigger patterns are created using the *patt_gen* utility.

In order to extend the trigger functionality of KauNet to external processes, a mechanism to transfer the trigger value of a fired trigger to a receiver is needed. This functionality is implemented by the trigger communication module. This module enables receiving triggers from a certain pipe using the UDP interface. The recipient of a trigger should be able to reside in either user space or kernel space, locally or on another host.

To conclude, opportunistic communication that depends on cross-layer information can be emulated by the pattern-driven triggers of the KauNet emulation system. The triggers can be controlled and synchronized with other emulation effects. The triggers are distributed to local and remote processes by the trigger communication module. The emulator allows fine-grained and repeatable control of bit errors, packet losses, and packet delays.

13.6 Research Directions

Efficient pattern handling inside the kernel of KauNet will improve the throughput of the emulator. The reduction in computation overhead to emulate delay, bit error, and variable bandwidth for an opportunistic network requires further research. Patterns are stored in the kernel using a compressed format and decompression occurs stepwise as the pattern information is consumed. Future research may be carried out to evolve better compression and decompression techniques.

Key Terminology & Definitions

Dynamic Source Routing (DSR): A routing protocol used in multi-hop wireless ad hoc networks of mobile nodes. DSR allows the network to be completely self-organizing and self-configuring, without the need for any existing network infrastructure or administration conditions. DSR forms a route only when a transmitting node requests one. However, it uses source routing instead of relying on the routing table at each intermediate device.

Network emulation: Experimental techniques that evaluate networking performance from the execution point of view. It provides different levels of abstractions and controls over the experiment along with a range of experimental conditions that can be reproduced. Emulation is done in real time and allows dynamic condition change options. Hence, network emulation can be done in a wide range of conditions depending on constraints like cost, time, etc. Performance evaluation for DTN using emulation is more effective.

Optimized Link State Routing (OLSR): An IP routing protocol optimized for mobile ad hoc networks, which can also be used on other wireless ad hoc networks. A few selected nodes called multi-point relays (MPRs) forward broadcast messages during the flooding process. This technique substantially reduces the message overhead as compared to a classical flooding mechanism, where every node retransmits each message when it receives the first copy of the message. Link state information is generated only by nodes elected as MPRs. Thus, a second optimization is achieved by minimizing the number of control messages flooded in the network.

References

Ahrenholz, J. (2010). Comparison of CORE network emulation platforms. *2010 – MILCOM Military Communications Conference*. http://dx.doi.org/10.1109/milcom.2010.5680218

Beuran, R. (2012). *QOMET User's Guide*. Retrieved from http://www.nict.go.jp

Beuran, R., Nakata, J., Tan, Y., & Shinoda, Y. (2012). Emulation testbed for IEEE 802.15.4 networked systems. *IEICE Transactions on Communications*, E95.B(9), 2892–2905.

Esmailpour, A. (2016). Interoperability among broadband wireless technologies in a testbed environment. *International Journal of Wireless Information Networks*, 23(2), 151–161.

Hahn, D., Lee, G. Walker, B. Beecher, M. & Mundur P. (2011). Using virtualization and live migration in a scalable mobile wireless testbed. *ACM Sigmetrics Performance Evaluation Review*, 38(3), 21.

Jin, D., Zheng, Y., & Nicol D. (2014). A parallel network simulation and virtual time-based network emulation testbed. *Journal of Simulation*, 8(3), 206–214.

Li, Y., Hui, P., Jin, D., & Chen, S. (2015). Delay-tolerant network protocol testing and evaluation. *IEEE Communications Magazine*, 53(1), 258–266.

Pérennou, T., Brunstrom, A., Hall, T., Garcia, J., & Hurtig, P. (2011). Emulating opportunistic networks with KauNet triggers. *EURASIP Journal on Wireless Communications and Networking*, 1–14.

14

Applications of DTN

Rahul Johari, Prachi Garg, Riya Bhatia, Kalpana Gupta, and Afreen Fatimah

CONTENTS

14.1 Applications 259
References 266

14.1 Applications

1.1 *Disaster-hit areas*: Delay-tolerant networks (DTNs) can be used in disaster-hit areas where wired and wireless communication is badly hampered due to natural or manmade calamities (Jonson, Pezeshki, Chao, Smith, & Fazio, 2008). Figure 14.1 depicts the DTN deployed in a disaster-hit area. It depicts a how ship stuck in a cyclone can be rescued with help from the Disaster Management Authority (DMA). The captain sends a message to the DMA to help the people on board. The message is communicated over the DTN because it doesn't allow a delay to vitiate the transmission. Similarly, Figure 14.2 represents a person stuck in blaze who sends an SOS message to his neighbor. The person sends a packet over the DTN so as to ensure the delivery of the message.

1.2 *Border regions/cross-border areas*: DTNs can be used in border regions or cross-border areas which are not easily accessible to human beings either due to poor geographical terrain or extremely low temperature, like in the Siachen Glacier. Figure 14.3 depicts a border region where troops are deployed, each having his own DTN node. The commanding officer, sitting at the control office, can send a message over the DTN to the troops. Troops can receive orders from the officer over the network.

2. *Assisting people with disabilities*: DTNs can be used to assist people with disabilities, like visually-challenged people, deaf-mute people, etc., in their daily life. Figure 14.4 depicts a disabled person on a wheelchair sending a message to a help desk to help him to climb up a staircase. DTN nodes are deployed to establish the network. Similarly, Figure 14.5 represents a person with a walking stick that has a DTN node installed in it. The node stores a map of the city and develops a force in the stick in the direction where the person is supposed to go. Also, it informs the person about pits on the road.

3. *Assisting people with disabilities*: DTNs can be applied in smart city informatics/rural informatics, wherein they can be used in the cleaning of clogged drains or in the disposal of garbage. Figure 14.6 depicts the DTN nodes being installed in the drainage pipes. These nodes detect the speed across the pipes. As soon as a node

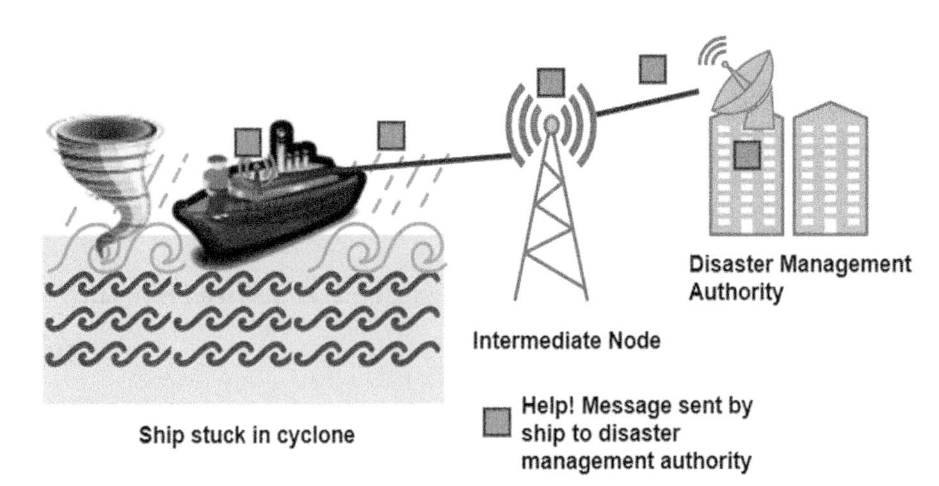

FIGURE 14.1
Delay-tolerant network in cyclone.

FIGURE 14.2
Disaster-hit area.

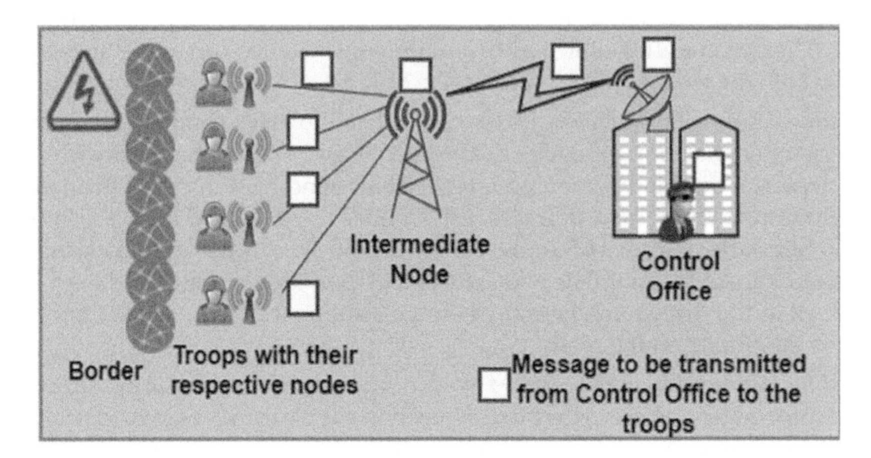

FIGURE 14.3
Border region.

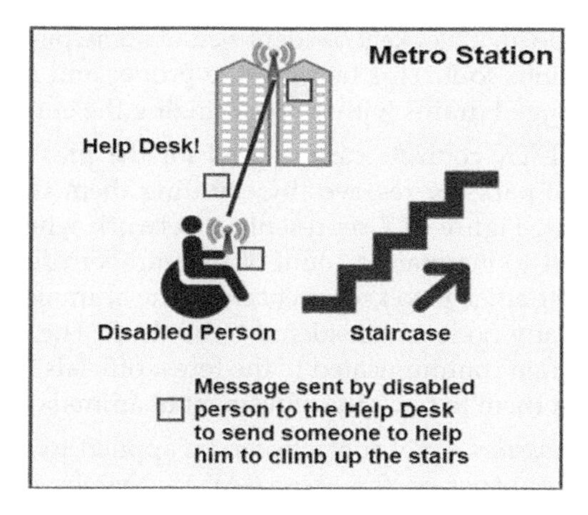

FIGURE 14.4
Delay-tolerant network deployed for disabled person.

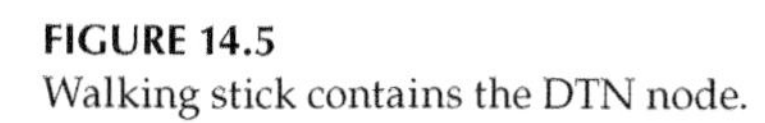

FIGURE 14.5
Walking stick contains the DTN node.

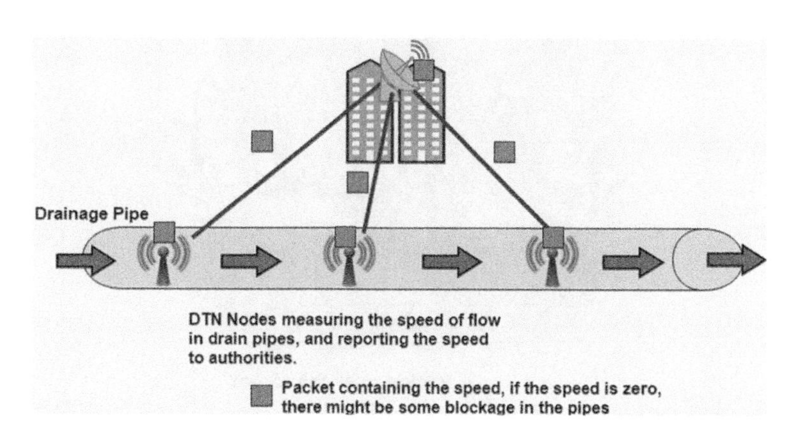

FIGURE 14.6
Delay-tolerant network deployed in drainage system.

reports a nought speed, a blockage is identified at some place between that node and the node previous to it. This helps in the proper and efficient identification and cleaning of clogged drains without scrutinizing the entire pipe.

4. *Preserving wildlife*: DTN contacts can be used for the preservation of birds and wildlife in national parks or reserves by counting them through keeping track of their movements. Figure 14.7 represents a network wherein DTN nodes are deployed in a forest to maintain a count of the number of animals in the forest. This helps the forest officials to keep a check on the animals. An animal not seen for many days by any node is considered to be dead. The count maintained by different nodes is then communicated to the forest officials. The DTN nodes have sensors installed in them to track the movement of animals around it.
5. *Predicting natural disasters*: DTN contacts can be applied to carrying out communication between countries on the ocean/seabed, thereby creating an underwater oceanic marine network. Figure 14.8 depicts the DTN established underwater. This underwater oceanic marine network is used to detect the speed of flowing water and currents to predict the occurrence of cyclones. Each node detects the speed and transmits it to officials, who interpret and analyse the speed to provide intimation of any natural calamity.
6. *Monitoring temperature/traffic*: By mounting them on the rooftops of city buses, DTN nodes can be used to sense the traffic and/or temperature of the regions in which they are operating (Hirakawa, Uchida, Arai, & Shibata, 2015). Figure 14.9 represents buses which have DTN nodes deployed on their rooftops. These sensors can be used to record the snarled-up traffic and notify the officials of traffic jams at those particular locations. Officials can then update the shortest route shown on Google Maps to make the people aware of different traffic jams in the city.
7. *Communication in difficult terrain*: DTNs can be used to provide communication facilities in difficult terrain where it is not feasible to set up wired and wireless communication, like in mountainous regions such as in the Himalayas, or in desert regions like the Thar desert in Rajasthan. Figure 14.10 represents the ease

FIGURE 14.7
Delay-tolerant network for wildlife protection.

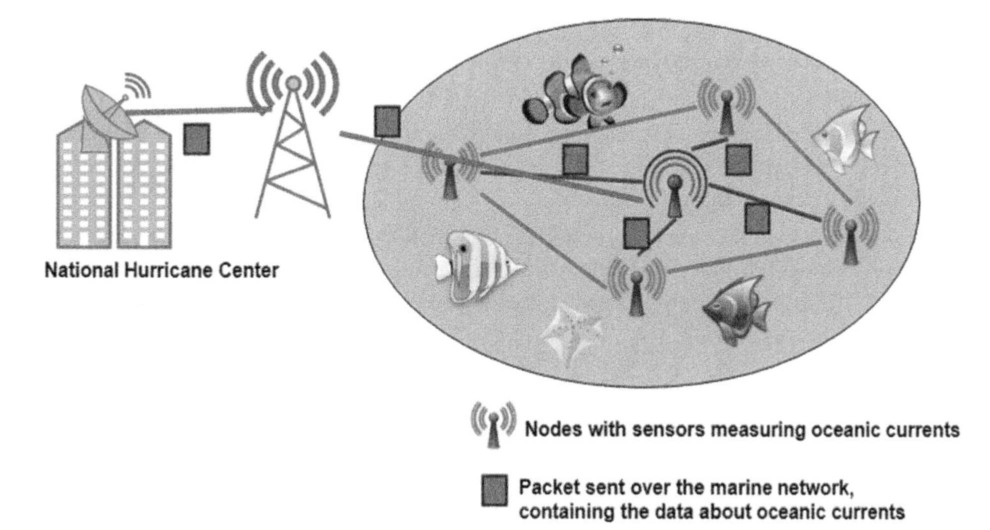

FIGURE 14.8
Oceanic bed network.

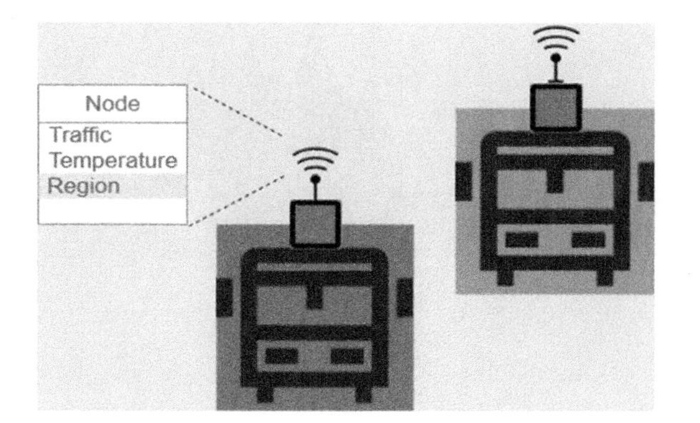

FIGURE 14.9
DTN nodes on the rooftop of the buses.

FIGURE 14.10
DTN deployed in mountainous regions.

of communication in areas that may lack continuous network connectivity. The nodes can be easily deployed in such areas and the communication network can be established.

8. *Preserving the biosphere*: DTNs can be used for the preservation of the plants and trees in forests, thereby preserving the biosphere. Figure 14.11 represents the DTN deployed to preserve the biosphere.
9. *Post office scenario*: The exchange/delivery of different messages by postmen to villagers or the exchange of different messages among different villages, e.g. Daknet. Figure 14.12 depicts messages being delivered by postmen to villagers across the delay-tolerant network.

FIGURE 14.11
Biological biosphere.

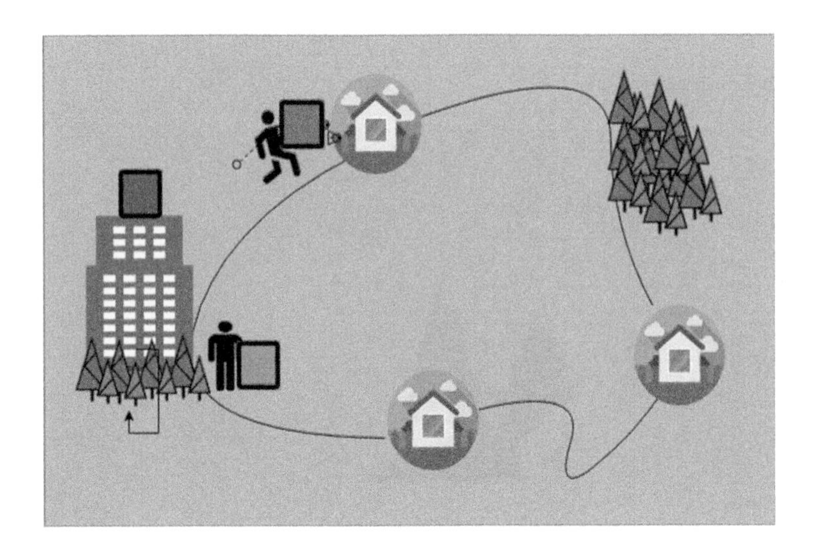

FIGURE 14.12
Post office scenario.

10. *Traffic sensors*: DTNs can be used to monitor traffic during peak hours. The purpose is to anticipate traffic jams and thus help travellers to decide their routes accordingly. This can be deployed by attaching sensors on each traffic signal and whenever a request is generated for a particular signal, it can respond. In this case, latency in communication is permissible as people generally enquire about such traffic issues before leaving from home. Figure 14.13 represents a sensor, deployed

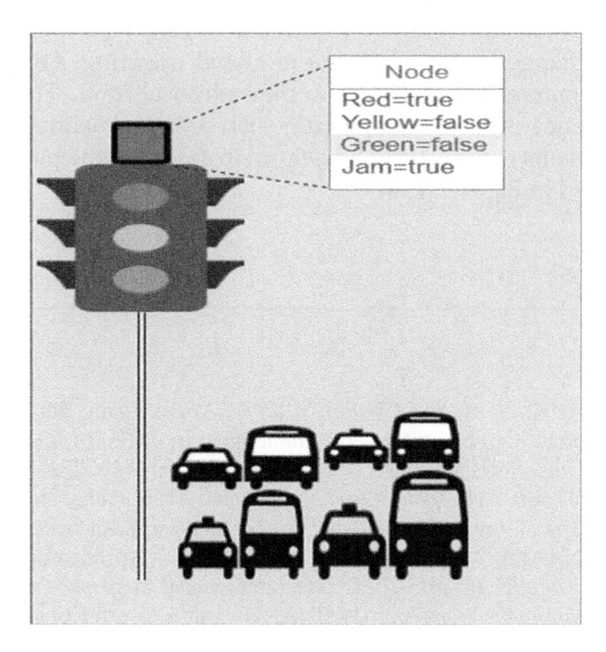

FIGURE 14.13
Traffic sensors.

FIGURE 14.14
DTN for preserving biological biosphere.

on a traffic signal, which records the traffic snarls at every signal and sends the report to officials, who in turn update the shortest route on the maps used by people to commute.

11. *Communication in deserted places*: Providing communication facilities in economically infeasible areas like remote areas/poor villages which are distant from the main cities and towns (Ntareme, Zennaro, & Pehrson, 2011). DTNs can also be deployed to gather information from very far-off and remote places, like Antarctica. The information can be acquired about any new species evolving there or topological changes happening due to global warming. This can be done by tracing some remarkable differences in their physical traits. This can be done by placing DTN nodes capable of capturing such data and sending back images or statistics. The results received will be helpful in studying the patterns of evolution there, as depicted in Figure 14.14.

References

Benhamida, F. Z., Bouabdellah, A., and Challal, Y. (2017, April). Using delay tolerant network for the Internet of things: Opportunities and challenges. In *2017 8th International Conference on Information and Communication Systems (ICICS)*, Irbid, Jordan, (pp. 252–257).

Fall, K. (2003). *Delay-tolerant Networking: Architecture and Applications*. Intel Research, Berkeley. Available from: https://www.cl.cam.ac.uk/~jac22/talks/sta/dtn-history.pdf.

Hirakawa, G., Uchida, N., Arai, Y., and Shibata, Y. (2015, March). Application of DTN to the vehicle sensor platform CoMoSE. In *2015 IEEE 29th International Conference on Advanced Information Networking and Applications Workshops (WAINA)*, Gwangiu, South Korea, (pp. 490–493).

Jonson, T., Pezeshki, J., Chao, V., Smith, K., and Fazio, J. (2008, November). Application of delay tolerant networking (DTN) in airborne networks. In *Military Communications Conference, 2008, MILCOM 2008 IEEE*, San Diego, CA, (pp. 1–7).

Ntareme, H., Zennaro, M., and Pehrson, B. (2011, September). Delay tolerant network on smartphones: Applications for communication challenged areas. In *Proceedings of the 3rd Extreme Conference on Communication: The Amazon Expedition*, Manaus, Brazil, (p. 14).

Raj, V. S. and Chezian, R. M. (2013). DELAY–Disruption Tolerant Network (DTN), its network characteristics and core applications. In *International Journal of Computer Science and Mobile Computing*, 2(9), (pp. 256–262).

Schildt, S. Wolf, L. Recent Trends: DTN Introduction and Applications. Available from: https://www.ibr.cs.tu-bs.de/courses/ss12/lnm/RecentTrendsDTNIntroduction.pdf.

Sun, W., Liu, C., and Wang, D. (2011, April). On delay-tolerant networking and its applications. In *International Conference on Computer Science and Information Technology (ICCSIT 2011)*, Venice, Italy, (pp. 238–244).

Vasilakos, A. V., Zhang, Y., and Spyropoulos, T. (Eds.). (2016). *Delay Tolerant Networks: Protocols and Applications*. Boca Raton, FL: CRC Press.

15

Performance Evaluation of Social-Aware Routing Protocols in an Opportunistic Network

Makshudur Rahman and Md. Sharif Hossen

CONTENTS

15.1 Introduction ... 267
15.2 Social-Aware Routing Protocols ... 268
15.2.1 Social-Aware Content-Based Opportunistic Routing Protocol (SCORP) ... 268
15.2.2 dLife ... 268
15.2.3 dLifeComm ... 268
15.3 Evaluation Methodology and Setting ... 269
15.3.1 The ONE Simulator ... 269
15.3.2 Simulation Environment Setup ... 269
15.4 Comparison Evaluation ... 270
15.4.1 Evaluation of TTL Impact ... 270
15.4.2 Evaluation of Node Impact ... 270
15.4.3 Evaluation of Simulation Time Impact ... 272
15.4.4 Evaluation on Buffer Impact ... 275
15.5 Conclusion ... 276
References ... 276

15.1 Introduction

A delay-tolerant network (DTN) (Fall, 2003) is a kind of challenged network where there is no direct path from source to destination. There are many real-life networks which follow this DTN paradigm, for example satellite communication (Prescott, Smith and Moe, 1999), wildlife tracking sensor networks (Juang et al., 2002), military networks, vehicular ad hoc networks (Ott and Kutscher, 2005), etc. In this scenario, network topology changes dynamically. Hence, routing in DTN uses a store and forward mechanism to enable successful communication. Various DTN routings apply different techniques to meet the target node based on particular routing metrics, such as estimated delivery probability, historical contact frequency, available network resources, or estimated delay (Schurgot, Comaniciu and Jaffres-Runser, 2012).

In this modern era, with the advent of powerful mobile devices, users crave connectivity while on the go. This leads to a networking scenario with heterogeneous, mobile, and power-constrained devices, as well as wireless networks with intermittent connectivity even in urban scenarios, due to the existence of shadowing, and costly internet services (Moreira, Mendes and Sargento, 2014).

We can get a better outcome of such a network having a direct path between source and destination (Costa et al., 2008; Boldrini, Conti and Passarella, 2010). Moreover, using the social interactions of a node with its corresponding structure (i.e., communities (Hui, Crowcroft and Yoneki, 2011), levels of social interaction (Moreira, Mendes and Sargento, 2012; Nguyen and Giordano, 2012)) we can ensure better performance of efficient routing.

Social-aware Content-based Opportunistic Routing Protocol (SCORP) considers the daily social interactions and interests of users to ensure better delivery in dense regions, i.e., urban areas (Moreira, Mendes and Sargento, 2014). The dLife routing protocol considers the dynamism of users' behavior (Moreira and Mendes, 2012) in their daily life routines. It takes the trace of social interaction for further data transmission. The dLifeComm routing protocol is the updated version of dLife, where it takes also the trace of social interaction and their interest for better performance than dLife.

In this chapter, we have evaluated the performance of the three aforementioned social-aware routings, namely, SCORP, dLife, and dLifeComm. This chapter is structured as follows. In Section 15.2, we present SCORP, dLife, and dLifeComm. Section 15.3 shows the simulation tool and environmental setup. Section 15.4 presents our evaluation study. In Section 15.5, we discuss the conclusions of this research study.

15.2 Social-Aware Routing Protocols

Here, we briefly discuss the SCORP, dLife, and dLifeComm social-aware routing protocols.

15.2.1 Social-Aware Content-Based Opportunistic Routing Protocol (SCORP)

The social-aware content-based opportunistic routing protocol that takes into account the social presence between nodes and the content knowledge that nodes have while taking ongoing decisions. SCORP is based on a utility function that reflects the probability of encountering nodes with a particular interest among the ones that have similar daily social functions. There are two reasons to use social appearance: first, nodes with the same daily habits have a higher probability of having similar (content) interest (Costa et al., 2008); second, social proximity metrics accommodate a faster dissemination of data, taking advantage of the more frequent and longer contacts between closer nodes (Moreira, Mendes and Sargento, 2014).

15.2.2 dLife

With dLife (Schurgot, Comaniciu and Jaffres-Runser, 2012; Moreira, Mendes and Sargento, 2014; Moreira and Mendes, 2012), the dynamism of users' behavior found in their daily life routines is calculated to aid routing. The goal is to keep track of the different levels of social interactions that nodes have throughout their daily tasks in order to conclude how well socially connected users are in different periods of the day.

15.2.3 dLifeComm

The dLifeComm (Moreira and Mendes, 2012) is community-based version of the dLife routing protocol. The dLifeComm social-aware routing protocol keeps track of the social interest and interaction history to gain higher performance in the intermittently connected network. It performs the same functionality as the dLife social-aware routing protocol except it focuses on community-based activities. It allows the structure of the Bubble Rap (Hui, Crowcroft and Yoneki, 2011) routing protocol.

15.3 Evaluation Methodology and Setting

Simulations are carried out using the Opportunistic Network Environment (ONE) simulator, program version 1.5.1. This section presents the ONE simulator and experimental settings.

15.3.1 The ONE Simulator

ONE is a Java-based simulator which is basically considered for analyzing the performance of ad hoc routing in opportunistic networks. This simulator is focused on mobility modeling, node tracing, message delivery, report generating, etc. (Keränen, Ott and Kärkkäinen, 2009; Netlab.tkk.fi, 2008).

15.3.2 Simulation Environment Setup

The following table depicts the simulation setting for analyzing the simulation time, time to live (TTL), and number of nodes, respectively. For varying the number of nodes, the simulation time is one day (defined in seconds, 86400 s) and TTL is 300 minutes, respectively. When varying the simulation time, the number of nodes is 150 (which are distributed equally in three groups), and TTL remains unchanged (which is 300 min). When varying the TTL, the number of nodes is 150 (50 per group), and the simulation time is one day (86400 s) (Table 15.1).

TABLE 15.1

Simulation Parameters

Parameters	Values
Simulation Time	3, 6, 12, 24, 48 (hours)
Update Interval	1 seconds
Number of nodes per Group	50, 100, 150, 200, 250
Interface	Bluetooth Interface
Interface Type	Simple Broadcast Interface
Transmit Speed	250 kbps
Transmit Range	10 m
Routing Protocols	SCORP, dLife, dLifeComm
Buffer Size	5, 50, 100, 200, 500 (MB)
Message TTL	50, 100, 150, 200, 250 (min)
Movement model	Shortest Path Map Based
Message Size	500 kB–1 MB
Simulation Area size	4500 m × 3400 m

15.4 Comparison Evaluation

In this section, we investigate the performance of three social-aware routing protocols, namely, SCORP, dLife, and dLifeComm using the ONE simulator in terms of three performance metrics. These are delivery probability (i.e., ratio between the total number of messages delivered to the destination over the total number of messages created at the source), average latency (i.e., time elapsed between message creation and delivery), and overhead ratio (i.e., how many redundant packets are relayed to successfully deliver a single packet). Hence, we have considered the impact of varying TTL, number of mobile nodes, simulation time, and buffer size.

15.4.1 Evaluation of TTL Impact

In this section, we see that delivery probability increases with the increase of TTL for all routings, as shown in Figure 15.1. Hence, SCORP has higher delivery than dLife and dLifeComm. In the case of average latency, we see that initially SCORP has much higher latency than dLife and dLifeComm, but with the increase of TTL the average latency of SCORP decreases and here we see from Figure 15.2 that the average latency of SCORP is lower than other routings when the TTL greater than or equal to 150 min. Again, we see from Figure 15.3 that the overhead ratio of SCORP initially is higher than others routings, but when the TTL is greater than 100 SCORP has lower delivery than dLife and dLifeComm. Therefore, we can consider the value of TTL is greater than or equal to 150. Therefore, SCORP shows the better delivery, lower latency, and lower overhead when TTL is greater than or equal to 150.

15.4.2 Evaluation of Node Impact

In this investigation, we observe the impact of mobile nodes (Hossen and Rahim, 2016) in the considered intermittently connected mobile network (ICMN) scenario. Here, we see that with increase of mobile nodes SCORP initially has much lower delivery, but when the

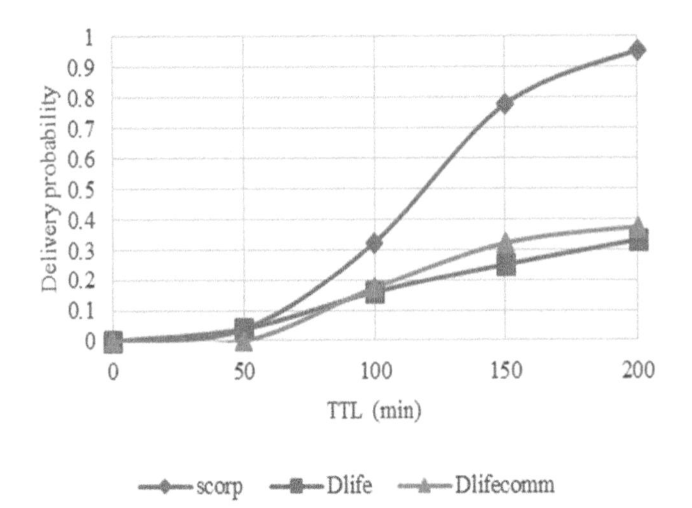

FIGURE 15.1
Delivery probability vs. TTL.

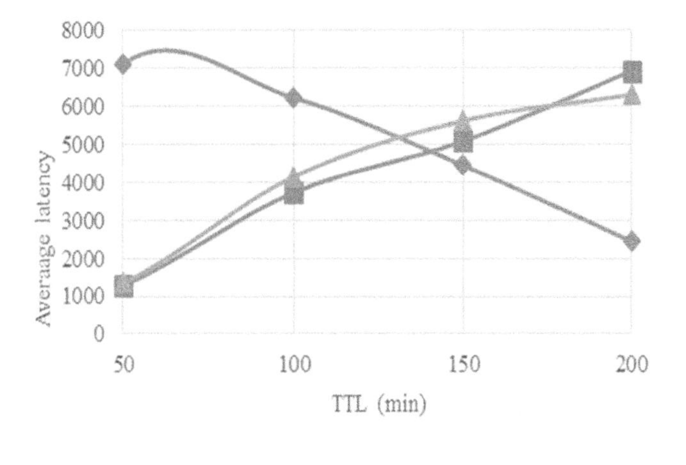

FIGURE 15.2
Average latency vs. TTL.

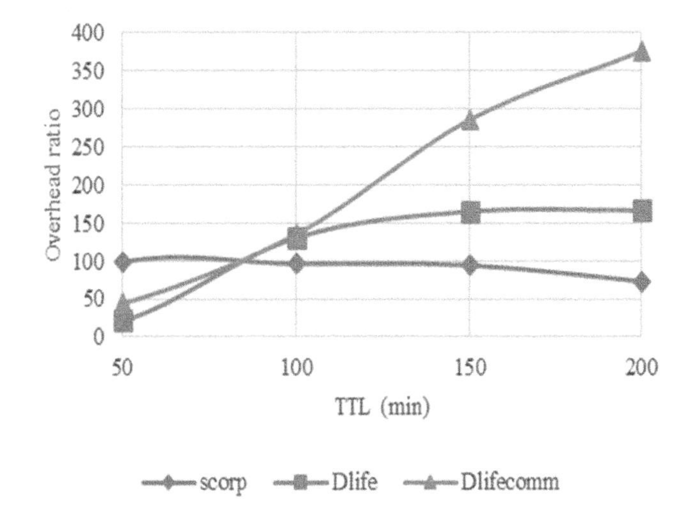

FIGURE 15.3
Overhead ratio vs. TTL.

number of nodes equals 150 SCORP has a higher delivery probability than dLife and dLife-Comm, as shown in Figure 15.4. Again, we see that the average latency of SCORP is lower when the number of nodes is between 150 and 200, as shown in Figure 15.5. In the case of considering overhead ratio, we see approximately constant overhead when the number of nodes is greater than 100, as shown in Figure 15.6. Therefore, we can say that SCORP routing exhibits better performance when the number of nodes is between 150 and 200 with the considerations of delivery probability, average latency, and overhead ratio.

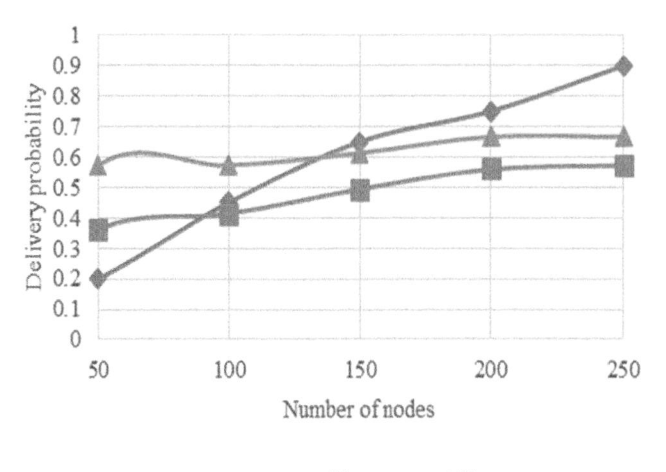

FIGURE 15.4
Delivery probability vs. number of nodes.

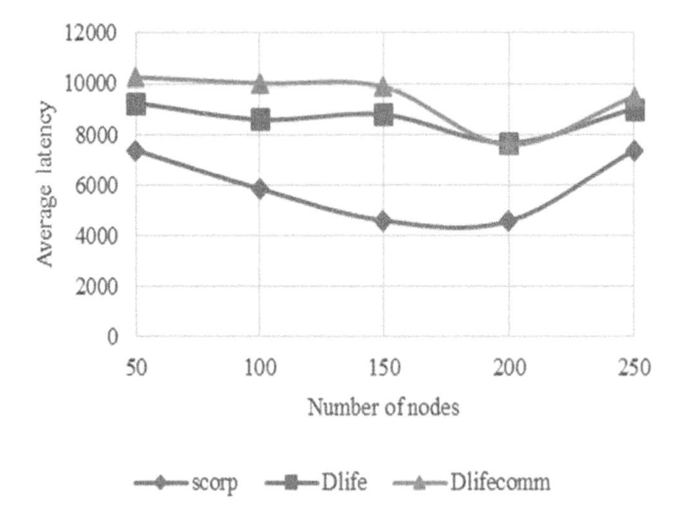

FIGURE 15.5
Average latency vs. number of nodes.

15.4.3 Evaluation of Simulation Time Impact

In this investigation, we see that with increase of simulation time the delivery probability of SCORP is much higher than dLife and dLifeComm, as shown in Figure 15.7. We see SCORP has a more or less constant average latency, as shown in Figure 15.8. Again, we see that SCORP has higher overhead, as shown in Figure 15.9. Therefore, we can say that with the consideration of simulation time, SCORP has much higher delivery probability.

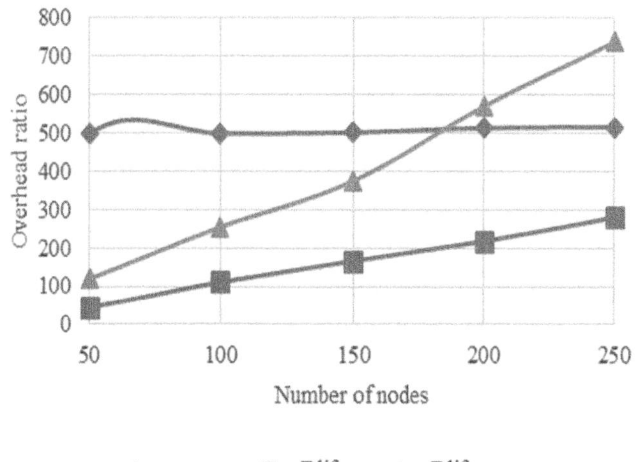

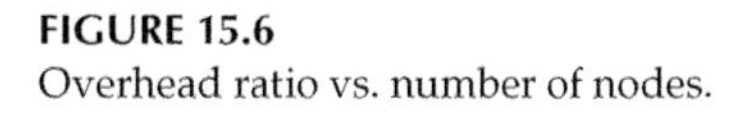

FIGURE 15.6
Overhead ratio vs. number of nodes.

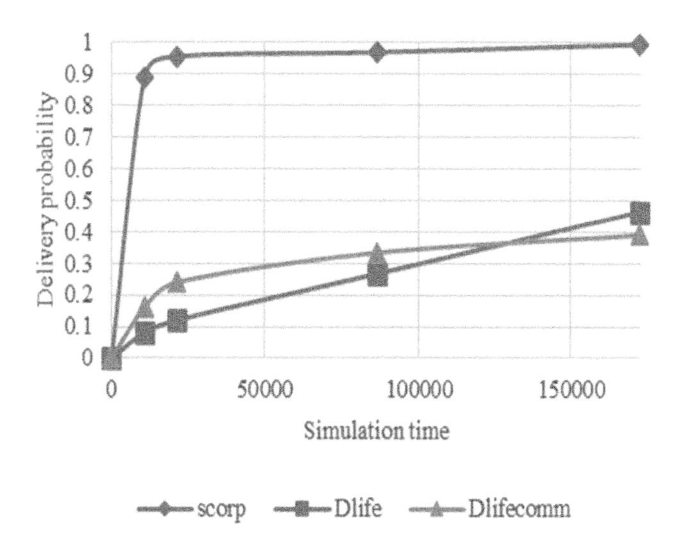

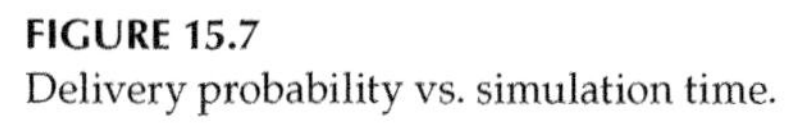

FIGURE 15.7
Delivery probability vs. simulation time.

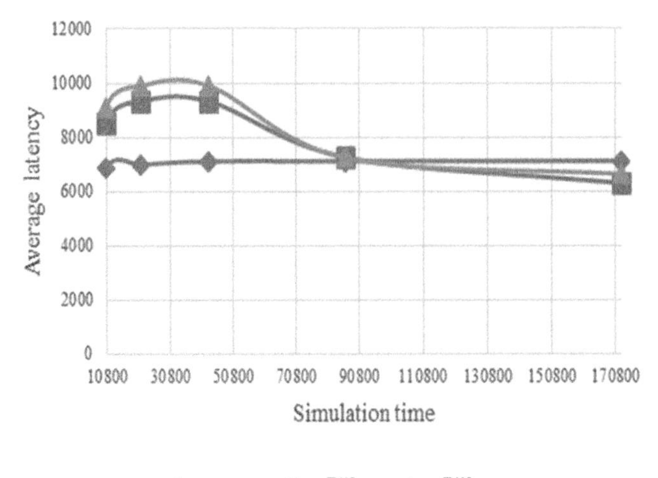

FIGURE 15.8
Average latency vs. simulation time.

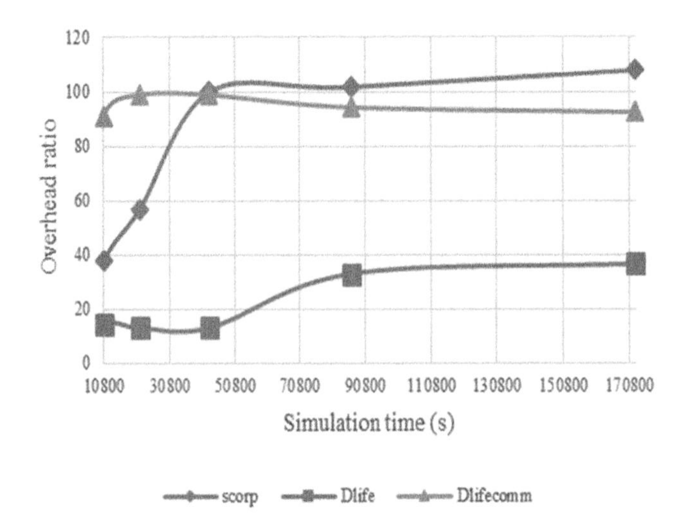

FIGURE 15.9
Overhead ratio vs. simulation time.

15.4.4 Evaluation on Buffer Impact

Here, with the increase of buffer size we see that SCORP routing shows higher delivery probability, lower latency, and lower overhead ratio. On the other hand, dLife routing shows lower delivery probability while dLifeComm exhibits higher latency and overhead ratio, as shown in Figures 15.10 through 15.12.

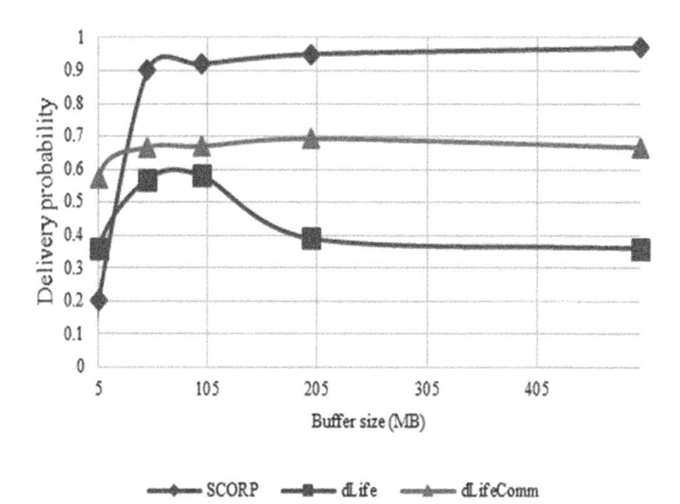

FIGURE 15.10
Delivery probability vs. buffer size.

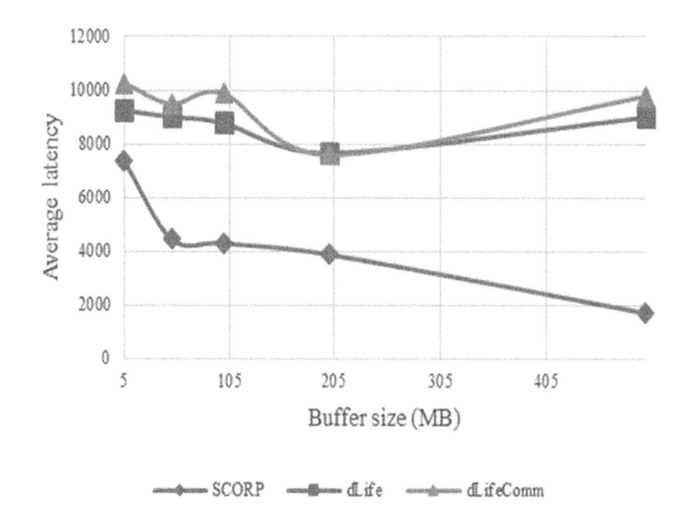

FIGURE 15.11
Average latency vs. buffer size.

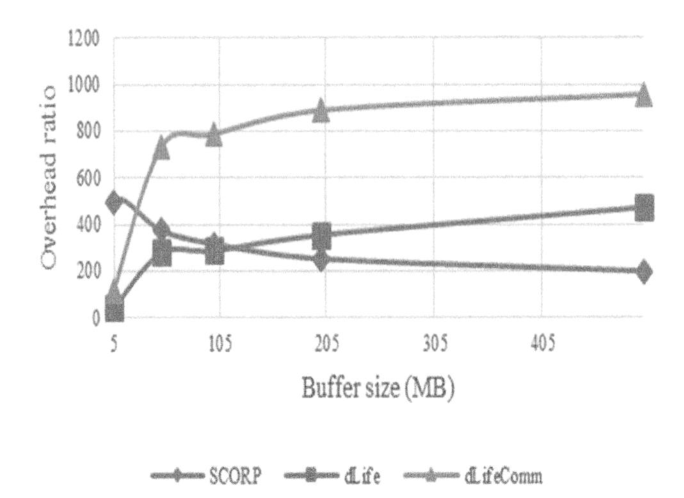

FIGURE 15.12
Overhead ratio vs. buffer size.

Therefore, we can see from this impact of TTL, number of mobile nodes, simulation time, and buffer size that SCORP has much higher delivery probability, lower average latency, and lower overhead compared to dLife and dLifeComm with the following considerations:

TTL ≥ 150 min

Number of Nodes = 150–200

Buffer ≥ 50 MB

15.5 Conclusion

In this chapter, we have investigated the performance of DTN social-aware routing protocols, i.e., SCORP, dLife, and dLifeComm (community-based version), in ICMNs scenario with the impact of TTL, number of mobile nodes, simulation time, and buffer size. Simulation results exhibit that SCORP shows better performance than dLife and dLifeComm in an opportunistic ICMN scenario in terms of three performance metrics, namely, message delivery probability, average latency, and overhead ratio with the considerations of TTL equal to or greater than 150 min, number of nodes between 150 and 200, and buffer size greater than 5 MB.

References

Boldrini, C., Conti, M. and Passarella, A. (2010). Design and performance evaluation of ContentPlace, a social-aware data dissemination system for opportunistic networks. *Computer Networks*, [online] 54(4), pp.589–604. Available at: https://doi.org/10.1016/j.comnet.2009.09.001 [Accessed 20 Jan. 2018].

Costa, P., Mascolo, C., Musolesi, M. and Picco, G. (2008). Socially-aware routing for publish-subscribe in delay-tolerant mobile ad hoc networks. *IEEE Journal on Selected Areas in Communications*, [online] 26(5), pp.748–760. Available at: https://doi.org/10.1109/JSAC.2008.080602 [Accessed 20 Jan. 2018].

Fall, K. (2003). A delay-tolerant network architecture for challenged internets. In: *Conference on Applications, Technologies, Architectures, and Protocols for Computer Communication*. [online] Karlsruhe, Germany: ACM SIGCOMM, pp.27–34. Available at: https://dl.acm.org/citation.cfm?id=863960 [Accessed 20 Jan. 2018].

Hossen, M. and Rahim, M. (2016). Impact of mobile nodes for few mobility models on delay-tolerant network routing protocols. In: *International Conference on Networking Systems and Security (NSysS)*. [online] Dhaka, Bangladesh: IEEE. Available at: https://doi.org/10.1109/NSysS.2016.7400704 [Accessed 20 Jan. 2018].

Hui, P., Crowcroft, J. and Yoneki, E. (2011). BUBBLE rap: Social-based forwarding in delay-tolerant networks. *IEEE Transactions on Mobile Computing*, [online] 10(11), pp.1576–1589. Available at: https://doi.org/10.1109/TMC.2010.246 [Accessed 20 Jan. 2018].

Juang, P., Oki, H., Wang, Y., Martonosi, M., Peh, L. and Rubenstein, D. (2002). Energy-efficient computing for wildlife tracking: Design tradeoffs and early experiences with zebranet. In: *International Conference on Architectural Support for Programming Languages and Operating Systems*. [online] San Jose, CA: ACM ACPLOS, pp.96–107. Available at: https://dl.acm.org/citation.cfm?id=605408 [Accessed 20 Jan. 2018].

Keränen, A., Ott, J., and Kärkkäinen, T. (2009). The ONE simulator for DTN protocol evaluation. In: *International Conference on Simulation Tools and Techniques (ICST)*. Rome, Italy: ICST (Institute for Computer Sciences, Social-Informatics and Telecommunications Engineering) [online] Available at: https://doi.org/10.4108/ICST.SIMUTOOLS2009.5674 [Accessed 20 Jan. 2018].

Moreira, W. and Mendes, P. (2012). *Opportunistic Routing Based on Users Daily Life Routine draft-moreira-dlife-00*. [online] Tools.ietf.org. Available at: https://tools.ietf.org/html/draft-moreira-dlife-00 [Accessed 20 Jan. 2018].

Moreira, W., Mendes, P. and Sargento, S. (2012). Opportunistic routing based on daily routines. *IEEE International Symposium on a World of Wireless, Mobile and Multimedia Networks (WoWMoM)*. [online], pp.1–6. Available at: https://doi.org/10.1109/WoWMoM.2012.6263749 [Accessed 20 Jan. 2018].

Moreira, W., Mendes, P. and Sargento, S. (2014). Social-aware opportunistic routing protocol based on user's interactions and interests. *Ad Hoc Networks*, [online] pp.100–115. Available at: https://link.springer.com/chapter/10.1007/978-3-319-04105-6_7 [Accessed 20 Jan. 2018].

Netlab.tkk.fi. (2008). *The ONE*. [online] Available at: http://www.netlab.tkk.fi/tutkimus/dtn/theone/ [Accessed 20 Jan. 2018].

Nguyen, H. and Giordano, S. (2012). Context information prediction for social-based routing in opportunistic networks. *Ad Hoc Networks*, [online] 10(8), pp.1557–1569. Available at: https://doi.org/10.1016/j.adhoc.2011.05.007 [Accessed 20 Jan. 2018].

Ott, J. and Kutscher, D. (2005). A disconnection-tolerant transport for drive-thru internet environments. In: *Annual Joint Conference of the IEEE Computer and Communications Societies*. [online] Miami, FL: IEEE. Available at: http://ieeexplore.ieee.org/abstract/document/1498464/ [Accessed 20 Jan. 2018].

Prescott, G., Smith, S. and Moe, K. (1999). Real-time information system technology challenges for NASAs earth science enterprise. In: *Real-Time Systems Symposium*. [online] Phoenix, AZ: IEEE. Available at: http://citeseerx.ist.psu.edu/viewdoc/summary?doi=10.1.1.25.7260 [Accessed 20 Jan. 2018].

Schurgot, M., Comaniciu, C. and Jaffres-Runser, K. (2012). Beyond traditional DTN routing: Social networks for opportunistic communication. *IEEE Communications Magazine*, [online] 50(7), pp.155–162. Available at: http://ieeexplore.ieee.org/document/6231292/ [Accessed 20 Jan. 2018].

16

Hands-On ONE Simulator: Opportunistic Network Environment

Anshuman Chhabra, Vidushi Vashishth, and Deepak Kumar Sharma

CONTENTS

16.1 Introduction 279
16.2 ONE Basics 280
16.2.1 Configuration File: default_settings.txt 280
16.2.1.1 Scenario Settings 280
16.2.1.2 Common Settings for all Groups 281
16.2.1.3 Settings Specific to a Particular Group 281
16.2.1.4 Message Creation Settings 282
16.2.1.5 Movement Model Settings 282
16.2.1.6 Reports Settings 282
16.2.2 Flow of the Code 283
16.3 Existing Routing Protocol Implementations in ONE 293
16.4 Implementation of a New Routing Protocol from Scratch in ONE 297
16.5 Conclusion 303
References 303

16.1 Introduction

The Opportunistic Network Environment (ONE) Simulator (Keränen et al., 2009) is a simulation tool widely used by researchers working on research related to Delay-Tolerant Networks (DTNs) and Opportunistic Networks. Its salient features are listed below -

- The ONE Simulator is a tool for researchers to carry out research-based simulations related to Delay-Tolerant Networks (DTNs) (Fall and Farrell, 2008), Wireless Sensor Networks (WSNs) and Opportunistic Networks.
- The different use cases of the ONE simulator are as follows:
  - Developing and testing different movement models for nodes
  - Adding message routing capabilities between nodes
  - Using JUnit and the GUI to debug the protocols being developed as well as visualize them
  - Generate results and statistics for tests in an effortless manner
- The ONE simulator is based in Java and can be downloaded from 'http://akeranen.github.io/the-one/'.

It is easy to set up and install the ONE after downloading the source code from the link above. The steps are fairly straightforward for setting up the ONE Simulator on Netbeans:

1. This requires working on a new Java Project with Existing Source(s). Therefore, choose 'New Project' from the drop-down menu and then choose the options corresponding to 'Java Project with Existing Sources'.
2. To add the source code, select 'Add Folder'. Give the path corresponding to where you downloaded the ONE source code.
3. Now also select 'Libraries' from the side window and choose 'Add Library'. Select JUnit 4.10 or greater and then add it.
4. Again from 'Libraries' select 'Add JAR/Folder'. You need to select the 'DTNConsoleConnection.jar' and 'ECLA.jar' from the 'lib' folder in the ONE source code.
5. Make sure the 'Working Directory' is the folder containing ONE source code after clicking on 'Run' in the side pane.
6. The last step is to select core.DTNSim as the main class after clicking on 'Run' again.

16.2 ONE Basics

16.2.1 Configuration File: default_settings.txt

This file is essential for running the simulation. It is provided as default with the ONE source code folder and contains some parameters with default values initially. For researchers and ONE developers, this file is going to be very useful – it will allow them to make changes and tweak parameters to carry out simulations corresponding to their new and proposed protocols.

The default_settings.txt file's utilities have been explained with code snippets below. The different tunable settings have been grouped differently so that it becomes easier for the reader to figure out where they are supposed to make changes or add lines when they want to specify characteristics of their simulations. As an illustration, the reader is shown how to make changes to the file if they had to see the variations in energy levels of the nodes throughout the simulation.

16.2.1.1 Scenario Settings

```
Scenario.name = default_scenario
Scenario.simulateConnections = true
Scenario.updateInterval = 0.1
Scenario.endTime = 43200
```

This corresponds to setting up a scenario. More of this will be discussed later while looking at the Java code running at the backend. However, it is apparent that the default_settings.txt file is read by the ONE and is used to initialize some variables. That is what the rest of the lines in the above code essentially depict. 'Scenario.endTime' is being set as 43200 signifying 43200 seconds which is equal to 12 hours.

Scenario.nrofHostGroups = 6

This is to specify the different types of hosts that exist in the system. The default has been selected as 6. There are settings which are universal to all the different groups and some which are only for certain group types. The formats corresponding to both types of settings in the default_settings.txt are shown in Section 16.2.1.2.

16.2.1.2 Common Settings for all Groups

Group.movementModel = RandomWaypoint
Group.router = ProphetRouter
Group.bufferSize = 10M
Group.waitTime = 0.120
Group.nrofInterfaces = 1
Group.interface1 = btInterface
Group.speed = 0.5,1.5
Group.msgTtl = 300
Group.nrofHosts = 58

16.2.1.3 Settings Specific to a Particular Group

Group1.groupID = p
 # group2 specific settings
Group2.groupID = c
 # cars can drive only on roads
Group2.okMaps = 1
 # 10-50 km/h
Group2.speed = 2.7, 13.9
 # another group of pedestrians
Group3.groupID = w
 # The Tram groups
Group4.groupID = t
Group4.bufferSize = 50M
Group4.movementModel = RandomWaypoint
Group4.routeFile = data/tram3.wkt Group4.routeType = 1
Group4.waitTime = 10, 30

The above code describes individual properties of different groups that are a part of the simulation. The users can define as many groups as they wish and can then set different parameters and variations for each group.

16.2.1.4 Message Creation Settings

Events.nrof = 1
Events1.class = MessageEventGenerator
Events1.interval = 25,35
Events1.size = 500k,1M
Events1.prefix = M

The above code deals with message generation. Parameters such as the frequency of new message generation, size of messages and ID of messages are set here for the MessageEventGenerator class. First, the number of events that are a part of the simulation are defined and then the class to be used is set as MessageEventGenerator. All the settings that follow are specific to MessageEventGenerator. The first line sets the message generation interval. This means messages will be sent out after every random interval chosen between 25 and 35 seconds. Similarly, the size of the messages is set to lie between 500 kB and 1 MB.

16.2.1.5 Movement Model Settings

MovementModel.rngSeed = 1
MovementModel.worldSize = 4500, 3400
MovementModel.warmup = 1000
MapBasedMovement.nrofMapFiles = 4
MapBasedMovement.mapFile1 = data/roads.wkt
MapBasedMovement.mapFile2 = data/main_roads.wkt
MapBasedMovement.mapFile3 = data/pedestrian_paths.wkt
MapBasedMovement.mapFile4 = data/shops.wkt

The map-based movement model requires that settings define the files being used to construct the movement models. It can be seen that the 4 map files used are essentially defining the paths and places of the real world that are being simulated. The reader can define their own map files in the *wkt* format. This will be covered later in more detail. One can also define the area of the world that they want for their simulations as well as the warm-up time, which is the time duration for which hosts are moved around in the defined area until the onset of the real simulation.

16.2.1.6 Reports Settings

Report.nrofReports = 1
Report.warmup = 0
Report.granularity = 43000
Report.reportDir = reports/
Report.report1 = MessageStatsReport

These settings determine where and how the simulation's reports will be saved. The first line specifies the number of reports one wants to generate. The user can also change the directory in which their reports are saved – which by default will be stored in the 'reports' folder in the current directory. The reports to be generated require the report class to be set as well. By default the report class is set to generate a report of only one type belonging to MessageStatsReport. To generate other types of reports, set the Report.nrofReports setting to the desired value and specify the types of reports to be generated (for example, EnergyLevelReport).

Q: Add settings to the default_settings.txt file to capture all energy levels for all possible node Groups and also save them as a report.

A: To specify energy settings in the default_settings.txt file for all, some extra settings will be added in two places. First, settings will be added to define the different energy levels for all the nodes. Second, an energy report will be generated using the EnergyLevelReport Report class.

The following lines have to be added to the file where common settings for all groups are present:

Group.initialEnergy = 4800
Group.scanEnergy = 0.06
Group.scanResponseEnergy = 0.08
Group.transmitEnergy = 0.08
Group.baseEnergy = 0.07
Group.charging_cofficient = 20
Group.threshold_energy = 3000

For report generation, it is first specified that there is a need for 2 reports to be generated (EnergyLevelReport along with the default MessageStatsReport):

Report.nrofReports = 2
Report.warmup = 0 #No change
Report.granularity = 43000 #No change
Report.reportDir = reports/ #No change
Report.report1 = MessageStatsReport
Report.report2 = EnergyLevelReport

This addition may be required in energy level-dependent protocol implementations (Lu and Hui, 2010; Ye et al., 2002; Chhabra et al., 2017a, b; 2018).

16.2.2 Flow of the Code

To be equipped with the skills to do hands-on development and research with the ONE, it is mandatory for the reader to understand how the ONE works internally. It is necessary to go through the various classes and Java files and get a brief understanding of the flow of the code. To reinforce the skills learnt by going through the code, implementation of routing protocols like Prophet routing (Lindgren et al., 2003) and Epidemic routing (Vahdat and Becker, 2000) in ONE will be looked at in Section 16.2. In Section 16.3 the process of writing a new routing algorithm from scratch using the ONE will be covered.

The flowchart in Figure 16.1 is a reference for how the code progresses through the ONE backend. Each rectangular box represents the function that is being called first along with the name of the class it belongs to. The circular boxes are an explanation of what the code does eventually – when it doesn't make any more calls to other functions. The graph is supposed to be read in a depth-first fashion. The reader should first go to the leftmost box starting at the top and then keep reading through all the calls straight down. They would then move to the next box on the right and continue the same way. The reader is recommended while going through the flow of the code to also simultaneously go through the code snippets that are being talked about in the ONE code base for a more thorough understanding of what is being discussed. Therefore, as an example of how to read the flowchart, one would start with the DTNSimGUI.start()function (belonging to DTNSim.java) and it can be seen that two calls are made – initModel() (belonging to DTNSimUI.java) and runSim() (belonging to DTNSimGUI.java). The call to initModel() progresses first. It can be seen that initModel() makes subsequent calls to four functions – Settings(), SimScenario(), addReport() and warmupMovementModel(). So, first, a new Settings() object is created to read parameters from the configuration file (default_settings.txt). Then the code progresses to SimScenario() (belonging to SimScenario.java) which makes three function calls of its own. These calls and their sub-calls will all be completed from left to right in the same way that has been discussed so far – EventQueueHandler(), then createHosts() and its sub-call to DTNHost() and then to World() with two sub-calls of its own to setNextEventQueue() and initSettings() in that order. After these, the flow will reach addReport() and then move to warmupMovementModel() at the end. Similarly, the reader would progress reading through the runSim() function. An understanding of what exactly is happening throughout all the function calls is now required.

The code starts with initModel() which does initializations for the ONE. It runs completely to set up a lot of the parameters and events. After initModel(), runSim() is run which handles the running of the simulation. These can be seen in Figure 16.2.

In initModel(), the first call is to Settings() to read all the parameter values set in default_settings.txt. The next call to SimScenario() (which belongs to SimScenario.java) is interesting – it first gets some parameter values from default_settings.txt and then makes three calls as can be seen in the code snippet in Figure 16.3. These are to EventQueueHandler (in EventQueueHandler.java), createHosts() (in SimScenario.java) and World() (in World.java).

EventQueueHandler initializes the handler for managing event queues like message creation. In this case, it initializes the event queue with MessageEventGenerator. The next function call is createHosts(). The main purpose of this is to create and initialize the hosts/nodes that populate the simulation. The createHosts() function makes another call to create a new list of DTNHost() objects via the DTNHost class as can be seen from Figure 16.4. The last call that SimScenario() makes is to create a new World() object. This constructor itself makes calls to setNextEventQueue() (in World.java) and initSettings() (also in World.java). This can be seen in Figure 16.5. The function setNextEventQueue() is used to set the next event in the queue to MessageEventGenerator again and initSettings() specifies the order in which the hosts are updated with settings throughout the simulation.

From initModel(), the last calls are made to addReport() (belonging to DTNSimUI.java) and warmupMovementModel(). The addReport() function just adds all the listeners to the simulation in order to aid report generation. These can be looked at in the DTNSimUI.java file by the reader. Finally, warmupMovementModel does two things – based on the warm-up time set in default_setttings.txt it sets the simulation clock and then it calls the moveHosts() function to facilitate node movements after the warm-up time has elapsed. This can be looked at by the reader in World.java.

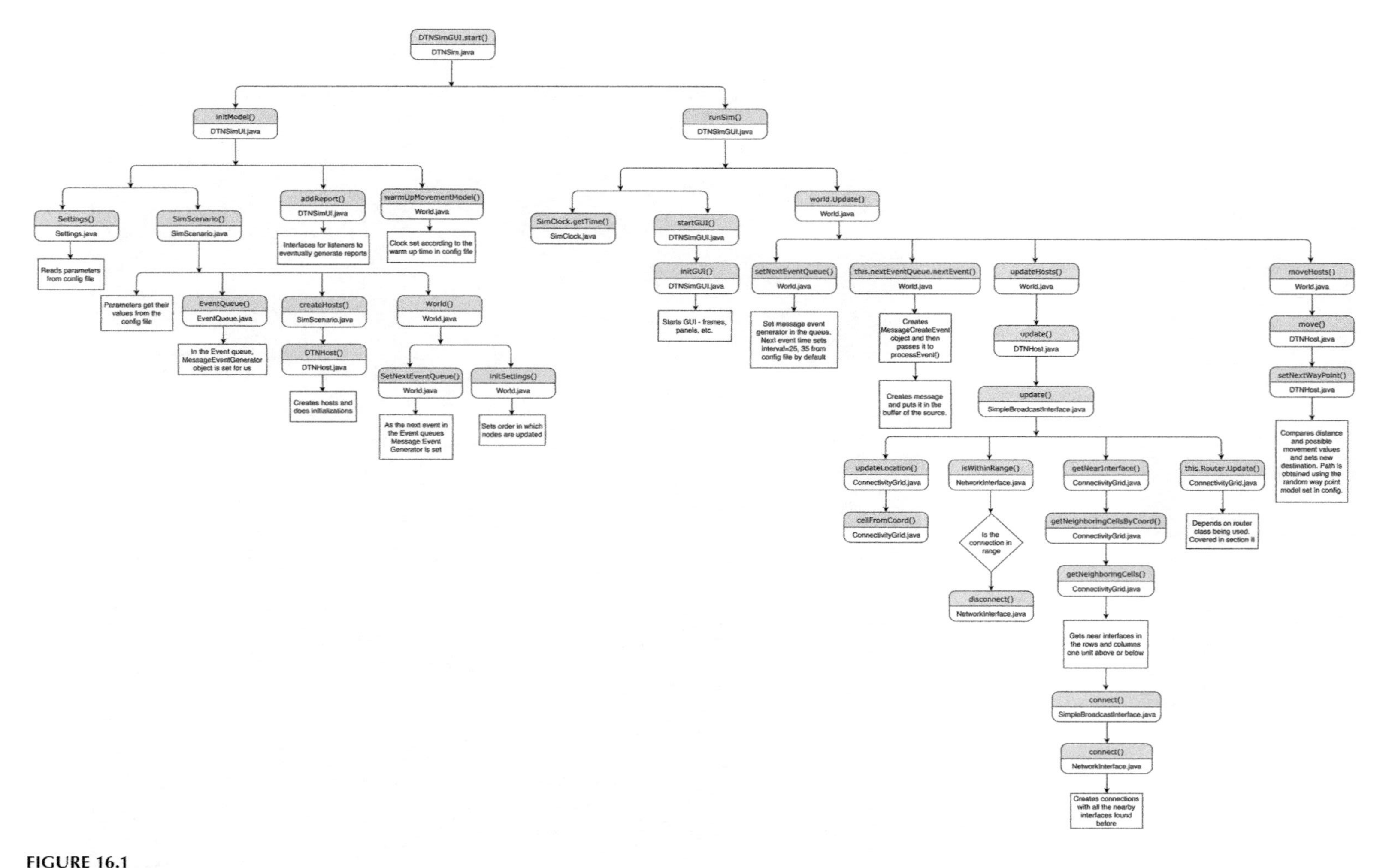

FIGURE 16.1
Class diagram giving a minimal overview of the ONE simulator's code.

```
72      /**
73       * Starts the simulation.
74       */
75      public void start() {
76          initModel();
77          runSim();
78      }
79
```

FIGURE 16.2
Code snippet - start().

```
/**
 * Creates a scenario based on Settings object.
 */
protected SimScenario() {
    Settings s = new Settings(SCENARIO_NS);
    nrofGroups = s.getInt(NROF_GROUPS_S);

    this.name = s.valueFillString(s.getSetting(NAME_S));
    this.endTime = s.getDouble(END_TIME_S);
    this.updateInterval = s.getDouble(UP_INT_S);
    this.simulateConnections = s.getBoolean(SIM_CON_S);

    s.ensurePositiveValue(nrofGroups, NROF_GROUPS_S);
    s.ensurePositiveValue(endTime, END_TIME_S);
    s.ensurePositiveValue(updateInterval, UP_INT_S);

    this.simMap = null;
    this.maxHostRange = 1;

    this.connectionListeners = new ArrayList<ConnectionListener>();
    this.messageListeners = new ArrayList<MessageListener>();
    this.movementListeners = new ArrayList<MovementListener>();
    this.updateListeners = new ArrayList<UpdateListener>();
    this.appListeners = new ArrayList<ApplicationListener>();
    this.eqHandler = new EventQueueHandler();

    /* TODO: check size from movement models */
    s.setNameSpace(MovementModel.MOVEMENT_MODEL_NS);
    int [] worldSize = s.getCsvInts(MovementModel.WORLD_SIZE, 2);
    this.worldSizeX = worldSize[0];
    this.worldSizeY = worldSize[1];

    createHosts();

    this.world = new World(hosts, worldSizeX, worldSizeY, updateInterval,
            updateListeners, simulateConnections,
            eqHandler.getEventQueues());
}
```

FIGURE 16.3
Code snippet SimScenario().

```
/**
 * Creates hosts for the scenario
 */
protected void createHosts() {
    this.hosts = new ArrayList<DTNHost>();

    for (int i=1; i<=nrofGroups; i++) {
        List<NetworkInterface> interfaces =
            new ArrayList<NetworkInterface>();
        Settings s = new Settings(GROUP_NS+i);
        s.setSecondaryNamespace(GROUP_NS);
        String gid = s.getSetting(GROUP_ID_S);
        int nrofHosts = s.getInt(NROF_HOSTS_S);
        int nrofInterfaces = s.getInt(NROF_INTERF_S);
        int appCount;
```

FIGURE 16.4
Code snippet - createHosts().

```
/**
 * Constructor.
 */
public World(List<DTNHost> hosts, int sizeX, int sizeY,
        double updateInterval, List<UpdateListener> updateListeners,
        boolean simulateConnections, List<EventQueue> eventQueues) {
    this.hosts = hosts;
    this.sizeX = sizeX;
    this.sizeY = sizeY;
    this.updateInterval = updateInterval;
    this.updateListeners = updateListeners;
    this.simulateConnections = simulateConnections;
    this.eventQueues = eventQueues;

    this.simClock = SimClock.getInstance();
    this.scheduledUpdates = new ScheduledUpdatesQueue();
    this.isCancelled = false;
    this.isConSimulated = false;

    setNextEventQueue();
    initSettings();
}
```

FIGURE 16.5
Code snippet - World().

Before moving on to the runSim() function, one can look at the definition of the initModel() function in Figure 16.6 for a summary of the above.

The runSim() function belongs to DTNSimGUI.java and is responsible for handling the running of the simulation. Its definition is referred to at the start as can be seen in Figure 16.7.

The function runSim() makes a number of interesting function calls. First, it sets the simulation time to a variable using the SimClock.getTime() function (which belongs to SimClock.java). Then it calls startGUI() (which belongs to DTNSimGUI.java). This function makes a sub-call to initGUI() (this also belongs to DTNSIMGUI.java) which initializes the GUI panels, frames and buttons prior to running the simulation. Next, as can be seen

```
/**
 * Initializes the simulator model.
 */
private void initModel() {
    Settings settings = null;

    try {
        settings = new Settings();
        this.scen = SimScenario.getInstance();

        // add reports
        for (int i=1, n = settings.getInt(NROF_REPORT_S); i<=n; i++){
            String reportClass = settings.getSetting(REPORT_S + i);
            addReport((Report)settings.createObject(REPORT_PAC +
                    reportClass));
        }

        double warmupTime = 0;
        if (settings.contains(MM_WARMUP_S)) {
            warmupTime = settings.getDouble(MM_WARMUP_S);
            if (warmupTime > 0) {
                SimClock c = SimClock.getInstance();
                c.setTime(-warmupTime);
            }
        }

        this.world = this.scen.getWorld();
        world.warmupMovementModel(warmupTime);
    }
    catch (SettingsError se) {
        System.err.println("Can't start: error in configuration file(s)");
        System.err.println(se.getMessage());
        System.exit(-1);
    }
    catch (SimError er) {
        System.err.println("Can't start: " + er.getMessage());
        System.err.println("Caught at " + er.getStackTrace()[0]);
        System.exit(-1);
    }
}
```

FIGURE 16.6
Code snippet - initModel().

from Figure 16.7, if the pause button is not pressed, the simulation world is updated via the world.update() method of the World class. The code snippet encapsulating the definition of the world.update() method can be seen in Figure 16.8.

The interesting function to observe here is setNextEventQueue(). This belongs to World.java and essentially sets the MessageEventGenerator as the next event to be processed in the event queue. Also, the nextQueueEventTime is the value of the time interval set in the default_settings.txt. In the earlier discussion on the default_settings.txt file in Message Creation Settings (Section 16.2.1.4), setting the message creation interval was elaborated upon. By default, it was set as Event1.interval = 25,35 which specified that messages would be created in an interval decided by a random number lying between 25 and 35 seconds. That interval time corresponds here to nextQueueEventTime and the while loop runs accordingly.

Inside the loop, another call is made to this.nextEventQueue.nextEvent(). Here the nextEventQueue is the MessageEventGenerator as had been seen previously. This function call creates a MessageCreateEvent object and then this event is passed to the processEvent() function in the next line. This finalizes creation of the message by putting it into the buffer of the source node.

```
98      protected void runSim() {
99          double simTime = SimClock.getTime();
100         double endTime = scen.getEndTime();
101
102         startGUI();
103
104         // Startup DTN2Manager
105         // XXX: Would be nice if this wasn't needed..
106         DTN2Manager.setup(world);
107
108         while (simTime < endTime && !simCancelled){
109             if (guiControls.isPaused()) {
110                 wait(10); // release CPU resources when paused
111             }
112             else {
113                 try {
114                     world.update();
115                 } catch (AssertionError e) {
116                     // handles both assertion errors and SimErrors
117                     processAssertionError(e);
118                 }
119                 simTime = SimClock.getTime();
120             }
121             this.update(false);
122         }
123
124         simDone = true;
125         done();
126         this.update(true); // force final GUI update
127
128         if (!simCancelled) { // NOT cancelled -> leave the GUI running
129             JOptionPane.showMessageDialog(getParentFrame(),
130                     "Simulation done");
131         }
132         else { // was cancelled -> exit immediately
133             System.exit(0);
134         }
135     }
```

FIGURE 16.7
Code snippet - runSim().

Next, updateHosts() will be looked at which updates all the hosts after the events. The function updateHosts() (belonging to World.java) then calls update() (belonging to DTNHost.java). The definition of this function can be seen in Figure 16.9.

There are two function calls worth mentioning that will also be elaborated upon – update() (belonging to SimpleBroadcastInterface.java) and this.router.update (specific to the settings of the router being used – ProphetRouter, EpidemicRouter, etc.). As the latter is specific to the type of routing being used it will be discussed in complete detail in Section 16.3. The former update() function of SimpleBroadcastInterface.java will also be discussed. The definition of this function is shown in Figure 16.10.

As can be seen, there are three main functions that are being called in this definition. The first is updateLocation() and it belongs to ConnectivityGrid.java. This function calls another function from the ConnectivityGrid.java Class called cellFromCoord(). First, it's imperative to understand that the ONE makes the entire simulation area into a grid of cells to judge position, movement and proximity of the node to other nodes. The cell is the simplest and smallest indivisible unit of the grid. The function cellFromCoord() gives the

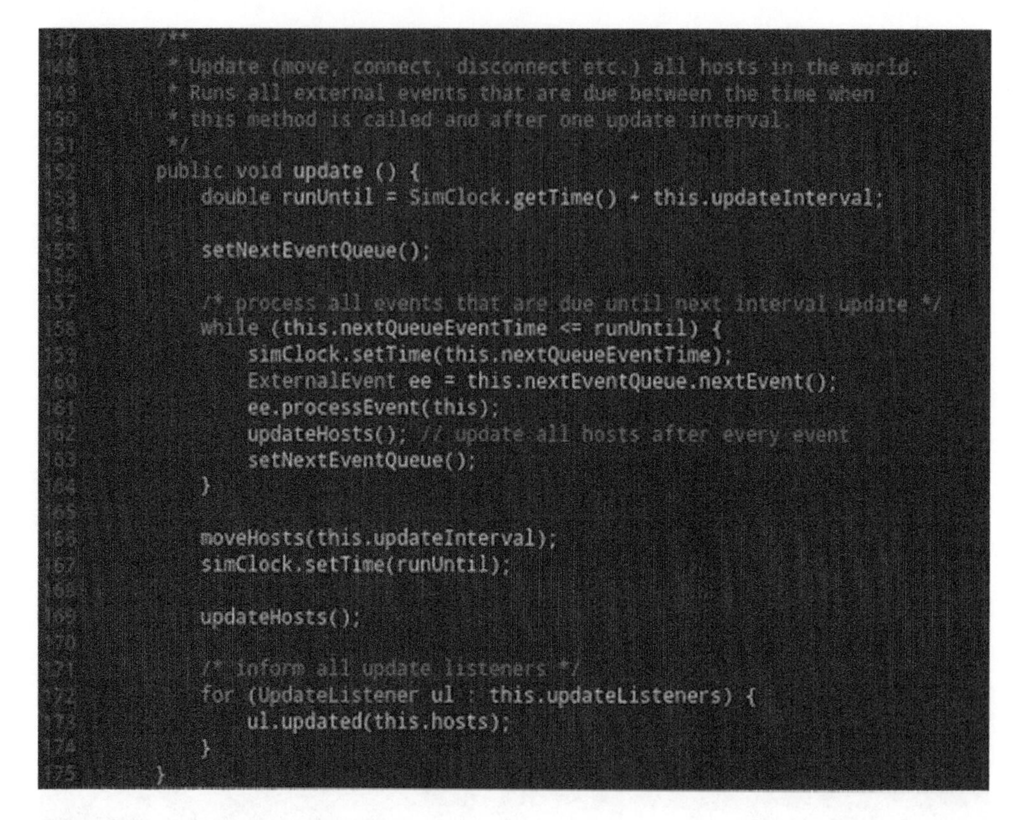

```
/**
 * Update (move, connect, disconnect etc.) all hosts in the world.
 * Runs all external events that are due between the time when
 * this method is called and after one update interval.
 */
public void update () {
    double runUntil = SimClock.getTime() + this.updateInterval;

    setNextEventQueue();

    /* process all events that are due until next interval update */
    while (this.nextQueueEventTime <= runUntil) {
        simClock.setTime(this.nextQueueEventTime);
        ExternalEvent ee = this.nextEventQueue.nextEvent();
        ee.processEvent(this);
        updateHosts(); // update all hosts after every event
        setNextEventQueue();
    }

    moveHosts(this.updateInterval);
    simClock.setTime(runUntil);

    updateHosts();

    /* inform all update listeners */
    for (UpdateListener ul : this.updateListeners) {
        ul.updated(this.hosts);
    }
}
```

FIGURE 16.8
Code snippet - update() of World.java.

```
/**
 * Updates node's network layer and router.
 * @param simulateConnections Should network layer be updated too
 */
public void update(boolean simulateConnections) {
    if (!isRadioActive()) {
        // Make sure inactive nodes don't have connections
        tearDownAllConnections();
        return;
    }

    if (simulateConnections) {
        for (NetworkInterface i : net) {
            i.update();
        }
    }
    this.router.update();
}
```

FIGURE 16.9
Code snippet - update() of DTNHost.java.

```
62
63      /**
64       * Updates the state of current connections (i.e. tears down connections
65       * that are out of range and creates new ones).
66       */
67      public void update() {
68          if (optimizer == null) {
69              return; /* nothing to do */
70          }
71
72          // First break the old ones
73          optimizer.updateLocation(this);
74          for (int i=0; i<this.connections.size(); ) {
75              Connection con = this.connections.get(i);
76              NetworkInterface anotherInterface = con.getOtherInterface(this);
77
78              // all connections should be up at this stage
79              assert con.isUp() : "Connection " + con + " was down!";
80
81              if (!isWithinRange(anotherInterface)) {
82                  disconnect(con,anotherInterface);
83                  connections.remove(i);
84              }
85              else {
86                  i++;
87              }
88          }
89          // Then find new possible connections
90          Collection<NetworkInterface> interfaces =
91              optimizer.getNearInterfaces(this);
92          for (NetworkInterface i : interfaces) {
93              connect(i);
94          }
95      }
```

FIGURE 16.10
Code snippet - update() of SimpleBroadcastInterface.java.

cell that the current node belongs to by using its current coordinates. Using this new cell value, the function updateLocation() updates the location to the current cell it belongs to. The next function being called by update() (of SimpleBroadcastInterface.java) is isWithinRange() which belongs to NetworkInterface.java. This function finds whether or not the connection is in range. If it is not, the disconnect() function which also belongs to NetworkInterface.java is called. The disconnect () function just calls connectionDown() of the DTNHost class to remove the current connection.

The next function that update() calls is getNearInterface() which belongs to ConnectivityGrid.java. The definition of this function can be seen in Figure 16.11. This function calls getNeighborCellsByCoord() which also belongs to ConnectivityGrid.java. This function makes another sub-call to getNeighborCells() which is also present in ConnectivityGrid.java. This finds all the neighboring interfaces which are the cells with rows and columns one unit in any direction from the current coordinates.

The neighboring interfaces are required for the connect() function of SimpleBroadcastInterface.java. This function calls another function called connect() which belongs to NetworkInterface.java. This creates connections with all the nearby interfaces that had been found earlier. The definition for the connect() function of SimpleBroadcastInterface.java is given in Figure 16.12.

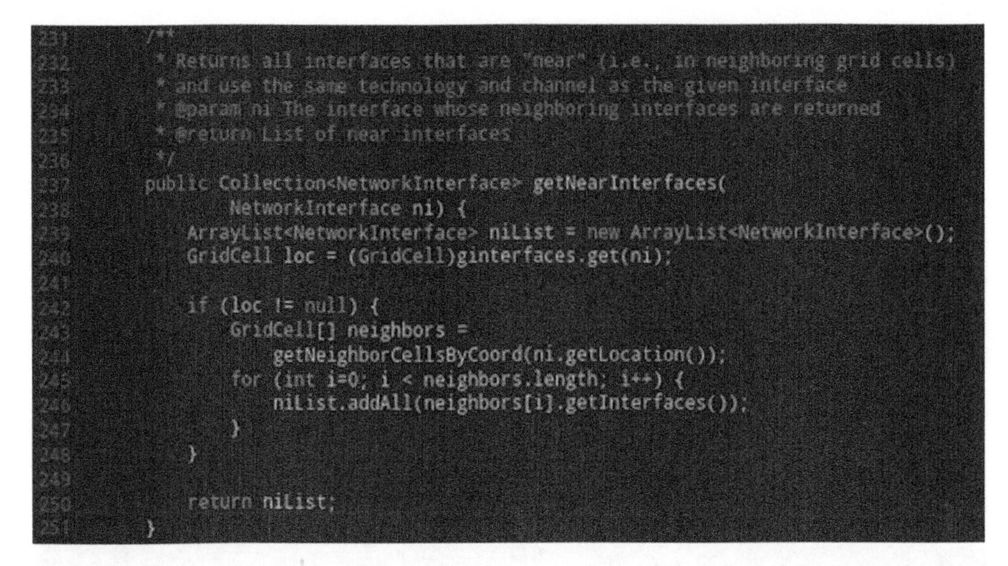

```
231    /**
232     * Returns all interfaces that are "near" (i.e., in neighboring grid cells)
233     * and use the same technology and channel as the given interface
234     * @param ni The interface whose neighboring interfaces are returned
235     * @return List of near interfaces
236     */
237    public Collection<NetworkInterface> getNearInterfaces(
238            NetworkInterface ni) {
239        ArrayList<NetworkInterface> niList = new ArrayList<NetworkInterface>();
240        GridCell loc = (GridCell)ginterfaces.get(ni);
241
242        if (loc != null) {
243            GridCell[] neighbors =
244                getNeighborCellsByCoord(ni.getLocation());
245            for (int i=0; i < neighbors.length; i++) {
246                niList.addAll(neighbors[i].getInterfaces());
247            }
248        }
249
250        return niList;
251    }
```

FIGURE 16.11
Code snippet - getNearInterfaces().

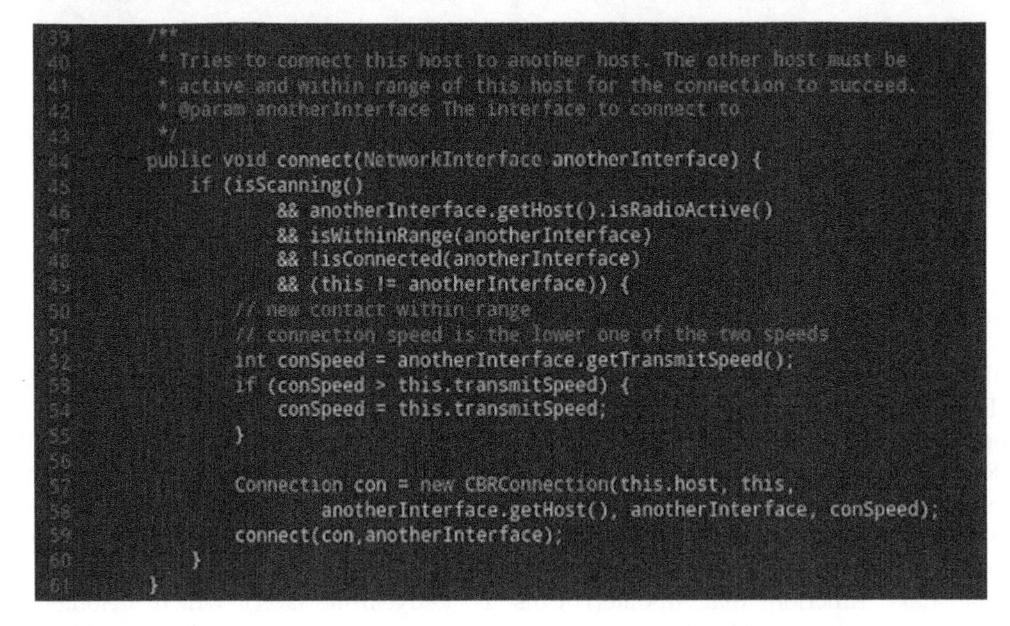

```
39    /**
40     * Tries to connect this host to another host. The other host must be
41     * active and within range of this host for the connection to succeed.
42     * @param anotherInterface The interface to connect to
43     */
44    public void connect(NetworkInterface anotherInterface) {
45        if (isScanning()
46                && anotherInterface.getHost().isRadioActive()
47                && isWithinRange(anotherInterface)
48                && !isConnected(anotherInterface)
49                && (this != anotherInterface)) {
50            // new contact within range
51            // connection speed is the lower one of the two speeds
52            int conSpeed = anotherInterface.getTransmitSpeed();
53            if (conSpeed > this.transmitSpeed) {
54                conSpeed = this.transmitSpeed;
55            }
56
57            Connection con = new CBRConnection(this.host, this,
58                    anotherInterface.getHost(), anotherInterface, conSpeed);
59            connect(con,anotherInterface);
60        }
61    }
```

FIGURE 16.12
Code snippet - connect().

Now that the updateHosts() function has been discussed, look back to the world.update() method in Figure 16.8, to observe that there is another important function called moveHosts() which also belongs to World.java. The definition of this function can be seen in Figure 16.13. It calls the move() function which further calls setNextWayPoint(). Both these functions belong to DTNHost.java. All these calls ensure that the next new destination is set for the node. The next point in the path is obtained using the RandomWayPoint movement model which had been set in default_settings.txt under the common settings for all groups.

```
209     /**
210      * Moves all hosts in the world for a given amount of time
211      * @param timeIncrement The time how long all nodes should move
212      */
213     private void moveHosts(double timeIncrement) {
214         for (int i=0,n = hosts.size(); i<n; i++) {
215             DTNHost host = hosts.get(i);
216             host.move(timeIncrement);
217         }
218     }
```

FIGURE 16.13
Code snippet - moveHosts().

With this, Section 16.2 comes to an end. In the next section, the functions responsible for routing messages throughout the network will be discussed. After the basics covered in Section 16.2, it is important to understand the specifics of routing that will be covered in Section 16.3. The main reason for this is that most research related with developing new routing algorithms can be implemented by making changes or adding extensions to an existing router class like ProphetRouter, EpidemicRouter, SprayAndWaitRouter (Spyropoulos et al., 2005) or the base routing class ActiveRouter. Implementation of a new routing algorithm will be discussed in Section 16.4.

16.3 Existing Routing Protocol Implementations in ONE

Going back to what has been covered in the previous section, this.router.update() method depends on the router class being used. Also, the this.router.update() method was called at the end of update() (belonging to DTNHost.java). This update() method was itself called by updateHosts() which was one of the sub-calls of world.update(). The world.update() function was being called by runSim().

Now that it has been summarized from the previous section where the router.update() function is being called, one can move on to look at the definitions. Moreover, the Prophet routing protocol implemented in ProphetRouter.java will be covered. One can refer to the definition of update() method in ProphetRouter.java in Figure 16.14.

The first call is made to the update() method of its superclass, ActiveRouter. The code snippet for this can be seen in Figure 16.15 and will be discussed subsequently.

This method makes its first call to update() method of its superclass MessageRouter.java. This update() method is used for updating the application layer and is not relevant to our discussions. Moving on, the next call is made to isMessageTransferred() in the loop to check whether the message has been transferred completely. The function isMessageTransferred() first calls getRemainingByteCount(). The total bytes deciding the size of the message are set using the default_settings.txt file as had been shown in Section 16.1. These are specified in the Message Creation Settings which by default is specified as Event1.size = 500k, 1M. This means that the message size will be randomly chosen to be between 500 kB and 1 MB. Therefore, the getRemainingByteCount() function gives us how many bytes of the message remain to be transferred. When this value reaches 0, the message has been transferred completely. Thus, isMessageTransferred() returns True when getRemainingByteCount() returns 0.

```
194      @Override
195      public void update() {
196          super.update();
197          if (!canStartTransfer() ||isTransferring()) {
198              return; // nothing to transfer or is currently transferring
199          }
200
201          // try messages that could be delivered to final recipient
202          if (exchangeDeliverableMessages() != null) {
203              return;
204          }
205
206          tryOtherMessages();
207      }
```

FIGURE 16.14
Code snippet - update() of ProphetRouter.java.

ActiveRouter update() method calls another interesting function, getMessage(). It returns the message that is currently being transferred between the connection that has been set up. The basic logic is that if the message was supposed to be transferred as given by the value in isMessageTransferred() but getMessage() returned a null, then the transfer was aborted due to some other reason. Therefore, this message needs to be removed from the buffer. Hence, the removeCurrent boolean variable is set to True and eventually is used to check for removal of the message from the buffer and the removal of the connection. If removeCurrent is not set to True, it moves on to the next connection. The other important conditional statement in the definition in Figure 16.15 checks for whether the Time-To-Live (TTL) has expired or not and then drops messages by calling dropExpiredMessages().

Moving on from the update() method of ActiveRouter back to the update() method of ProphetRouter given in Figure 16.14. There are two functions – canStartTransfer() and isTransferring() which are being called. The canStartTransfer() function ensures that there is at least one message and a connection to ensure transmission. The isTransferring() function returns a boolean value depending on whether or not a transfer is happening at the current time.

The next important function is exchangeDeliverableMessages(). Here, this function requests to transfer all messages that are supposed to be delivered to this node from all the other nodes. The function stops this process as soon as a transfer is initiated. The function definition is given in Figure 16.16. Looking at the definition in Figure 16.16, it can be seen that exchangeDeliverableMessages() first gets a list of all the possible connections using getConnections() for the particular host. Then it creates a tuple of relevant messages and connections after shuffling them according to the current queue mode.

If this does not yield any transfers, the function requestDeliverableMessages() is called. The definition of this function is shown in Figure 16.17. It basically requests all other nodes in the present connections with the current node to transfer messages.

The last function in the update() method of ProphetRouter is called is tryOtherMessages(). The definition of this function can be seen in Figure 16.18. The tryOtherMessages() function is what defines ProphetRouter. It first gets all the messages for this particular node. After all the initial checks, it then filters messages by checking whether the other node in the connection has a higher chance (probability) of delivering messages and then adds it to a list consisting of tuples of such messages and connections. This list is then

```
@Override
public void update() {
    super.update();

    /* in theory we can have multiple sending connections even though
      currently all routers allow only one concurrent sending connection */
    for (int i=0; i<this.sendingConnections.size(); ) {
        boolean removeCurrent = false;
        Connection con = sendingConnections.get(i);

        /* finalize ready transfers */
        if (con.isMessageTransferred()) {
            if (con.getMessage() != null) {
                transferDone(con);
                con.finalizeTransfer();
            } /* else: some other entity aborted transfer */
            removeCurrent = true;
        }
        /* remove connections that have gone down */
        else if (!con.isUp()) {
            if (con.getMessage() != null) {
                transferAborted(con);
                con.abortTransfer();
            }
            removeCurrent = true;
        }

        if (removeCurrent) {
            // if the message being sent was holding excess buffer, free it
            if (this.getFreeBufferSize() < 0) {
                this.makeRoomForMessage(0);
            }
            sendingConnections.remove(i);
        }
        else {
            /* index increase needed only if nothing was removed */
            i++;
        }
    }

    /* time to do a TTL check and drop old messages? Only if not sending */
    if (SimClock.getTime() - lastTtlCheck >= TTL_CHECK_INTERVAL &&
            sendingConnections.size() == 0) {
        dropExpiredMessages();
        lastTtlCheck = SimClock.getTime();
    }

    if (energy != null) {
        /* TODO: add support for other interfaces */
        NetworkInterface iface = getHost().getInterface(1);
        energy.update(iface, getHost().getComBus());
    }
}
```

FIGURE 16.15
Code snippet - update() of ActiveRouter.java.

passed through the tryMessagesForConnected function to send them across. This is the process through which messages are routed through the network. Moreover, even though this process was covered for ProphetRouter, it remains much the same for the other routing protocols albeit with some minor changes. The reader is encouraged to read through the other implementations and understand them with the help of the concepts covered in this section. In the next section, a new routing algorithm based on EpidemicRouter will be implemented but with additional functionality.

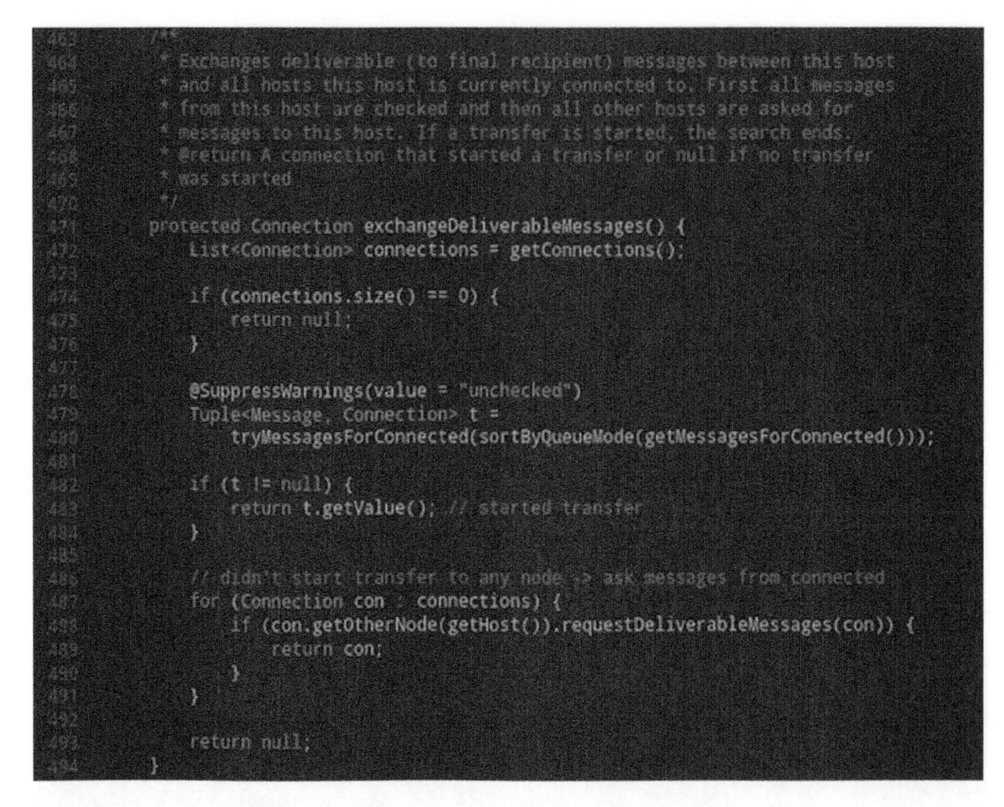

```
463    /**
464     * Exchanges deliverable (to final recipient) messages between this host
465     * and all hosts this host is currently connected to. First all messages
466     * from this host are checked and then all other hosts are asked for
467     * messages to this host. If a transfer is started, the search ends.
468     * @return A connection that started a transfer or null if no transfer
469     * was started
470     */
471    protected Connection exchangeDeliverableMessages() {
472        List<Connection> connections = getConnections();
473
474        if (connections.size() == 0) {
475            return null;
476        }
477
478        @SuppressWarnings(value = "unchecked")
479        Tuple<Message, Connection> t =
480            tryMessagesForConnected(sortByQueueMode(getMessagesForConnected()));
481
482        if (t != null) {
483            return t.getValue(); // started transfer
484        }
485
486        // didn't start transfer to any node -> ask messages from connected
487        for (Connection con : connections) {
488            if (con.getOtherNode(getHost()).requestDeliverableMessages(con)) {
489                return con;
490            }
491        }
492
493        return null;
494    }
```

FIGURE 16.16
Code snippet - exchangeDeliverableMessages().

```
102    @Override
103    public boolean requestDeliverableMessages(Connection con) {
104        if (isTransferring()) {
105            return false;
106        }
107
108        DTNHost other = con.getOtherNode(getHost());
109        /* do a copy to avoid concurrent modification exceptions
110         * (startTransfer may remove messages) */
111        ArrayList<Message> temp =
112            new ArrayList<Message>(this.getMessageCollection());
113        for (Message m : temp) {
114            if (other == m.getTo()) {
115                if (startTransfer(m, con) == RCV_OK) {
116                    return true;
117                }
118            }
119        }
120        return false;
121    }
```

FIGURE 16.17
Code snippet - requestDeliverableMessages().

```
209     /**
210      * Tries to send all other messages to all connected hosts ordered by
211      * their delivery probability
212      * @return The return value of {@link #tryMessagesForConnected(List)}
213      */
214     private Tuple<Message, Connection> tryOtherMessages() {
215         List<Tuple<Message, Connection>> messages =
216             new ArrayList<Tuple<Message, Connection>>();
217
218         Collection<Message> msgCollection = getMessageCollection();
219
220         /* for all connected hosts collect all messages that have a higher
221            probability of delivery by the other host */
222         for (Connection con : getConnections()) {
223             DTNHost other = con.getOtherNode(getHost());
224             ProphetRouter othRouter = (ProphetRouter)other.getRouter();
225
226             if (othRouter.isTransferring()) {
227                 continue; // skip hosts that are transferring
228             }
229
230             for (Message m : msgCollection) {
231                 if (othRouter.hasMessage(m.getId())) {
232                     continue; // skip messages that the other one has
233                 }
234                 if (othRouter.getPredFor(m.getTo()) > getPredFor(m.getTo())) {
235                     // the other node has higher probability of delivery
236                     messages.add(new Tuple<Message, Connection>(m,con));
237                 }
238             }
239         }
240
241         if (messages.size() == 0) {
242             return null;
243         }
244
245         // sort the message-connection tuples
246         Collections.sort(messages, new TupleComparator());
247         return tryMessagesForConnected(messages);   // try to send messages
248     }
```

FIGURE 16.18
Code snippet - tryOtherMessages().

16.4 Implementation of a New Routing Protocol from Scratch in ONE

As a part of this section, a new routing algorithm based on EpidemicRouter will be implemented. The features of this protocol will be:

1. A scoring mechanism for all nodes will be introduced.
2. This scoring mechanism will be based on how often a node comes in contact with another node and a connection is made.
3. Unlike EpidemicRouter where all possible connections are tried for all messages, this protocol will only try to transfer messages to connections where the other node in the connection has a higher score.
4. The default score for every node will be 100 and for every next encounter it will be incremented by a value of 5.

This new routing protocol will be called EpidemicRouterWithScore. After writing the protocol code, reports will be taken with both EpidemicRouterWithScore and EpidemicRouter. Upon examination, it will be seen how a simple addition such as the scoring mechanism gives quite an increase in delivery probability.

For creating a new Java file with the name EpidemicRouterWithScore, the first step is to import all the relevant Java packages.

After this, most of the functionality provided by the ActiveRouter class will be intelligently used as it will make writing new functionalities much easier. One could also have opted for using EpidemicRouter but the epidemic routing protocol doesn't have many features that cannot be re-written. The code will be made simpler by just extending ActiveRouter. Moreover, since there is a need for the router class to have a score attribute for the scoring mechanism a HashMap will be declared that will be able to keep the score for a particular node. These steps can be seen in Figure 16.19. Next, the constructors will be written. These are almost identical to the EpidemicRouter Class except for the fact that the score will have to be declared as a new HashMap. The code for this step can be seen in Figure 16.20.

Now, the function that will describe the scoring mechanism has to be written. For this, the changedConnection function of ActiveRouter will be overridden to add the scoring functionality to it. It will first be checked when the connection is made. When a connection is made it will be first seen if the other node in the connection already has a score value from a previous encounter. If so, its score is increased by 5. Otherwise, if it is establishing a connection for the first time (first contact) it will be given the default score of 100 and then the score is increased by 5, since it came in contact with another node. The code for achieving this is fairly straightforward to implement and can be seen in Figure 16.21. The con.getOtherNode(getHost()).getAddress() function gives us the address for the other node in the connection. First con.getOtherNode(getHost()) gives us the other node in the connection of the DTNHost class. Then the getAddress() method of DTNHost is called to get its integer address. The rest of the code can be understood via the explanation provided above.

The next step in writing the protocol is to write the update() method that was discussed in detail in the previous section. Here most of the nodes will have their messages tried after establishing connections. This method will be almost identical to the EpidemicRouter

```
14  public class EpidemicRouterWithScore extends ActiveRouter {
15
16      protected HashMap<Integer,Integer> score;
```

FIGURE 16.19
Extending ActiveRouter class.

```
18      public EpidemicRouterWithScore(Settings s) {
19      super(s);
20      score = new HashMap<Integer, Integer>();
21      }
22
23      public EpidemicRouterWithScore(EpidemicRouterWithScore r) {
24      super(r);
25      score = new HashMap<Integer, Integer>();
26      }
```

FIGURE 16.20
Writing constructors for EpidemicRouterWithScore.

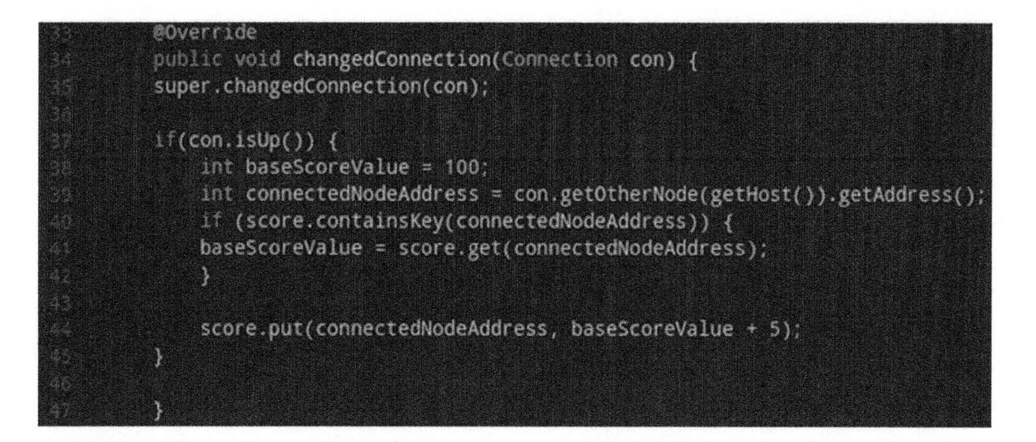

```
@Override
public void changedConnection(Connection con) {
super.changedConnection(con);

if(con.isUp()) {
    int baseScoreValue = 100;
    int connectedNodeAddress = con.getOtherNode(getHost()).getAddress();
    if (score.containsKey(connectedNodeAddress)) {
    baseScoreValue = score.get(connectedNodeAddress);
    }

    score.put(connectedNodeAddress, baseScoreValue + 5);
}

}
```

FIGURE 16.21
Writing the changedConnection() function.

update() method but for one difference. Instead of specifying tryAllMessagesToAllConnections (a function belonging to ActiveRouter) as the last function call, a new function called tryAllMessagesToAllConnectionsWithGreaterScore will be specified. The definition of this function has to be written in the code to give it the functionality which is desired – which is to allow for sending messages to only those nodes in the connection that have a greater score than the host node. The update() method can be seen in Figure 16.22.

Just before writing the code for tryAllMessagestoAllConnectionsWithGreaterScore another function will be written. To make it easier to get the score for a current node, a getter function called getCurrentScore is written which will take the node's address and give its score value. The code for this function can be seen in Figure 16.23.

Now we will write the code for tryAllMessagesToAllConnectionsWithGreaterScore. The definition of this function can be seen in Figure 16.24. Each line of code will be explained for better understanding.

First, a new empty list is created that will later be populated with connections in which the other node has a greater score than the current node. This list is called connectionsWithGreaterScore. Then, a list of all the connections can be acquired using the getConnections() function. Like the function tryAllMessagesToAllConnections we will first check whether these connections are empty or not and whether they have any messages. After

```
@Override
public void update() {
super.update();

if (isTransferring() || !canStartTransfer()) {
    return; // The node is transferring right now , cannot try any other connections yet
}

// First the messages that can be delivered to final recipient are tried
if (exchangeDeliverableMessages() != null) {
    return;
}

// then try all messages but only to specific connections with greater scores
this.tryAllMessagesToAllConnectionsWithGreaterScore();
}
```

FIGURE 16.22
Writing the update() method for EpidemicRouterWithScore.

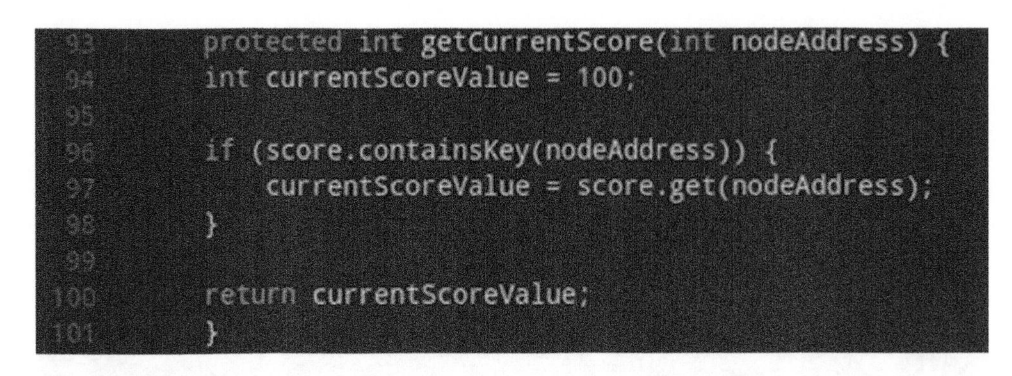

FIGURE 16.23
Writing the getCurrentScore() function.

FIGURE 16.24
Writing the tryAllMessagesToAllConnectionsWithGreaterScore() function.

this is done, all messages are compiled into a list. Now the routing based on the scoring mechanism will be implemented. It iterates over all possible connections one by one. For each connection, the address of the other node in the connection is extracted. Then the EpidemicRouterWithScore object for the other node is created to be able to call the getter function to get its score. The getter function getCurrentScore is then used to do the comparison and if the other node has a greater score, the connection is added to the originally empty list connectionsWithGreaterScore. At the end of this, tryMessagesToConnections is called and all the messages and connectionsWithGreaterScore are passed to it. In EpidemicRouter this function would have called tryMessagesToConnections but would have passed all the connections and messages without doing any comparisons.

The implementation is nearly finished. Only the replicate method is to be defined so that every time a node gets the EpidemicRouterWithScore Class object it does not retain any past history from a previous call. This replicate method is very common and is present in the EpidemicRouter implementation as well. The replicate function for the implementation can be seen in Figure 16.25.

```
28 @Override
29 public EpidemicRouterWithScore replicate() {
30 return new EpidemicRouterWithScore(this);
31 }
```

FIGURE 16.25
Writing the replicate() method.

The code for the scoring-based EpidemicRouter routing protocol is now complete. First, without using the protocol and just running the default simulation with EpidemicRouter, the default_scenario_MessageStatsReport.txt is examined in the reports directory. This can be seen in Figure 16.26. Out of 1463 messages created, only 350 were delivered leading to a delivery probability of 23.92%.

Now the protocol will be simulated. To do this, the group settings common to all the groups in default_settings.txt will be changed. This can be seen done in Figure 16.27. The router will be specified as EpidemicRouterWithScore (corresponding to the protocol's class name) without changing any other settings. Again the simulation is run using Netbeans IDE and we wait to examine the results. After the completion of the simulation default_scenario_MessageStatsReport.txt is checked in the reports directory. This is shown in Figure 16.28. We can see much better results! Out of the 1463 messages created

```
1  Message stats for scenario default_scenario
2  sim_time: 43200.1000
3  created: 1463
4  started: 60425
5  relayed: 33273
6  aborted: 27149
7  dropped: 33164
8  removed: 0
9  delivered: 350
10 delivery_prob: 0.2392
11 response_prob: 0.0000
12 overhead_ratio: 94.0657
13 latency_avg: 4684.2286
14 latency_med: 3548.1000
15 hopcount_avg: 4.3771
16 hopcount_med: 4
17 buffertime_avg: 1381.6837
18 buffertime_med: 870.9000
19 rtt_avg: NaN
20 rtt_med: NaN
```

FIGURE 16.26
Results for EpidemicRouter.

```
53  # Common settings for all groups
54  Group.movementModel = ShortestPathMapBasedMovement
55  Group.router = EpidemicRouterWithScore
56  Group.bufferSize = 5M
57  Group.waitTime = 0, 120
58  # All nodes have the bluetooth interface
59  Group.nrofInterfaces = 1
60  Group.interface1 = btInterface
61  # Walking speeds
62  Group.speed = 0.5, 1.5
63  # Message TTL of 300 minutes (5 hours)
64  Group.msgTtl = 300
```

FIGURE 16.27
Selecting router as EpidemicRouterWithScore.

```
1   Message stats for scenario default_scenario
2   sim_time: 43200.1000
3   created: 1463
4   started: 1301
5   relayed: 518
6   aborted: 783
7   dropped: 928
8   removed: 0
9   delivered: 518
10  delivery_prob: 0.3541
11  response_prob: 0.0000
12  overhead_ratio: 0.0000
13  latency_avg: 6956.7878
14  latency_med: 6425.4000
15  hopcount_avg: 1.0000
16  hopcount_med: 1
17  buffertime_avg: 15395.9154
18  buffertime_med: 17945.6000
19  rtt_avg: NaN
20  rtt_med: NaN
```

FIGURE 16.28
Results for EpidemicRouterWithScore.

just like before, 518 were successfully delivered. This increased the delivery probability to 35.41%, which is more than a 10% increase. Such a simple change and addition to the EpidemicRouter protocol was able to give much better results.

16.5 Conclusion

This tutorial on development using the ONE simulator, started off with the basics – covering how to add or change parameters in the configuration file and the way routing protocols are implemented in the ONE. Lastly, using all this knowledge, a new routing protocol was implemented from scratch which outperformed EpidemicRouter. This chapter served the purpose of helping researchers get familiar with the ONE. There is much more to explore in the ONE simulator, including using real-world mobility traces and data to writing tests using JUnit which are left to the reader.

References

Chhabra, A., Vashishth, V., and Sharma, D. K. (2017a). SEIR: A Stackelberg game based approach for energy-aware and incentivized routing in selfish Opportunistic Networks. *2017 51st Annual Conference on Information Sciences and Systems (CISS)*, Baltimore, MD. doi:10.1109/ciss.2017.7926113

Chhabra, A., Vashishth, V., and Sharma, D. K. (2017b). A game theory based secure model against Black hole attacks in Opportunistic Networks. *2017 51st Annual Conference on Information Sciences and Systems (CISS)*, Baltimore, MD, 2017, pp. 1–6. doi: 10.1109/CISS.2017.7926114

Chhabra, A., Vashishth, V., and Sharma, D. K. (2018). A fuzzy logic and game theory based adaptive approach for securing opportunistic networks against black hole attacks. *Int J Commun Syst*, 31, e3487. https://doi.org/10.1002/dac.3487

Fall, K. and Farrell, S. (2008). DTN: An architectural retrospective. *IEEE Journal on Selected Areas in Communications*, 26 (5), 828–836. doi:10.1109/jsac.2008.080609

Keränen, A., Ott, J., and Kärkkäinen, T. (2009). The ONE simulator for DTN protocol evaluation. *Proceedings of the Second International ICST Conference on Simulation Tools and Techniques*, Rome, Italy. doi:10.4108/icst.simutools2009.5674

Lindgren, A., Doria, A., and Schelén, O. (2003). Probabilistic routing in intermittently connected networks. *ACM SIGMOBILE Mobile Computing and Communications Review*, 7 (3), 19. doi:10.1145/961268.961272

Lu, X. and Hui, P. (2010). An energy-efficient n-epidemic routing protocol for delay tolerant networks. *2010 IEEE Fifth International Conference on Networking, Architecture, and Storage*, Macau, China. doi:10.1109/nas.2010.46

Spyropoulos, T., Psounis, K., and Raghavendra, C. S. (2005). Spray and wait: An efficient routing scheme for intermittently connected mobile networks. *Proceedings of the 2005 ACM SIGCOMM workshop on Delay-tolerant networking - WDTN 05*, Philadelphia, PA. doi:10.1145/1080139.1080143

Vahdat, A. and Becker, D (2000). Epidemic routing for partially connected ad-hoc networks. *(2000) Tech. Rep. Duke CS-2000-06*

Ye, W., Heidemann, J., and Estrin, D. (2002). An energy-efficient MAC protocol for wireless sensor networks. *Proceedings of the Twenty-First Annual Joint Conference of the IEEE Computer and Communications Societies*, New York, NY. doi:10.1109/infcom.2002.1019408

Index

A

Adaptive routing, 119–125
neighborhood determination and message forwarding, 120–123
congestion-aware adaptive routing protocol (CAARP), 122
performance indicators and evaluation, 123–125
message delivery ratio (MDR), 123
protocol with congestion avoidance, 125–128
congestion-aware adaptive routing protocol (CAARP), 125–128
addReport(), 284
AFSnW, *see* Adaptive Fuzzy Spray and Wait (AFSnW)
AONC, *see* Adaptive Opportunistic Network Coding (AONC)
Auto-Adjustable Opportunistic Acknowledgment/Timer-Based Routing, 105
Better Approach To Mobile Ad hoc Networking (BATMAN), 105

B

BAB, *see* Bundle Authentication Block (BAB)
BIONETS, *see* bio-inspired next-generation networks (BIONETS)
Black hole and selective forwarding security attacks, 198–199
Blackmail attack, 207–208
Bogus information security attacks, 202–203
Buffer impact, 275–276
Buffer management in delay-tolerant networks
fuzzy-based techniques, 43
Adaptive Fuzzy Spray and Wait (AFSnW), 43
Enhanced Fuzzy-based Spray and Wait Routing (EFSnWR), 43
Fuzzy Logic Controller (FLC), 43
global policies, 44–49
history-based dropping and scheduling (HSBD), 45
optimal buffer management (OBM), 45
Resource Allocation Protocol for Intentional DTN (RAPID), 45
issues in DTN routing, 27–28
dropping policy, 28
forwarding policy, 27
replication policy, 27–28
scheduling policy, 28
local policies, 29–43
buffer replacement policy, 41
comprehensive-integrated buffer management (CIM), 41
congestion control, 42
drop-least-encountered, 29
drop-least-recently-received, 29
drop-oldest, 29
drop-random, 29
Enhanced Buffer Management Policy (EBMP), 37
first in first out (FIFO), 31
High Weight Message List (HWML), 40
law of diminishing marginal utility, 39
least probable (LEPR), 31
Least Recently Forwarded (LRF), 32
Less Probable Sprayed (LPS), 32
Low Weight Message List (LWML), 40
Message Transmission Status Based Buffer Management Scheme (MTSBS), 39
most forwarded (MOFO), 31
most probable (MOPR), 31
Prioritized Epidemic (PREP), 33
Probabilistic Routing Protocol using History of Encounters and Transitivity (PRoPHET), 30
redundant deletion, 42
shortest lifetime (SHLI), 31
weight-based buffer management policy (WBD), 40
overview, 25–27
social-based routing algorithms, 43–45
pocket switched networks (PSNs), 44
Social Congestion Metric (SCM), 44
Socially Aware Congestion Control algorithm (SACC), 44
traffic differentiation schemes, 49–52
differentiated services (DiffServ), 49
integrated services (IntServ), 49
Bundle Authentication Block (BAB), 252

C

CAARP, *see* congestion-aware adaptive routing protocol (CAARP)
Campus Waypoint model, 65
CBF, *see* Contention-Based Forwarding (CBF)
cellFromCoord(), 289
CIM, *see* comprehensive-integrated buffer management (CIM)
City Section mobility model, 74
CMM, *see* Community-based Mobility Model (CMM)
Colluding misrelay attack, 206
Column mobility model, 71
Congestion-aware adaptive routing
  adaptive routing, 119–125
    neighborhood determination and message forwarding, 120–123
    performance indicators and evaluation, 123–125
    protocol with congestion avoidance, 125–128
  overview, 117–119
    content distribution, 118
    research initiatives in opportunistic networks, 118–119
    technical background, 117–118
connect(), 291
connectionDown(), 291
ConnectivityGrid.java, 289
Context-aware, 97–101
  fully context-aware, 99–101
    Context-Aware Routing (COR), 99–100
    History-Based Opportunistic Routing Protocol (HiBOp), 101
    SimBet, 101
  partially context-aware, 97–99
    Bubble Rap, 98–99
    MaxProp, 98
    MobySpace routing, 98
    probabilistic routing protocol using history of encounters and transitivity (PRoPHET), 97
    PRoPHET+, 99
    Resource Allocation Protocol for Intentional DTN Routing (RAPID), 99
CORMAN, *see* Cooperative Opportunistic Routing in Mobile Ad Hoc Networks (CORMAN)
  cooperative communication, 93
CPS, *see* cyber-physical system (CPS)
createHosts(), 284
Cross-layer, 108–112
  MAC Aware, 109–111
    Opportunistic Routing for Low Power and Lossy Networks (ORPL), 110
    Opportunistic Routing for Wireless Sensor Networks (ORW), 110–111
    Opportunistic Routing with Congestion Diversity (ORCD), 110
    QoS Oriented Opportunistic Routing (QOR), 110
  PHY and MAC Aware, 111–112
    Cross-Layer Aided Energy-Efficient Opportunistic Routing (CL EE), 111–112
    Maximizing Transmission Opportunities in Wireless Multi-Hop Network (MTOP), 111
    Protocol for Retransmitting Opportunistically (PRO), 111
    Sensor Context-Aware Adaptive Duty-Cycled Beaconless Opportunistic Routing (SCAD), 112
  Physical-Layer-Aware (PHY-Aware), 108–111
    Energy Efficient Opportunistic Routing (EEOR), 109
    High-Speed Opportunistic Routing (HS OR), 109
    interference limited opportunistic relaying (ILOR), 108
    Parallel Opportunistic Routing (Parallel OR), 108–109
    Simple and Practical Opportunistic Routing (SPOR), 109

D

DakNet, 241–243
  information and communication technology (ICT), 241
Delay-tolerant networking (DTN)
  applications, 259–266
    communication in deserted places, 266
    post office scenario, 264
    traffic sensors, 265–266
Desynchronization attack, 209
DiffServ, *see* differentiated services (DiffServ)
disconnect(), 291
Distributed denial of service (DDoS), 203–204
DOC, *see* decentralized oppnet controller (DOC)
DTNHost.java, 292
DTNSimGUI.java, 287
DTNSIMGUI.java, 287
DTRP, *see* Directed Transmission Routing Protocol (DTRP)

E

E-ACK, *see* end-to-end acknowledgement (E-ACK)
EAR, *see* expected advancement rate (EAR)
EBR, *see* encounter-based routing (EBR)
EEOR, *see* Energy Efficient Opportunistic Routing (EEOR)
EFSnWR, *see* Enhanced Fuzzy-based Spray and Wait Routing (EFSnWR)
EMT, *see* expected medium time (EMT)
Emulation of cross-layer feedback
  design aspects, 250–252
    bundle protocols for overlay network, 251–252
    DTN architecture, 250–251
    pervasive computing, 250
    security, 252
  network basics, 248–250
    advantages, 249
    challenges, 249–250
    simulation and real world experiments, 248–249
  overview, 247–248
  testbeds, 252–254
    overview, 253
    quality observation and mobility experiment tools (QOMET), 253–254
  time and data-driven triggering, 254–256
    KauNet, 255–256
    pattern and scenario files, 255
    pattern generation, 255
Energy management in oppnets
  consevation methodologies, 163–165
    encounter frequency, 163–164
    intelligent infrastructure-based energy conservation, 165
    interest in the type of content, 165
    regularity of encounter, 164–165
  constraints, 162
    affecting factors, 162
  customized routing protocols, 165–167
    flooding-based approach, 167
    forwarding-based approach, 166
  opportunistic mobile networks (OMNs), 160–161
    categories, 160
    comparison of MANETs and WSNs, 161
  performance analysis of existing routing protocols, 167
EOT, *see* expected one-hop throughput (EOT)
EventQueueHandler, 284
ExOR, *see* Extremely Opportunistic Routing (ExOR)

F

FIFO, *see* first in first out (FIFO)
FLC, *see* Fuzzy Logic Controller (FLC)
former update(), 289
Foundation
  delaydisruption-tolerant networking (DTN), 8–9
    network topology classification, 8
    replication and semantic classification, 8–9
    routing strategy classification, 8
  Freenet, 9
  migration from MANET, 2–3
    mobile ad hoc network (MANET), 2
    OppNet, 2
  Mobile Ad Hoc Network (MANET), 5–6
    advantages, 6
  Mobile Social Networks (MSNs), 6
    human–computer interaction (HCI), 6
  Mobile Ubiquitous LAN Extensions (MULEs), 4
    ultra-wideband (UWB), 4
  overview, 1–2
  sensor networks, 3
    applications, 3
    cyber-physical system (CPS), 3
    Internet of Things (IoT), 3
    Wireless sensor networks (WSNs), 3
  Shared Wireless Infostation Model (SWIM), 5
  Vehicular Ad Hoc Networks (VANETs), 7–8
    applications, 7
    characteristics, 7–8
  ZebraNet, 4–5
    Global Positioning System (GPS), 4
    Impala middleware layer, 4
FPOR, *see* Fixed Point Opportunistic Routing (FPOR)
FreeNet, 239–240
Future networks
  bio-inspired next-generation networks (BIONETS), 230–231
  DakNet, 241–243
    information and communication technology (ICT), 241
  FreeNet, 239–240
  Haggle, 243–244
  mobile ubiquitous LAN extensions (MULE), 231–232
  overview, 229–230
    intermittently connected mobile networks (ICMNs), 229–230

shared wireless infostation model (SWIM), 233–234
Social Networking for Pervasive Adaptation (SOCIALNETS), 235–236
Future and Emerging Technologies (FET), 235
Social-Aware Mobile and Pervasive Computing (SCAMPI), 236–238
vehicular delay-tolerant network (VDTN), 234–235
ZebraNet, 238–239

G

Gauss–Markov mobility model, 69
Geographical-restriction-based mobility model, 72–74
City Section mobility model, 74
Manhattan mobility model, 74
Obstacle mobility model, 73–74
Pathway mobility model, 73
GeRaF, *see* Geographic Random Forwarding (GeRaF)
getNearInterface(), 291
getNeighborCells(), 291
getNeighborCellsByCoord(), 291
GOR, *see* Geographic Opportunistic Routing (GOR)
expected one-hop throughput (EOT), 89
GPS spoofing security attacks, 201–202
GPS, *see* Global Positioning System (GPS)

H

H-ACK, *see* hop-to-hop acknowledgement (H-ACK)
Haggle, 243–244
Hands-on ONE simulator
code flow, 283–293
configuration file, 280–283
existing routing protocol implementations, 293–297
new routing protocol implementation, 297–303
overview, 279–280
Opportunistic Network Environment (ONE), 279
HAPs, *see* high-speed access points (HAPs)
HCI, *see* human-computer interaction (HCI)
HCM, *see* Home-cell Community-Based Mobility (HCM)
HS OR, *see* High-Speed Opportunistic Routing (HS OR)
HSBD, *see* history-based dropping and scheduling (HSBD)
HWML, *see* High Weight Message List (HWML)
Hybrid networks, 160

I

ICMNs, *see* intermittently connected mobile networks (ICMNs)
ICT, *see* information and communication technology (ICT)
ILOR, *see* interference limited opportunistic relaying (ILOR)
Impala middleware layer, 4
Impersonation security attacks, 202
initModel(), 284
Inter-session network coding, 175–186
Adaptive Opportunistic Network Coding (AONC), 184–186
overview, 184–185
packet header, 186
performance, 186
protocol mechanism, 185–186
BEND, 181–184
overview, 181–182
packet decoding, 183
packet header, 183–184
packet mixing and queuing strategy, 182–183
performance, 184
traffic concentration, 181
COPE
Medium Access Control (MAC), 175
overview, 175–177
packet coding, 177–178
packet decoding, 178
packet structure, 178–179
performance, 179–181
Intra-session network coding, 187–190
CodeOR, 187–189
motivation, 187–188
overview, 187
performance, 188–189
protocol mechanism, 187
sending window and acknowledgment, 188
SlideOR, 189–190
encoding and decoding mechanism, 189
overview, 189
performance, 190
sliding window process, 190
IntServ, *see* integrated services (IntServ)
IoT, *see* Internet of Things (IoT)
isWithinRange(), 291

J

Jellyfish security attacks, 200–201

L

LAOR, *see* Location Aided Opportunistic Routing (LAOR)
one-hop throughput (OEOT), 90
LCAR, *see* Least Cost Anypath Routing (LCAR)
LEPR, *see* least probable (LEPR)
LinGo, *see* Link Quality and Geographical Aware Opportunistic Routing (LinGo)
Link spoofing attack, 206–207
LOR, *see* Localized Opportunistic Routing (LOR)
close-node-sets (CNS's), 108
minimum transmission selection (MTS-B), 108
LPS, *see* Less Probable Sprayed (LPS)
LRF, *see* Least Recently Forwarded (LRF)
LWML, *see* Low Weight Message List (LWML)

M

Man-in-the-middle attack, 204–205
MANET, *see* mobile ad hoc network (MANET)
Manhattan mobility model, 74
MAP, *see* Multi-constrained Anypath Routing (MAP)
NP-hard problem, 107
MessageEventGenerator, 284
MGOR, *see* multi-rate geographic opportunistic routing (MGOR)
expected advancement rate (EAR), 90
expected medium time (EMT), 90
Mobility models
impact of mobility models, 79
map-based mobility models, 75–77
evaluation, 77
Route-based Map Movement (RBMM), 75
Rush Hour Traffic model, 75–76
Shortest Path Map-based Movement (SPMBM), 76
Working Day Movement model, 76
overview, 59–63
Social-network-based mobility models, 77
Community-based Mobility Model (CMM), 77
social network models, 77
stochastic mobility model, 65–68
random-based mobility models, 65–68
synthetic models, 68–75
evaluation, 72–74
geographical-restriction-based mobility model, 72–74
spatial-dependency-based mobility models, 70–72
temporal-dependency-based mobility models, 68–70
testing tools, 78–79
trace-based models, 64–65
analysis, 64–65
TRansportation ANalysis SIMulation System (TRANSIMs), 78
Modified Random Direction mobility model, *see also* Random Direction Model, 67
MOFO, *see* most forwarded (MOFO)
MOPR, *see* most probable (MOPR)
MORE, *see* MAC Independent Opportunistic Routing (MORE)
MSNs, *see* Mobile Social Networks (MSNs)
MTSBS, *see* Message Transmission Status Based Buffer Management Scheme (MTSBS)
MULE, *see* mobile ubiquitous LAN extensions (MULE)
MULEs, *see* Mobile Ubiquitous LAN Extensions (MULEs)
Multicast communication, 139
geographical multicast routing protocol (GeMuRo), 139
group header-based multicasting (GHM), 139
on-demand multicast routing protocol (ODMRP), 139

N

Network coding schemes
inter-session, 175–186
Adaptive Opportunistic Network Coding (AONC), 184–186
BEND, 181–184
COPE, 175–181
intra-session, 187–190
overview, 172–175
Network emulation basics, 248–250
simulation and real world experiments, 248–249
Common Open Research Emulator (CORE), 248
graphical user interface (GUI), 248
Transmission Control Protocol (TCP), 249
User Datagram Protocol (UDP), 249
Network optimization
graph-based, 106–108
Least Cost Anypath Routing (LCAR), 107

Localized Opportunistic Routing (LOR), 108
Multi-constrained Anypath Routing (MAP), 107
PLASMA, 107–108
Polynomial Time Algorithm for Multi-rate Anypath Routing (PTAS MRA), 107
shortest multi-rate anypath first (SMAF), 106–107
learning-based, 105–106
Adapt Opportunistic Routing (d-Adapt OR), 106
Opportunistic Routing with Learning Algorithm (ORL), 106
selfish aware, 106
utility-based, 104–105
Auto-Adjustable Opportunistic Acknowledgment/Timer-Based Routing, 105
Dice, 105
JOKER, *see also* Auto-Adjustable Opportunistic Acknowledgment/ Timer-Based Routing, 105
Node-Constrained Opportunistic Routing (Consort), 104
Opportunistic Residual Expected Network Utilities (OpRENU), 104–105
Optimized Multi-Path Network Coding (OMNC), 104
Time Sensitive Utility-Based Opportunistic Routing (TOUR), 105
Network topology classification, 8
coding-based schemes., 8
deterministic, 8
stochastic, 8
Node impact, 270–272
Node replication attack, 208–209
Nomadic Community mobility model, 72

O

OAPF, *see* Opportunistic AnyPath Forwarding (OAPF)
expected anypath transmission (EAX), 97
expected transmission count (ETX), 97
OBM, *see* optimal buffer management (OBM)
Obstacle mobility model, 73–74
OEOT, *see* one-hop throughput (OEOT)
OMNC, *see* Optimized Multi-Path Network Coding (OMNC)
OOF, *see* optimal opportunistic forwarding (OOF)
OPF, *see* optimal probabilistic forwarding (OPF)
OppNet, 2
Opportunistic mobile networks (OMNs), 160–161
categories, 160
human-centered mobile phones, 160
hybrid networks, 160
VANETs, 160
comparison of MANETs and WSNs, 161
density of nodes, 161
energy conservation, 161
mobility of nodes, 161
OPRAH, *see* Opportunistic Routing in Dynamic Ad hoc Networks (OPRAH)
interference, 92
OpRENU, *see* Opportunistic Residual Expected Network Utilities (OpRENU)
ORCD, *see* Opportunistic Routing with Congestion Diversity (ORCD)
Lyapunov, 110
ORL, *see* Opportunistic Routing with Learning Algorithm (ORL)
ORW, *see* Opportunistic Routing for Wireless Sensor Networks (ORW)
expected number of duty-cycled wakeups (EDC), 111
OVM, *see* Oppnet Virtual Machine (OVM)

P

Pathway mobility model, 73
Payload Confidentiality Block (PCB), 252
Payload Integrity Block (PIB), 252
PCB, *see* Payload Confidentiality Block (PCB)
Pervasive trust foundation for security and privacy
Oppnets, 214–217
helpers and categories, 216–217
security and privacy challenges, 217
structure and operation, 214–216
overview, 213–214
decentralized oppnet controller (DOC), 214
PTF-based oppnet architecture, 220–224
POA operations, 222–224
POA structure, 221–222
PIB, *see* Payload Integrity Block (PIB)
Ping-pong effect, 64
POR, *see* Position-Based Opportunistic Routing (POR)
PREP, *see* Prioritized Epidemic (PREP)
PRO, *see* Protocol for Retransmitting Opportunistically (PRO)

Probabilistic issues, 101–104
delegation forwarding, 102–103
encounter-based routing (EBR), 103
epidemic routing, 101–102
Fixed Point Opportunistic Routing (FPOR), 102
MaxOPP, 103
opportunistic flooding (OR Flood), 102
wireless sensor networks (WSNs), 102
optimal opportunistic forwarding (OOF), 104
optimal probabilistic forwarding (OPF), 103–104
OR tree, 103
Spray and Wait, 102
PRoPHET, *see* Probabilistic Routing Protocol using History of Encounters and Transitivity (PRoPHET)
PSNs, *see* pocket switched networks (PSNs)
PTAS MRA, *see* Polynomial Time Algorithm for Multi-rate Anypath Routing (PTAS MRA)
PTF-based oppnet architecture, 220–224
POA operations, 222–224
POA structure, 221–222
Pursue mobility model, 71

Q

QOR, *see* QoS Oriented Opportunistic Routing (QOR)

R

Random Direction Model, 67
Random Waypoint (RWP), 66
Random-based mobility models, 65–68
Random Walk Mobility Model, *see also* Brownian motion, 65–66
RAPID, *see* Resource Allocation Protocol for Intentional DTN (RAPID)
RBMM, *see* Route-based Map Movement (RBMM)
Reference Point Group mobility model (RPGM), 70–71
Resource utilization and related technologies
applications, 14–16
healthcare and wellness monitor, 14–16
Oppnet Virtual Machine (OVM), 13–14
overview, 11–12
structure and operation, 12–13
ROMER, *see* Resilient Opportunistic Mesh Routing (ROMER)
Extremely Opportunistic Routing (ExOR), 93
high-speed access points (HAPs), 91
Routing strategy classification, 8
RPGM, *see* Reference Point Group mobility model (RPGM)
runSim(), 287
Rushing attacks, 205
RWP, *see* Random Waypoint (RWP)

S

SACC, *see* Socially Aware Congestion Control algorithm (SACC)
SCM, *see* Social Congestion Metric (SCM)
setNextEventQueue(), 284
setNextWayPoint(), 292
SHLI, *see* shortest lifetime (SHLI)
Shortest multi-rate anypath first (SMAF), 106–107
expected anypath transmission time (EATT), 107
SimScenario(), 284
Simulation time impact, 272–274
Sinkhole security attacks, 199–200
SlideOR, 189–190
encoding and decoding mechanism, 189
end-to-end acknowledgement (E-ACK), 188
hop-to-hop acknowledgement (H-ACK), 188
SM, *see* Smooth mobility model (SM)
SMAF, *see* shortest multi-rate anypath first (SMAF)
Smooth mobility model (SM), 67
Smooth random mobility model, 69–70
Snare attack, 207
Social-aware routing protocols
comparison evaluation, 270–276
buffer impact, 275–276
node impact, 270–272
simulation time impact, 272–274
TTL impact, 270
dLifeComm, 268–269
evaluation methodology and setting, 269
ONE simulator, 269
overview, 267–268
simulation environment setup, 269
social-aware content based-opportunistic routing protocol (SCORP), 268–269
Social-network-based mobility models
Community-based Mobility Model (CMM), 77
Home-cell Community-Based Mobility (HCM), 77
SOCIALNETS, *see* Social Networking for Pervasive Adaptation (SOCIALNETS)

Spatial-dependency-based mobility models, 70–72
Column mobility model, 71
Nomadic Community mobility model, 72
Pursue mobility model, 71
Reference Point Group mobility model (RPGM), 70–71
SPOR, *see* Simple and Practical Opportunistic Routing (SPOR)
startGUI(), 287
Stochastic mobility model, 65–68
Modified Random Direction mobility model, *see also* Random Direction Model, 67
Random Direction Model, 67
Random Waypoint (RWP), 66
random-based mobility models, 65–68
border effect, 65
Random Walk Mobility Model, 65–66
Smooth mobility model (SM), 67
SWIM, *see* Shared Wireless Infostation Model (SWIM)
Sybil security attacks, 197–198
Synthetic models, 68–75
geographical-restriction-based mobility model, 72–74
City Section mobility model, 74
Manhattan mobility model, 74
Obstacle mobility model, 73–74
Pathway mobility model, 73
spatial-dependency-based mobility models, 70–72
Column mobility model, 71
Nomadic Community mobility model, 72
Pursue mobility model, 71
Reference Point Group mobility model (RPGM), 70–71
temporal-dependency-based mobility models, 68–70
Gauss–Markov mobility model, 69
smooth random mobility model, 69–70

T

Taxonomy of routing protocols
context-aware, 97–101
fully context-aware, 99–101
partially context-aware, 97–99
cross-layer, 108–112
MAC Aware, 109–111
PHY and MAC Aware, 111–112
Physical-Layer-Aware (PHY-Aware), 108–111
geographic-based, 88–93
Contention-Based Forwarding (CBF), 89
Cooperative Opportunistic Routing in Mobile Ad Hoc Networks (CORMAN), 93
Cross-Layer Link Quality and Geographical-Aware Beaconless Opportunistic Routing (XLinGo), 93
Directed Transmission Routing Protocol (DTRP), 92
Geographic Opportunistic Routing (GOR), 89–90
Geographic Random Forwarding (GeRaF), 89
Link Quality and Geographical Aware Opportunistic Routing (LinGo), 91
Location Aided Opportunistic Routing (LAOR), 90
multi-rate geographic opportunistic routing (MGOR), 90
Opportunistic Routing in Dynamic Ad hoc Networks (OPRAH), 92
Position-Based Opportunistic Routing (POR), 91
Resilient Opportunistic Mesh Routing (ROMER), 91–92
Topology- and Link-Quality-Aware Geographical Opportunistic Routing (TLG OR), 91
topology-assisted geographic opportunistic routing (To Go), 93
link state aware, 94–97
Code OR, 95
Cumulative Coded Acknowledgment (CCACK), 96
economy, duplicate free opportunistic routing, 94
MAC Independent Opportunistic Routing (MORE), 95
O3 (Optimized Overlay-Based Opportunistic Routing), 96
Opportunistic AnyPath Forwarding (OAPF), 97
Opportunistic Multi-Hop Routing for Wireless Networks (ExOR), 94
Simple Opportunistic Adaptive Routing Protocol for Wireless Mesh Networks (SOAR), 96
Slide OR: Online Opportunistic Network Coding in Wireless Mesh Networks, 95
XCOR (Synergistic Interflow Network Coding and Opportunistic Routing), 95

optimization, 104–108
graph-based, 106–108
learning-based, 105–106
utility-based, 104–105
overview, 87–88
probabilistic issues, 101–104
delegation forwarding, 102–103
encounter-based routing (EBR), 103
epidemic routing, 101–102
Fixed Point Opportunistic Routing (FPOR), 102
MaxOPP, 103
opportunistic flooding (OR Flood), 102
optimal opportunistic forwarding (OOF), 104
optimal probabilistic forwarding (OPF), 103–104
Spray and Wait, 102
Taxonomy of security attacks
attributes, 195–196
authentication, 195
availability, 196
data integrity, 196
nonrepudiation, 195
privacy, 196
authentication schemes, 196–197
cryptography-based, 196–197
signature-based, 196
black hole and selective forwarding, 198–199
artificial neural networks (ANNs), 199
Network Simulator 3 (NS-3), 199
Simulator of Urban MObility (SUMO), 199
VANET Car Mobility Manager (VaCaMobil), 199
blackmail, 207–208
bogus information attack, 202–203
classification, 195
active *vs.* passive, 195
insider *vs.* outsider, 195
malicious *vs.* rational, 195
colluding misrelay, 206
desynchronization, 209
distributed denial of service (DDoS), 203–204
GPS spoofing, 201–202
impersonation, 202
jellyfish, 200–201
throughput-feedback routing (TUF), 200
link spoofing, 206–207
man-in-the-middle, 204–205
node replication, 208–209
overview, 194
rushing, 205
rushing attack prevention (RAP), 205
sinkhole, 199–200
multipoint relay (MPR), 199
optimized link state routing protocol (OLSR), 199
snare, 207
Sybil, 197–198
priority batch verification algorithm (PBVA), 198
roadside unit (RSU), 197
wormhole, 200
Temporal-dependency-based mobility models, 68–70
Gauss–Markov mobility model, 69
smooth random mobility model, 69–70
Time and data-driven triggering, 254–256
KauNet, 255–256
pattern and scenario files, 255
pattern generation, 255
TLG OR, *see* Topology- and Link-Quality-Aware Geographical Opportunistic Routing (TLG OR)
To Go, *see* topology-assisted geographic opportunistic routing (To Go)
forwarding set selection (FSS), 93
next hop prediction algorithm (NPA), 93
TOUR, *see* Time Sensitive Utility-Based Opportunistic Routing (TOUR)
cyclic-based mobile social networks (MSNs), 105
Deadline-Sensitive Opportunistic Utility-Based Routing (DOUR), 105
Trace-based models, 64–65
analysis, 64–65
Campus Waypoint model, 65
ping-pong effect, 64
TRANSIMs, *see* TRansportation ANalysis SIMulation System (TRANSIMs)
TTL impact, 270

U

Unicast communication, 138
distance routing effect algorithm for mobility (DREAM), 138
greedy perimeter stateless routing (GPSR), 138
update(), 289
updateHosts(), 289
updateLocation(), 289
UWB, *see* ultra-wideband (UWB)

V

VANETs, *see* Vehicular Ad Hoc Networks (VANETs)
VDTN, *see* vehicular delay-tolerant network (VDTN)
Vehicular ad hoc networks
  applications, 136–137
    commercial, 137
    non-safety, 136
    safety, 136
  broadcaste protocols comparisons, 152–153
  broadcasting protocols for connected and fragmented, 147–152
    Acknowledgement parameterless broadcast in static to highly dynamic mobile (Ack-PBSM), 150–151
    Density-aware reliable broadcasting for vehicular ad hoc networks (DECA), 148–149
    distributed robust geocast(DRG), 147–148
    DV-Cast, 148
    EDB, 149
    edge aware epidemic protocol (EAEP), 150
    inter-vehicle geocast (IVG), 147
    mobicast, 148
    Position-aware reliable broadcasting protocol (POCA), 149
    Simple and robust dissemination (SRD), 149–150
    streetcast, 152
    Urban vehicular broadcast (UV-CAST), 151
  dissemination protocols for connected, 141–144
    broadcast storm problem., 141
    heuristic, 141–142
    topology-based, 143–144
  homogeneous and heterogeneous, 137–139
    broadcast, 140
    broadcast communication, 139
    multicast communication, 139
    unicast communication, 138
  overview, 134–136
    dedicated short range communication (DSRC) standard, 135–136
  packet forwarding and dissemination, 137
  routing approaches for fragmented, 145–147
    epidemic, 145
    MobySpace, 146–147
    spray and wait, 146
    Vehicle-assisted data delivery (VADD), 145–146

W

warmupMovementModel(), 284
WBD, *see* weight-based buffer management policy (WBD)
world.update(), 288
Wormhole security attacks, 200
WSNs, *see* Wireless sensor networks (WSNs)

X

XLinGo, *see* Cross-Layer Link Quality and Geographical-Aware Beaconless Opportunistic Routing (XLinGo)
  flying ad hoc networks (FANETs), 93

Z

ZebraNet, 4–5, 238–239